AGRICULTURAL NONPOINT SOURCE POLLUTION

Watershed Management and Hydrology

Edited by
William F. Ritter
Adel Shirmohammadi

LEWIS PUBLISHERS

Boca Raton London New York Washington, D.C.

Library of Congress Cataloging-in-Publication Data

Agricultural nonpoint source pollution : watershed management and hydrology / edited by William F. Ritter, Adel Shirmohammadi

   p. cm.

Includes bibliographical references.

ISBN 1-56670-222-4 (alk. paper)

   1. Agricultural pollution--Environmental aspects--United States. 2. Nonpoint source pollution--United States. 3.Watershed management--United States. 4. Water quality management--United States. I. Ritter, William F. II. Shirmohammadi, Adel, 1952-

TD428.A37 A362 2000

628.1'.684—dc21

                                                      00-046349

                                                           CIP

Visit the CRC Press Web site at www.crcpress.com

© 2001 by CRC Press LLC

Lewis Publishers is an imprint of CRC Press LLC
No claim to original U.S. Government works
International Standard Book Number 1-56670-222-4
Library of Congress Card Number 0046349
Printed in the United States of America   3 4 5 6 7 8 9 0
Printed on acid-free paper

Preface

Despite the tremendous progress that has been achieved in water pollution, almost 40% of the U.S. waters that have been assessed by states do not meet water quality goals. About 20,000 water bodies are impacted by siltation, nutrients, bacteria, oxygen depletion substances, metals, habitat alterations, pesticides, and toxic organic chemicals. With pollution from point sources being dramatically reduced, nonpoint source pollution is the major cause of most water that does not meet water quality goals. About 50 to 70% of the assessed surface waters are adversely affected by agricultural nonpoint source pollution caused by soil erosion from cropland and overgrazing and from pesticide and fertilizer applications. States have identified almost 500,000 kilometers of rivers and streams and more than two million hectares of lakes that do not meet state water quality goals. In 1998, about one-third of the 1062 beaches reporting to the U.S. Environmental Protection Agency had at least one health advisory or closing. More than 2500 fish consumption advisories or bans were issued by states in areas where fish were too contaminated to eat.

Clean water is important for the nation's economy. A third of Americans visit coastal areas each year, generating new jobs and billions of dollars. Closed beaches and fish advisories result in lost revenue. Water used for irrigating crops and raising livestock helps American farmers produce and sell $197 billion worth of food and fiber each year. Manufacturers use thirty-five trillion liters of fresh water annually.

This book is intended to give a comprehensive overview of agricultural nonpoint source pollution and its management on a watershed scale. The first chapter provides background information on watershed hydrology, with a discussion on each phase of the hydrologic cycle. The second chapter is on soil erosion and sedimentation. The basic processes of soil erosion as it occurs in upland areas are discussed, most of it focused on rill and interrill erosion. Process-based soil erosion models and cropping and management effects on erosion are treated and contrasted in some detail.

Chapters 3, 4, and 5 take up the nonpoint source pollutants nitrogen, phosphorus, and pesticides in detail. Both surface and subsurface processes are discussed in each chapter. Chapters 3 and 4 begin with nitrogen and phosphorus cycles, respectively. Management practices to control nonpoint source pollution from nitrogen, phosphorus, and pesticides are discussed.

Chapter 6 discusses nonpoint source pollution from the livestock industry. Surface water and groundwater quality effects from feedlots, manure storage and treatment systems, and land application of manures are presented, along with nonpoint source pollution control practices for each of these sources.

Chapter 7 addresses the impact of irrigated agriculture on water quality. The nonpoint source pollutants nitrates, pesticides, salts, trace elements, and suspended sediments are discussed, along with management practices for reducing nonpoint source pollution from irrigation. Chapter 8 is focused on the impact of agricultural drainage on water quality. Both conventional drainage and water-table management are discussed.

Chapter 9 provides an overview of water quality models. Different types of water quality models are discussed along with model development, sensitivity analysis, model validation and verification, and the role of geographic information systems in water quality modeling. Chapter 10 provides a treatment of best management practices (BMPs) to control nonpoint source pollution and the framework for the design of a monitoring system for BMP impact assessment. Fourteen BMPs are discussed in detail.

The final chapter discusses monitoring, including monitoring system design, data needs and collection, and implementation strategies, along with methods to monitor edge-of-field overland flow, bottom of root zone, soil, groundwater, and surface water.

The editors thank all authors for their valuable contribution to this book. We hope it will give people a better insight into the issues involved in agricultural nonpoint source pollution and its control.

William F. Ritter
Adel Shirmohammadi

Editors

William F. Ritter, Ph.D. is Professor of Bioresources and Civil and Environmental Engineering at the University of Delaware and a Senior Policy Fellow in the Center for Energy and Environment Policy.

In 1965 Dr. Ritter received his B.S.A. in agricultural engineering from the University of Guelph, and in 1966 received a B.A.S. in civil engineering from the University of Toronto. He obtained his M.S. in 1968 in water resources and his Ph.D. in 1971 in sanitary and agricultural engineering from Iowa State University. He was a research associate at Iowa State University from 1966 to 1971 and joined the Agricultural Engineering Department at the University of Delaware as an assistant professor in 1971. He served as department chair of the Agricultural Engineering Department from 1992 to 1998.

Dr. Ritter is a registered professional engineer in Delaware, Maryland, Pennsylvania, and New Jersey and is a fellow of the American Society of Agricultural Engineers and American Society of Civil Engineers. He is also a member of the American Water Works Association, Water Environment Federation, Canadian Society of Agricultural Engineers, and American Society of Engineering Education. He has taught courses on hydrology, soil erosion, irrigation, drainage, soil physics, solid waste management, wastewater treatment, and land application of wastes. He has conducted research on irrigation water management, livestock waste management, surface and groundwater quality, and land application of wastes. He has served as a consultant to government and industry on wastewater management, water quality, land application of wastes, and livestock waste management.

Dr. Ritter is the author of more than 270 papers, reports, and book contributions and has presented over 140 papers at regional, national, and international conferences. He has also received numerous awards that include the College of Agriculture Outstanding Research Award (1990), ASAE Gunlogson Countryside Engineering Award (1989), ASCE Outstanding News Correspondent (1997), and ASCE Delaware Section Civil Engineer of the Year (1999).

Dr. Adel Shirmohammadi, Ph.D. is Professor of Biological Resources Engineering at the University of Maryland, College Park campus.

In 1974, Dr. Shirmohammadi received his B.S. in agricultural engineering from the University of Rezaeiyeh in Iran. He obtained an M.S. in 1977 in agricultural engineering from the University of Nebraska and a Ph.D. in 1982 in biological and agricultural engineering from North Carolina State University. From 1982 to 1986 he was a post-doctoral agricultural research engineer and assistant research scientist in the Agricultural Engineering Department at the University of Georgia Coastal Plains Experiment Station at Tifton. In 1986, he joined the Agricultural Engineering Department at the University of Maryland as an assistant professor.

Dr. Shirmohammadi is a member of the American Society of Agricultural Engineers, Soil and Water Conservation Society of America, and American

Geophysical Union. He has taught courses in hydrology, soil and water conservation engineering, water quality modeling, flow-through porous media, and nonpoint source pollution. He has conducted research in hydrologic and water quality modeling, drainage, and nonpoint source pollution. He has developed an international reputation in water quality modeling for his work with CREAMS, GLEAMS, DRAINMODE, and ANSWERS.

Dr. Shirmohammadi has received numerous competitive grants and has served as a consultant to industry and government. He is the author of more than 100 refereed publications, conference proceedings, papers, and book contributions.

Contributors

Lars Bergstrom, Ph.D.
Professor
Swedish University of Agricultural
  Sciences
Division of Water Quality Research
Uppsala, Sweden
lars.bergstrom@mv.slu.se

Kevin M. Brannan, M.S.
Research Associate
Biological Systems Engineering
  Department
Virginia Polytechnic and State
  University
Blacksburg, VA
kbrannan@vt.edu

Adriana C. Bruggeman, Ph.D.
Research Associate
Biological Systems Engineering
  Department
Virginia Polytechnic and State
  University
Blacksburg, VA

Kenneth L. Campbell, Ph.D.
Professor
Agricultural and Biological Engineering
  Department
University of Florida
Gainesville, FL
klc@agen.ufl.edu

Theo A. Dillaha III, Ph.D.
Professor
Biological Systems Engineering
  Department
Virginia Polytechnic and State
  University
Blacksburg, VA
dillaha@vt.edu

Dwayne R. Edwards, Ph.D.
Associate Professor
Biosystems and Agricultural
  Engineering Department
University of Kentucky
Lexington, KY

Blaine R. Hanson, Ph.D.
Irrigation and Drainage Specialist
Department of Land, Air and Water
  Resources
University of California
Davis, CA
brhanson@ucdavis.edu

Walter G. Knisel, Jr., Ph.D.
Retired Hydraulic Engineer of USDA-
  ARS and Affiliate Professor
Biological and Agricultural Engineering
  Department
Coastal Plains Experiment Station
University of Georgia
Tifton, GA
wknisel@planttel.net

William L. Magette, Ph.D.
Lecturer
Agricultural and Food Engineering
  Department
University College Dublin
Dublin, Ireland
william.magette@ucd.ie

Hubert J. Montas, Ph.D.
Assistant Professor
Biological Resources Engineering
  Department
University of Maryland
College Park, MD
hm66@umail.umd.edu

Saied Mostaghimi, Ph.D.
H. E. and Elizabeth Alphin Professor
Biological Systems Engineering
  Department
Virginia Polytechnic and State
  University
Blacksburg, VA
smostagh@vt.edu

Mark A. Nearing, Ph.D.
Scientist
USDA-ARS National Soil Erosion
  Research Laboratory
West Lafayette, IN
nearing@ech.perdue.edu

L. Darrell Norton, Ph.D.
Scientist
USDA-ARS National Soil Erosion
  Research Laboratory
West Lafayette, IN

Adel Shirmohammadi, Ph.D.
Biological Resources Engineering
  Department
University of Maryland
College Park, MD
adel.shir@mv.slu.se

William F. Ritter, Ph.D.
Bioresources Engineering
  Department
University of Delaware
Newark, DE
william.ritter@mvs.udel.edu

Thomas J. Trout, Ph.D.
Agricultural Engineer
USDA-ARS Water Management
  Research Laboratory
Fresno, CA

Mary Leigh Wolfe, Ph.D.
Associate Professor
Biological Systems Engineering
  Department
Virginia Polytechnic and State
  University
Blacksburg, VA
mlwolfe@vt.edu

Xunchang Zhang, Ph.D.
Scientist
USDA-ARS Soil Erosion Research
  Laboratory
West Lafayette, IN

Table of Contents

1 Hydrology

M. L. Wolfe

CONTENTS

1.1 INTRODUCTION

Sources of water pollution can be classified broadly into two categories: point sources and nonpoint sources. Point sources are most readily identified with industrial sources such as manufacturing, processing, power generation, and waste treatment facilities where pollutants are delivered through a pipe (discharge point). In contrast, nonpoint, or diffuse, sources include areas such as agricultural fields, parking lots, and golf courses.

Nonpoint pollutants such as sediment, nutrients, pesticides, and pathogens are transported across the land surface by runoff and through the soil by percolating water. Nonpoint source (NPS) pollution is intermittent, associated very closely with rainfall runoff. Nonpoint source pollution is a function of climatic factors and site-specific land characteristics such as soil type, land management, and topography.

This chapter focuses on the hydrologic processes that strongly influence NPS pollution. First, an overview of the hydrologic cycle is given, with emphasis on the interaction of the processes. Interaction of hydrologic processes is highlighted throughout the chapter because it is difficult, if not impossible, to describe one

1-56670-222-4/01/$0.00+$.50
© 2001 by CRC Press LLC

process without mentioning others. The sections that follow include qualitative descriptions of each process, presentations of estimation techniques, and discussions of the relationship of each process to NPS pollution. Information related to measurement of each process is included in Chapter 11.

1.2   HYDROLOGIC CYCLE

Nonpoint source pollution is tied closely to the hydrologic cycle (Figure 1.1). Falling rain can be followed to several fates. Some rain evaporates as it falls and returns to the atmosphere. Some rainfall is intercepted by vegetation. Intercepted rainfall then either evaporates or drips to the soil surface. Some rainfall reaches the soil surface, where some of it infiltrates into the soil, some ponds on the soil surface, and some runs off. Ponded rainfall can evaporate, infiltrate into the soil, or run off. Rainfall that infiltrates can be used by plants, remain in the soil profile, or percolate to groundwater. The proportions of rainfall that reach the various fates depend on dynamic site-specific conditions such as vegetative cover, soil moisture content, soil texture, and slope. Similar to rainfall, snowmelt can run off or infiltrate.

Nonpoint pollutants are transported by runoff to surface water and by leaching to groundwater. In addition, groundwater feeds streams, so pollutants can also reach surface water via groundwater. In the following sections, hydrologic processes that are particularly important with respect to NPS pollution are described.

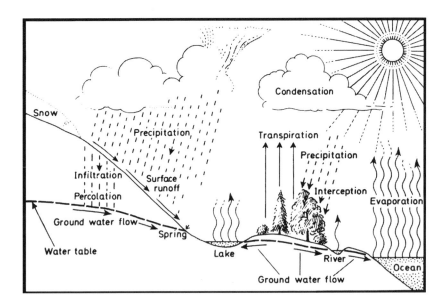

FIGURE 1.1   The hydrologic cycle. (From Shaw, E. M., Hydrology—a multidisciplinary subject, in *Environment, Man and Economic Change,* Phillips, A. D. M. and Turton, B. J., Eds., Longman, London and New York, 1975, 164. ©Longman Group Limited 1975. With permission.)

1.2.1 Precipitation

1.2.1.1 Description

Precipitation occurs in a number of different forms, including drizzle, mist, rain, snow, sleet, hail, and dew (Brooks et al.[1]). Drizzle consists of drops less than 0.5 mm in diameter. Rain consists of drops 0.5 to 7 mm in diameter. Mist describes a rate of less than one mm/h. Snow is precipitation that changes directly from water vapor to ice. Sleet refers to frozen raindrops cooled to ice while falling through air at sub-freezing temperatures. Hail is formed by alternate freezing and melting as raindrops are carried up and down in a turbulent air current. Dew is caused by condensation of moisture in air on cooler surfaces.

The relationship among atmospheric moisture, temperature, and vapor pressure determines the occurrence and amounts of precipitation. Precipitation occurs when three conditions are met (Eagleson[2]): (1) saturation conditions in the atmosphere, (2) phase change of water content from vapor to liquid or solid state, and (3) growth of the small water droplets or ice crystals to precipitable size. Detailed descriptions of these phenomena are presented in many sources (e.g., Eagleson,[2] Brooks et al.[1]).

Rain is the precipitation of primary importance to NPS pollution. Rainfall varies both temporally (Figure 1.2) and spatially (Figure 1.3), which means that NPS pollution varies temporally and spatially. Characteristics of rainfall that are important to NPS pollution include rainfall intensity, duration, amount, drop size distribution,

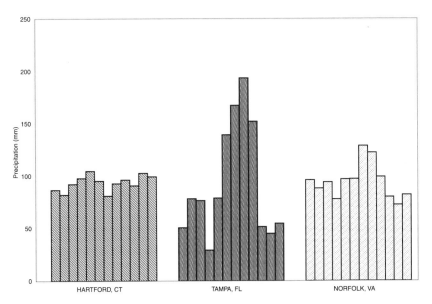

FIGURE 1.2 Distribution of mean (1961–1990) monthly precipitation (mm) for three locations that receive about 1120 mm total annual precipitation. (Based on data from National Climatic Data Center, http://www.ncdc.noaa.gov/ol/climate/online/ccd/nrmlprcp.html)

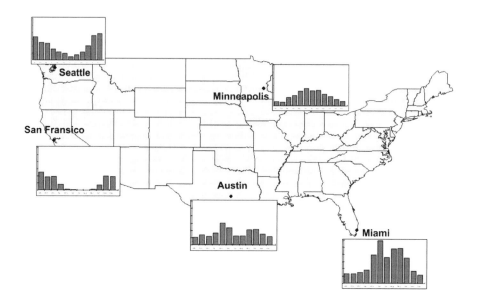

FIGURE 1.3   Mean (1961–1990) annual precipitation for selected locations in the United States. (Based on data from National Climatic Data Center, http://www.ncdc.noaa.gov/ol/climate/online/nrmlprcp.html)

raindrop energy, and frequency of occurrence. Intensity and duration determine the total amount of rainfall. Both total amount and intensity of rainfall are important influences on NPS pollution. For example, in general, a short-duration, high-intensity rainfall will cause more runoff than a long-duration, low-intensity rainfall of the same amount.

Drop size and velocity determine raindrop energy ($KE = 1/2 \, mv^2$, KE = kinetic energy, m = mass, v = velocity), which influences infiltration and, therefore, runoff and erosion. Drop size distribution is related to rainfall intensity (Laws and Parsons[3]). As rainfall intensity increases, the range of drop sizes increases and there are more drops of large diameter. Higher energy has the potential to decrease infiltration through surface sealing and to increase soil erosion through increased soil detachment. Terminal velocity ranges from about 5 m/s for a 1-mm drop to about 9 m/s for a 5-mm drop (Laws[4]).

Frequency of rainfall and other hydrologic events is typically described in terms of a return period, or recurrence interval. Return period is the average number of years within which a given event will be equaled or exceeded. A rainfall event is described fully in terms of its depth and duration. For example, a 25-year, 24-hour rainfall is the amount of rainfall during a 24-hour duration that is equaled or exceeded on the average once every 25 years. It does not mean that an exceedance occurs every 25 years, but that the average time between exceedances is 25 years. Depth-duration-frequency relationships have been developed for the United States for durations of

30 minutes to 24 hours and return periods of 1 to 100 years (Hershfield[5]). Frequency of rainfall events is important in designing some management practices and structures for NPS pollution control.

1.2.1.2 Rainfall Estimation

Daily rainfall is a complex process and therefore difficult to model (Richardson[6]). The randomness of rainfall occurrence and characteristics must be represented. Stochastic modeling of rainfall has often used the approach of first estimating the occurrence of rainfall and then modeling the rainfall event characteristics of depth and duration. For example, Mills[7] modeled occurrence of rainfall using a Poisson distribution and then estimated duration using a Weibull marginal probability density function (PDF) and depth using a log-normal conditional PDF given duration. Monte Carlo simulation (Mills[7]) and Markov type rainfall models (Jimoh and Webster[8]) are often used to describe the occurrence of daily rainfall occurrence (i.e., wet day/dry day sequences). Jimoh and Webster[8] investigated the optimum order of Markov models for simulating rainfall occurrence.

A second approach to simulating rainfall combines occurrence and depth of rainfall. Khaliq and Cunnane[9] described cluster-based models and a three-state continuous Markov process occurrence model (Hutchinson[10]). Cluster-based models represent rainfall events as clusters of rain cells. Each cell is considered to be a pulse with a random duration and random intensity that is constant throughout the cell duration. Cells are distributed in time according to the Neyman-Scott cluster process or the Bartlett-Lewis cluster process (Rodriguez-Iturbe et al.[11]).

Efforts continue to improve estimation of rainfall occurrence and event characteristics. The increasing availability of space-time rainfall data from radar and satellite is contributing to the effort (Mellor[12]). Detailed information on estimating rainfall events can be found in a number of publications (e.g., Singh[13] and O'Connell and Todini[14]).

1.2.2 SURFACE RUNOFF

1.2.2.1 Description

Surface runoff occurs when the infiltration capacity of the soil is exceeded by the rainfall rate. Excess rain (in excess of infiltration) accumulates on the soil surface and runs off when the depth of ponding and other surface conditions cause the water to flow. Runoff travels across the land surface, increasing and decreasing in flow velocity and changing course depending on slope, vegetation, surface roughness, and other surface characteristics. Some runoff can infiltrate as it flows (transmission losses). Previously infiltrated water can reemerge (interflow or shallow subsurface flow) to join the surface flow.

The amount of runoff depends on other components of the hydrologic cycle such as infiltration, interception, evapotranspiration (ET), and surface storage. If the rate of rainfall does not exceed the rate of infiltration, there is no runoff. The amount of interception is a function of the type and growth stage of vegetation and wind velocity. There is little information available about amount of interception by agricultural crops, but there has been considerable work done on interception by forests. Interception by a well-developed forest canopy is about 10 to 20% of the annual rainfall (Linsley et al.[15]). Evapotranspiration affects soil moisture conditions, which in turn affect infiltration capacity of the soil. Rainfall that reaches the soil surface but does not immediately infiltrate becomes part of surface retention or surface detention. Surface retention is water retained on the land surface in micro-depressions. Retained water will eventually evaporate or infiltrate. Surface detention is water temporarily detained on the land surface prior to running off. Microtopography, or surface roughness, and surface macroslope affect both retention and detention. In addition, detention is influenced by vegetation and rainfall excess distribution (Huggins and Burney[16]).

Runoff transports NPS pollutants in dissolved forms and in forms adsorbed to sediment. The detachment and transport capacity of runoff are dependent on the velocity and depth of flow. The velocity and depth of flow both change with time and space as runoff flows over a land surface. Sometimes the flow can be characterized as shallow sheet flow across the surface. Often the flow will be concentrated into small channels called rills on an agricultural field. The temporal distribution of runoff at a location is described graphically by a hydrograph (Figure 1.4) with runoff plotted on the y-axis and time on the x-axis. Runoff can be expressed in units of volume per time (cfs or m^3/s) or stage (L) of flow. Hydrographs can show surface runoff, direct runoff or total runoff. The time of concentration refers to the time required for runoff to reach the watershed outlet from the farthest hydraulic distance from the outlet. The time of concentration is a function of topography, surface cover, and distance of flow.

The amount and rate of runoff depend on rainfall and watershed characteristics. Important rainfall characteristics include duration, intensity, and areal distribution.

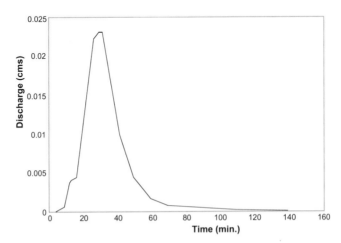

FIGURE 1.4 Hydrograph for Watershed W-1, Moorefield, WV, May 23, 1962. (Based on data from Agricultural Research Service Water Database, *http://hydrolab.arsusda.gov/arswater.html*)

Watershed characteristics that influence runoff include soil properties, land use, vegetation cover, moisture condition, size, shape, topography, orientation, geology, cultural practices, and channel characteristics. Larger watersheds generally produce larger volumes and rates of runoff. Long, narrow watersheds have longer times of concentration compared with compact watersheds. Storms moving upstream cause lower runoff rates at the watershed outlet than storms moving downstream. In the upstream case, rain stops at the lower end of the watershed before the upper end of the watershed contributes to runoff at the outlet. In the downstream case, runoff from the upper parts of the watershed reach the outlet while runoff is being contributed by the lower part of the watershed as well. Steeper slopes generally have higher runoff rates. The geology of a watershed affects runoff through its effect on infiltration. Vegetation in general retards overland flow and increases infiltration. Different vegetation types affect runoff differently. Close-growing plants such as sod retard flow more than woody plants that do not have much ground cover.

1.2.2.2 Estimating Runoff

Runoff is clearly a complex, variable process, influenced by many factors. Runoff calculations typically include estimating the amount of runoff, or rainfall excess, and then translating that amount of runoff into a hydrograph. Common approaches for estimating rainfall excess and runoff hydrographs are described in the following sections.

1.2.2.3 Rainfall Excess

Rainfall excess is determined as the total amount of rainfall minus infiltration and interception. Rainfall excess is typically estimated in two ways. In one approach, infiltration is estimated directly and then subtracted from rainfall. Methods of estimating infiltration are described later in this chapter.

The second approach is the USDA Soil Conservation Service (SCS) (now Natural Resources Conservation Service, NRCS) method of estimating runoff volume, commonly called the curve number approach. The SCS method correlates the difference between rainfall and runoff with antecedent soil moisture (ASM), or antecedent moisture condition (AMC), soil type, vegetative cover, and cultural practices. Rainfall excess is computed using the following relationship (SCS[17]):

$$Q = \frac{(P - 0.2S)^2}{P + 0.8S} \qquad (1.1)$$

$$S = \frac{25,400}{CN} - 254 \qquad (1.2)$$

where Q is the direct storm runoff volume (mm), P is the storm rainfall depth (mm), S is the maximum potential difference between rainfall and runoff starting at the time the storm begins (mm), and CN is the runoff curve number (Table 1.1), which

TABLE 1.1
Runoff Curve Numbers for Hydrologic Soil-Cover Complexes (Antecedent Moisture Condition II and $I_a = 0.2S$) (From SCS, *Hydrology, Section 4. National Engineering Handbook,* **U.S. Soil Conservation Service, GPO, Washington, DC, 1972)**

Land Use Description/Treatment/Hydrologic Condition			A	B	C	D
Residential:[a]						
Average Lot Size	Average % Impervious[b]					
0.05 ha or less	65		77	85	90	92
0.10 ha	38		61	75	83	87
0.13 ha	30		57	72	81	86
0.20 ha	25		54	70	80	85
0.40 ha	20		51	68	79	84
Paved parking lots, roofs, driveways, etc.[c]			98	98	98	98
Street and roads:						
paved with curbs and storm sewers[c]			98	98	98	98
gravel			76	85	89	91
dirt			72	82	87	89
Commercial and business areas (85% impervious)			89	92	94	95
Industrial districts (72% impervious)			81	88	91	93
Open Spaces, lawns, parks, golf courses, cemeteries, etc.						
good condition: grass cover on 75% or more of the area			39	61	74	80
fair condition: grass cover on 50% to 75% of the area			49	69	79	84
Fallow	Straight row	—	77	86	91	94
Row crops	Straight row	Poor	72	81	88	91
	Straight row	Good	67	78	85	89
	Contoured	Poor	70	79	84	88
	Contoured	Good	65	75	82	86
	Contoured & terraced	Poor	66	74	80	82
	Contoured & terraced	Good	62	71	78	81
Small grain	Straight row	Poor	65	76	84	88
		Good	63	75	83	87
	Contoured	Poor	63	74	82	85
		Good	61	73	81	84
	Contoured & terraced	Poor	61	72	79	82
		Good	59	70	78	81
Close–seeded	Straight row	Poor	66	77	85	89
legumes[d]	Straight row	Good	58	72	81	85
or	Contoured	Poor	64	75	83	85
rotation	Contoured	Good	55	69	78	83
meadow	Contoured & terraced	Poor	63	73	80	83
	Contoured & terraced	Good	51	67	76	80

TABLE 1.1 (cont'd.)

Land Use Description/Treatment/Hydrologic Condition			Hydrologic Soil Group			
Pasture		Poor	68	79	86	89
or range		Fair	49	69	79	84
		Good	39	61	74	80
	Contoured	Poor	47	67	81	88
	Contoured	Fair	25	59	75	83
	Contoured	Good	6	35	70	79
Meadow		Good	30	58	71	78
Woods or		Poor	45	66	77	83
Forest land		Fair	36	60	73	79
		Good	25	55	70	77
Farmsteads		—	59	74	82	86

[a] Curve numbers are computed assuming the runoff from the house and driveway is directed toward the street with a minimum of roof water directed to lawns where additional infiltration could occur.

[b] The remaining pervious areas (lawn) are considered to be in good pasture condition for these curve numbers.

[c] In some warmer climates of the country, a curve number of 95 may be used.

[d] Close-drilled or broadcast.

---

represents runoff potential of a surface. Rainfall depth, P, must be greater than 0.2 S for the equation to be applicable.

The CN indicates the runoff potential of a surface based on soil characteristics and land use conditions and ranges from 1 to 100 (Table 1.1), increasing with increasing CN. Required information to use the table includes the hydrologic soil group (defined in Table 1.2), the vegetal and cultural practices of the site, and the AMC (defined in Table 1.2). The CN obtained from Table 1.1 for AMC II can be converted to AMC I or III using the values in Table 1.3.

Curve numbers can be determined from rainfall runoff data for a particular site. Investigations have been conducted to determine CN values for conditions not included in Table 1.1 or similar tables. Examples include exposed fractured rock surfaces (Rasmussen and Evans[18]), animal manure application sites (Edwards and Daniel[19]), and dryland wheat-sorghum-fallow crop rotation in the semi-arid western Great Plains (Hauser and Jones[20]).

The CN approach is widely used for estimating runoff volume. Because the CN is defined in terms of land use treatments, hydrologic condition, AMC, and soil type, the approach can be applied to ungaged watersheds. Errors in selecting CN values can result from misclassifying land cover, treatment, hydrologic conditions, or soil type (Bondelid et al.[21]). The magnitude of the error depends on the size of the area misclassified and the type of misclassification. In a sensitivity analysis of runoff estimates to errors in CN estimates, Bondelid et al.[21] found that effects of variations in CN decrease as design rainfall depth increases and confirmed Hawkins'[22] conclusion that errors in CN estimates are especially critical near the threshold of runoff.

TABLE 1.2
Hydrologic Soil Group Descriptions and Antecedent Rainfall Conditions for Use with the SCS Curve Number Method (From SCS, *Hydrology, Section 4.* *National Engineering Handbook,* **U.S. Soil Conservation Sservice, GPO, Washington, DC, 1972)**

Soil Group	Description
A	*Lowest Runoff Potential.* Includes deep sands with very little silt and clay, also deep, rapidly permeable loess.
B	*Moderately Low Runoff Potential.* Mostly sandy soils less deep than A, and loess less deep or less aggregated than A, but the group as a whole has above-average infiltration after thorough wetting.
C	*Moderately High Runoff Potential.* Comprises shallow soils and soils containing considerable clay and colloids, though less than those of group D. The group has below-average infiltration after presaturation.
D	*Highest Runoff Potential.* Includes mostly clays of high swelling percentage, but the group also includes some shallow soils with nearly impermeable subhorizons near the surface.

Condition	General Description	5-Day Antecedent Rainfall (mm)	
		Dormant Season	Growing Season
I	Optimum soil condition from about lower plastic limit to wilting point	<6.4	<35.6
II	Average value for annual floods	6.4 − 27.9	35.6–53.3
III	Heavy rainfall or light rainfall and low temperatures within 5 days prior to the given storm	>27.9	>53.3

The CN approach is used in a number of NPS pollution models. Bingner[23] found that although most of the five models he evaluated use the CN approach, it is not implemented in the same way in each model. Bingner thus cautions that a user must understand the purpose for which a model was developed to avoid improper use of the model. Sensitivity analyses (e.g., Ma et al.,[24] Chung et al.[25]) have demonstrated the sensitivity of runoff estimates to CN in those models.

Additional concerns have been raised about the CN method. It is not clear whether the data from which the relationship was developed were ever presented. The method was developed only for estimating runoff volume from storms of long duration medium to large watersheds (5–50 km^2).

1.2.2.4   Runoff Hydrographs

Runoff, or overland flow, can be visualized as sheet-type flow (as opposed to channel flow) with small depths of flow and slow velocities (less than 0.3 m/sec). Considerable volumes of water can move through overland flow. In routing overland

TABLE 1.3
Conversion Factors for Converting Runoff Curve
Numbers AMC II to AMC I and III ($I_a = 0.2S$) (From
SCS, *Hydrology, Section 4. National Engineering*
***Handbook*, U.S. Soil Conservation Sservice, GPO,**
Washington, DC, 1972)

Curve Number for Condition II	Factor to Convert Curve Number for Condition II to	
	Condition I	Condition III
10	0.40	2.22
20	0.45	1.85
30	0.50	1.67
40	0.55	1.50
50	0.62	1.40
60	0.67	1.30
70	0.73	1.21
80	0.79	1.14
90	0.87	1.07
100	1.00	1.00

flow (i.e., determining the flow hydrograph), travel time needs to be considered. Overland flow is spatially varied, usually unsteady, nonuniform (i.e., the velocity and flow depth vary in both time and space). Input (rainfall) to the flow is distributed over the flow surface.

Overland flow can be described mathematically by theoretical hydrodynamic equations attributed to St. Venant (Huggins and Burney[16]). These equations are based on the fundamental laws of conservation of mass (continuity) and conservation of momentum applied to a control volume or fixed section of channel with the assumptions of one-dimensional flow, a straight channel, and a gradual slope. With these assumptions, a uniform velocity distribution and a hydrostatic pressure distribution can be assumed, resulting in quasi linear partial differential equations. Detailed derivations of continuity and momentum equations as they apply to unsteady, nonuniform flow can be found in Strelkoff.[26]

Lighthill and Whitham,[27] cited by Huggins and Burney,[16] proposed that the dynamic terms in the momentum equation had negligible influence in cases in which backwater effects were absent. Neglecting these terms yields a quasi steady approach known as the kinematic wave approximation. The kinematic approximation is composed of the continuity equation

$$\frac{\delta y}{\delta t} + \frac{\delta Q}{\delta x} = q - f \qquad (1.3)$$

and a flow (depth-discharge) equation of the general form

$$Q = ay^m \qquad (1.4)$$

where α and m are parameters. The flow equation can be one describing laminar or turbulent channel flow, with the overland flow plane represented by a wide channel. Overton[28] analyzed 200 hydrographs for relatively long, impermeable planes and found that flow was turbulent or transitional. Foster et al.[29] concluded that both Manning and Darcy-Weisbach flow equations were satisfactory for describing overland flow on short erodible slopes.

The most commonly used flow equation for overland flow is the Manning equation, which can be written for overland flow as

$$Q = \frac{1}{n} y^{5/3} S^{1/2} \qquad (1.5)$$

where Q is the discharge (m^3/s/m of width), n is the roughness coefficient, y is the flow depth (m), and S is the slope of energy gradeline, usually taken as surface slope (decimal). Values of Mannings n factor vary from 0.02 for smooth pavement to 0.40 for average grass cover. Mannings n values are tabulated in a variety of sources (e.g., Novotny and Olem[30] and Linsley et al.[15]).

Woolhiser and Liggett[31] developed an accuracy parameter to assess the effect of neglecting dynamic terms in the momentum equation

$$k = \frac{S_o L}{HF^2} \qquad (1.6)$$

where k is a dimensionless parameter, S_o is the bed slope, L is the length of bed slope, H is the equilibrium flow depth at the outlet, and F is the equilibrium Froude number for flow at the outlet. For values of k greater than 10, very little advantage in accuracy is gained by using the momentum equation in place of a depth-discharge relationship. Because k is usually much greater than 10 in virtually all overland flow conditions, the kinematic wave equations generally provide an adequate representation of the overland flow hydrograph (Huggins and Burney[16]).

Another approach to translating rainfall excess into a hydrograph is the unit hydrograph (UH) approach, proposed by Sherman.[32] The UH results from one unit (e.g., cm, mm) of rainfall excess generated uniformly over a watershed at a uniform rate during a specified period of time. The following assumptions are inherent in the UH technique (Huggins and Burney[16]): (1) excess is applied with a uniform spatial distribution over the watershed during the specified time period, (2) excess is applied at a constant rate, (3) time base of the hydrograph of direct runoff is constant, (4) discharge at any given time is directly proportional to the total amount of direct runoff, and (5) the hydrograph reflects all combined physical characteristics of the watershed.

A UH is typically developed through analysis of measured rainfall-runoff data but can also be generated synthetically when rainfall-runoff data are not available. In developing a UH from measured data, an average UH from several storms of the same duration rather than a single storm should be developed (Linsley et al.[15]). The average UH should be determined by computing an average peak discharge and time to peak and then giving the UH a shape that is similar to the measured hydrographs.

One common method for developing synthetic UHs is to use formulas that relate hydrograph features, such as time of peak, peak flow, and time base, to watershed characteristics. For example, the SCS synthetic hydrograph is triangular. There are equations for computing time to peak, peak discharge, and time base of the hydrograph. Detailed information about developing unit hydrographs is included in many hydrology books.

The usefulness of unit hydrographs with respect to NPS pollution applications is limited. One assumption of UH theory is that the hydrograph reflects all combined physical characteristics of the watershed. Most NPS pollution applications are concerned with evaluating the potential of alternative management schemes to control NPS pollution on a watershed or land unit. Changing management practices in a watershed changes physical characteristics of the watershed that will, in most cases, affect the runoff hydrograph, thus changing the UH.

1.2.3   SOIL WATER MOVEMENT

Water moves into the soil profile through infiltration and through capillary movement from groundwater. Water moves out of the soil profile through leaching into groundwater, through plant uptake, and through evaporation at the soil surface. Three useful terms in describing the continuum of soil moisture content are saturation, field capacity, and wilting point. Saturation refers to the condition in which all soil pores are filled with water. This condition does not occur in the field because, typically, some air is trapped in the soil pores. Field saturation of agricultural soils varies between $0.8\theta_s$ and $0.9\theta_s$ (Slack[33]), where θ_s is saturated moisture content. Field saturation varies with initial moisture content and rainfall intensity as well as soil texture (Slack and Larson[34]). When soil is saturated, matric potential is zero and water moves because of gravity.

The term field capacity is used to describe the moisture content at which free drainage from gravity ceases, traditionally considered to occur 2–3 days after rain or irrigation. Factors that affect redistribution of moisture, and thus field capacity, include the following (Hillel[35]): soil texture, type of clay, organic matter content, depth of wetting and antecedent moisture, presence of impeding layers, and evapotranspiration. Field capacity is more identifiable in coarse-textured soils than in medium- or fine-textured soils because clayey soils hold more water longer than sandy soils. Well-graded soils, with a wide distribution of pore sizes, also allow moisture movement for some time. Field capacity may vary from about 4% (mass basis) in sands to about 45% in heavy clay soils, and up to 100% or more in some organic soils (Hillel[35]).

Permanent wilting point was traditionally considered to be the soil water content below which plant activity ceases. Wilting point was traditionally associated with a matric potential of -1500 kPa. The water held by a soil between field capacity and permanent wilting was considered as available water for plants. In recent years, the dynamic nature of the soil-plant-atmosphere system has been more fully recognized and investigated, leading to replacement of the traditional view that field capacity, wilting point, and available water are soil constants. The traditional view is still helpful in providing a general understanding of soil moisture.

Soil moisture content and movement are important concepts for NPS pollution for two reasons. Soil moisture content is a major factor in determining how much precipitation infiltrates into the soil and how much is available for runoff. The role of runoff in NPS pollution was described earlier. In addition, soil moisture movement influences groundwater contamination. Potential contaminants that are water-soluble, such as phosphorus, nitrate and pesticides, dissolved in percolating soil water, can move through the root zone and potentially to groundwater.

In agricultural settings, leaching is usually defined as water movement beyond the root zone. It is not typically equivalent to movement into an aquifer. Leaching occurs most often when soil moisture is above field capacity and water is moving primarily because of gravitational forces. Leaching is a concern for NPS pollution because dissolved constituents, such as nitrate and pesticide residues, are transported with leachate. Leaching is also used to refer to downward movement of liquid from runoff and waste storage ponds and lagoons, another potential source of groundwater contamination.

Soil water varies in the energy with which it is retained in the soil. Total soil water potential describes the work required to move an incremental volume of water from some reference state. Total soil water potential, Ψ, is the sum of other potentials

$$\Psi = \Psi_g + \Psi_p + \Psi_o + \Psi_n \tag{1.7}$$

where Ψ_g is the gravitational potential, Ψ_p is the matric or pressure potential, Ψ_o is the osmotic potential, and Ψ_n is the pneumatic potential. Potentials are expressed in units of pressure (e.g., kPa) or units of head (e.g., cm).

Gravitational potential is due to gravitational forces and is determined by position. Matric, or pressure, potential is due to the attraction of soil surfaces for water as well as to the influence of soil pores and the curvature of the soil-water interface. Osmotic potential is a function of solutes in the soil water. The presence of solutes decreases the potential energy of pure soil water. This has an important impact on plant uptake of water through roots but does not influence soil water flow appreciably because solutes can move with the water. Pneumatic potential refers to air pressure. It is usually considered to be uniform throughout the soil profile and is ignored in characterizing soil water flow. For cases where these assumptions are not justified, solutions for two-phase flow have been developed by a number of authors (e.g., McWhorter,[36] Brustkern and Morel-Seytoux[37]).

Soil moisture movement, or flux, is directly proportional to the hydraulic gradient (also called total potential gradient) and can be described by Darcy's equation

$$q_s = -K \frac{\delta H}{\delta s} \tag{1.8}$$

where q_s is the flux or volume of water moving through the soil in the s-direction per unit area per unit time ($L^3 L^{-2} T^{-1}$), K is the hydraulic conductivity (L/T), and $\delta H/\delta s$ is the hydraulic gradient in the s-direction. Hydraulic head, H, is the same as total soil water potential, except it is expressed in units of head of water. If osmotic and pneumatic potentials are assumed negligible, as discussed earlier, the hydraulic head, H, is the sum of the pressure head, h, and the elevation (or gravitational) head, z. If the datum is taken at the soil surface, then

$$H = h - z \qquad (1.9)$$

where z is the distance measured positively downward from the surface.

Hydraulic conductivity is a function of moisture content. The matric potential is also a function of moisture content, described by the soil water characteristic curve (Fig. 1.5). Matric potential is considered to be a continuous function of water content so that it is positive in a saturated soil below the water table and negative in an unsaturated soil. Matric potential becomes less negative as soil moisture content increases. The water content in a soil at a given potential depends upon the wetting and drying history of the soil (Figure 1.5). The difference between the drying curve, also called desorption, water retention, or water release, and the wetting curve, also called sorption or imbibition, is caused by hysteresis. The moisture content during drying is

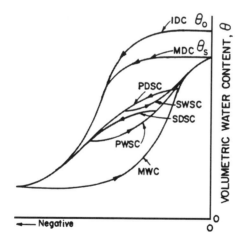

FIGURE 1.5 Soil water characteristic curve, indicating typical hysteresis curves, where IDC is the initial drainage curve, MWC and MDC are main wetting and drainage curves, respectively, and PWSC and PDSC are primary wetting and drainage scanning curves, and SWCS and SCSC are secondary wetting and drainage scanning curves. (From Skaggs, R. W. and Khaleel, R., Infiltration, in *Hydrologic Modeling of Small Watersheds,* Haan, C. T., Johnson, H. P., and Brakensiek, D. L., Eds., ASAE, St. Joseph, MI, 1982, 119. With permission.)

greater than during wetting in hysteretic soils. The change in volumetric water content per unit change in matric potential, $d\theta/dh$, is termed the soil water capacity, $C(h)$.

The continuity, or conservation of mass, equation for soil water flow in the vertical direction can be written as (Skaggs and Khaleel[38])

$$\frac{\delta\theta}{\delta t} = -\frac{\delta q_z}{\delta_z} \tag{1.10}$$

where θ is the volumetric moisture content (L^3/L^3), t is time (T), and q_z is water flux in the z-direction. Combining Darcy's equation with the continuity equation yields the general equation of flow in porous media, known as the Richards[39] equation, written for the vertical direction:

$$C(h)\frac{\delta h}{\delta t} = \frac{\delta}{\delta z}\left[K(h)\frac{\delta h}{\delta z}\right] - \frac{\delta K}{\delta z} \tag{1.11}$$

This equation was developed with the assumptions of no resistance to soil air movement and constant air pressure throughout the soil profile. With appropriate boundary and initial conditions, Richards' equation can be solved to describe moisture movement in porous media as a function of space and time. Richards' equation can be written in terms of h, as above, or in terms of moisture content, θ. The h-based equation includes two soil parameters, $C(h)$ and $K(h)$, whereas the θ-based equation includes the soil water diffusivity, $D(\theta)$, and $K(\theta)$. These soil parameters are related for unsaturated soil by $D = K/C$. For most soils, all three parameters vary markedly with water content or pressure head (Skaggs and Khaleel[38]).

1.2.4   INFILTRATION

Infiltration is defined as the entry of water from the surface into the soil profile. From a ponded surface or a rainfall situation, infiltration rate decreases over time and asymptotically approaches a final infiltration rate (Figure 1.6). The final infiltration rate is approximately equal to the saturated hydraulic conductivity, K_s, of the soil. The amount and rate of infiltration depend on infiltration capacity of the soil and the availability of water to infiltrate. Infiltration capacity is influenced by soil properties that govern water movement in soil, including $K(h)$, $C(h)$, and $D(\theta)$. Soil structure or pore size affects infiltration capacity, particularly during early stages of infiltration. The wider the range of pore sizes, the more gradual the change in the infiltration rate. Soil texture influences infiltration capacity with coarser soils having higher capacity (Figure 1.7) than finer-textured soils. Initial soil moisture content influences infiltration rate strongly at the beginning of an infiltration event (Figure 1.8) and less as the event continues. Lower initial soil moisture corresponds to a higher initial infiltration rate because of higher hydraulic gradients and more available storage volume. After the soil becomes wetted during the infiltration event, the effect of initial soil moisture virtually disappears from the infiltration rate but influences the cumulative infiltration because of higher initial rates.

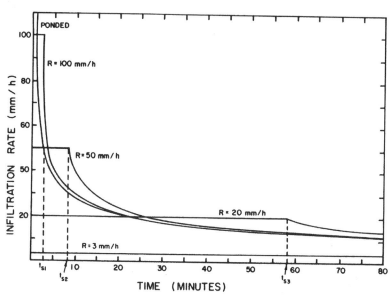

FIGURE 1.6 Predicted infiltration rates for a deep homogeneous Geary silt loam profile for constant surface application rates and for a shallow ponded surface. The initial water contant was uniform at $\theta_i = 0.26$ which corresponds to $h_i = -750$ cm of water. (From Skaggs, R. W. and Khaleel, R., Infiltration, in *Hydrologic Modeling of Small Watersheds,* Haan, C. T., Johnson, H. P., and Brakensiek, D. L., Eds., ASAE, St. Joseph, MI, 1982, 119. With permission.)

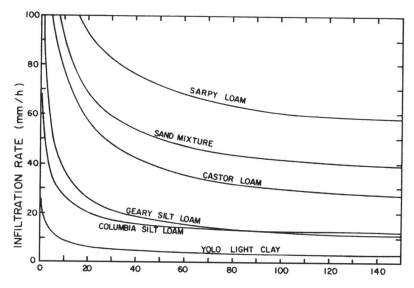

FIGURE 1.7 Predicted infiltration rates from numerical solutions to the Richards equation for deep soils with a shallow ponded surface. (From Skaggs, R. W. and Khaleel, R., Infiltration, in *Hydrologic Modeling of Small Watersheds,* Haan, C. T., Johnson, H. P., and Brakensiek, D. L., Eds., ASAE, St. Joseph, MI, 1982, 119. With permission.)

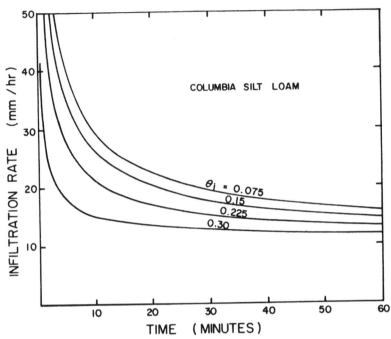

FIGURE 1.8   Predicted infiltration rates for a deep Columbia silt loam with different initial water contents. Saturated volumetric water content for this soil is $\theta_s = 0.34$. (From Skaggs, R. W. and Khaleel, R., Infiltration, in *Hydrologic Modeling of Small Watersheds*, Haan, C. T., Johnson, H. P., and Brakensiek, D. L., Eds., ASAE, St. Joseph, MI, 1982, 119. With permission.)

The actual infiltration rates and volumes that occur are also a function of the amount of water available to be infiltrated (i.e., precipitation or ponded water). Rainfall intensity affects infiltration rate (Figure 1.6). If the infiltration capacity of the soil is exceeded by the rainfall intensity (L/T), water will pond on the soil surface and the infiltration rate will equal the infiltration capacity. If the rainfall rate is less than the saturated hydraulic conductivity of the soil, the infiltration rate will equal the rainfall rate and ponding will not occur.

Surface conditions, including roughness, vegetation characteristics, and surface sealing, affect infiltration rates. Standing vegetation can intercept rainfall, which can then evaporate or drip to the soil surface. Residue on the soil surface can also intercept rainfall and affect infiltration rates. Roots of vegetation can affect the macroporosity of the soil and, thus, infiltration rates.

Surface seals form as wet soil aggregates and are broken down by raindrop impact and slaking (McIntyre[40]). Surface seals reduce infiltration rates because they reduce the hydraulic conductivity of the surface layer of soil (Figure 1.9). Examples of measured reductions in infiltration rates caused by surface sealing include 25 to 35% for sandy loam to silty clay loam and 75% for a clay loam (Duley[41]), 20 to 30% (Mannering[42]), and up to 50% (Edwards and Larson[43]).

The equations for computing infiltration are those that govern soil moisture movement (Darcy's, continuity, and Richards). The pronounced nonlinear variation of the soil parameters K, C, and D with water content and the surface boundary condition are sources of difficulty in solving the Richards equation for infiltration (Skaggs and Khaleel[38]). In addition, variations in soil properties from point to point and with depth make it difficult to describe field conditions adequately.

In practice, approximate equations rather than the governing partial differential equations are used. Often, approximate equations are tested against results obtained through use of the Richards equation to determine validity of the equations. Approximate equations have been developed based on simplified concepts to express infiltration rate, f, and cumulative infiltration, F, in terms of time and certain soil properties (parameters). All approximate infiltration equations have the characteristic that for a ponded surface, the infiltration rate decreases rapidly with time during the early part of an infiltration event. Some approximate equations have been developed by applying the principles governing soil water movement for simplified boundary and initial conditions. The parameters in such models can be determined from soil water properties when they are available. Other models are strictly empirical and the parameters must be obtained from measured infiltration data or estimated using more approximate procedures.

For NPS pollution applications, physically-based equations with measurable parameters are usually the most appropriate because the objective in many NPS

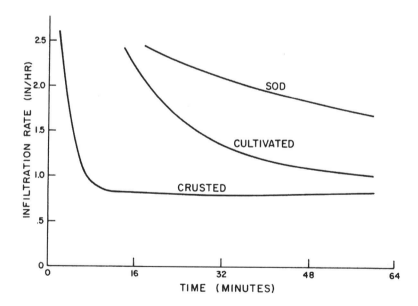

FIGURE 1.9 Effect of surface sealing and crusting due to rainfall impact on infiltration rate for a Zanesville silt loam. (From Skaggs, R. W. and Khaleel, R., Infiltration, in *Hydrologic Modeling of Small Watersheds,* Haan, C. T., Johnson, H. P., and Brakensiek, D. L., Eds., ASAE, St. Joseph, MI, 1982, 119. With permission.)

applications is to determine the impact of different management practices. Because those practices have not been installed, no data are available for applying an empirical equation. Two infiltration equations that have been used in NPS models are those of Holtan[44] and Green and Ampt.[45]

Holtan[44] developed an empirical equation based on a storage concept. After several modifications, the equation for infiltration capacity was presented as (Holtan and Lopez[46])

$$f_p = GI \cdot a \cdot SA^{1.4} + f_c \tag{1.12}$$

where f_p is the infiltration capacity (cm/hr), SA is the available storage in the surface layer (cm), GI is a crop growth index (percent maturity), a is an index of surface connected porosity which is a function of surface conditions and the density of plant roots (cm/hr/cm$^{1.4}$), and f_c is constant or steady-state infiltration rate (cm/hr). The available storage in the surface layer is determined as the difference between initial and final (field saturation) moisture content multiplied by the control depth.

Skaggs and Khaleel[38] reviewed the use of the Holtan equation; they found that its advantages include the relative ease of use for rainfall infiltration, and the input parameters can be obtained from a rather general description of the soil type and crop conditions. A major difficulty with the Holtan equation is the determination of the control depth on which to base SA. Holtan and Creitz[47] (cited by Skaggs and Khaleel[38]) suggested using the depth of the plow layer or the depth to the first impeding layer. Huggins and Monke[48] found that the effective control depth was highly dependent on both the surface condition and cultural practices used in preparing the seedbed. Experience with the Holtan equation indicates that, because of the generality of the inputs, its accuracy is questionable on a local or point-by-point basis in the watershed. Smith[49] argued that the infiltration curves are physically related to gradients and hydraulic conductivity far more than to soil porosity and that the Holtan equation should not be expected to describe the process adequately.

The Green and Ampt[45] approach, although approximate, has a theoretical basis and uses measurable parameters. The original equation was derived for infiltration from a ponded surface into a deep, homogeneous soil profile with uniform initial water content. Water is assumed to enter the soil as slug flow resulting in a sharply defined wetting front that separates a zone that has been wetted from an unwetted zone. Mein and Larson[50] applied the Green-Ampt equation to rainfall conditions by determining cumulative infiltration at the time of surface ponding, F_p. The Green-Ampt equation with the Mein-Larson modification is a two-stage model. First, the time of ponding is estimated using the following equations

$$F_p = \frac{S_f M}{\dfrac{R}{K_s} - 1} \tag{1.13}$$

$$t_p = F_p/R \text{ for constant } R \tag{1.14}$$

where F_p is the cumulative infiltration at time of ponding (L), S_f is the wetting front suction, M is the initial moisture deficit (decimal), R is the rainfall intensity (L/T), K_s is the saturated hydraulic conductivity (L/T), and t_p is the time of ponding (T). If R is less than K_s, surface ponding will not occur, providing the profile is deep and homogeneous, and f will be equal to R.

The infiltration rate prior to time of ponding is equal to the rainfall rate. After ponding, the infiltration rate is computed as

$$f = f_p = K_s \left(\frac{MS_f}{F} \right) \text{ for } t > t_p \qquad (1.15)$$

where f is the infiltration rate (L/T), f_p is the infiltration capacity under ponded conditions, and F is the cumulative infiltration (L).

The Green-Ampt-Mein-Larson (GAML) infiltration model has been used increasingly in recent years in NPS models. It has replaced other more empirical infiltration equations as well as being a choice over the curve number approach for computing rainfall excess. Researchers, e.g., Rawls et al.,[51] Brakensiek and Rawls,[52] Rawls and Brakensiek,[53] have developed improved estimates of the parameters in the GAML model.

1.2.5 GROUNDWATER

A cross-section of the subsurface profile (Figure 1.10) illustrates a series of subsurface zones through which water can move. The vadose zone is composed of the root zone and the unsaturated zone extending to the saturated zone. The root zone is usually unsaturated, except during periods of high infiltration of rainfall or irrigation. The thickness of the unsaturated zone varies due to geology, season, and other factors. Below the vadose zone is the saturated zone, or groundwater, in which all pores are filled with water. The upper bound of the saturated zone is the water table.

There are several different types of geologic formations that may contain water. The following descriptions are drawn from Novotny and Olem,[30] Shaw,[54] and Serrano.[55] An aquifer is a geologic formation saturated by water that yields appreciable quantities of water that can be economically used and developed. If the upper boundary of an aquifer is the water table, the aquifer is classified as unconfined, or phreatic (Figure 1.11). The water level in a well in an unconfined aquifer will rise to the level of the surrounding water table. Confined aquifers, also known as artesian or pressure aquifers, are bounded above and below by formations with significantly lower hydraulic conductivity than the aquifer. The confining layers cause a confined aquifer to be under pressure. The water level in a well in a confined aquifer will rise to the level of the hydraulic head at the upstream end of the confined aquifer. If the hydraulic head is higher than the ground surface, the well will be artesian, or free-flowing. Aquitards are geologic formations that are not permeable enough for economic development as a groundwater source. An aquiclude is a formation that stores water but is incapable of transmitting, (e.g., clays).

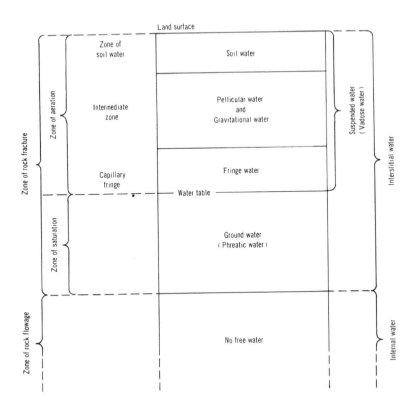

FIGURE 1.10 Divisions of subsurface water. (From SCS, *Groundwater, Section 18. National Engineering Handbook,* U.S. Soil Conservation Service, GPO, Washington, DC, 1968.)

Aquifers and aquitards can exist in layers with an unconfined aquifer on top and underlain by one or more confined zones. The top unconfined aquifer, often called a shallow aquifer, is most susceptible to NPS pollution and contamination.

Flow in groundwater systems is usually slow. Typical velocities may range from less than 1 cm/yr in tight clays to more than 100 m/yr in permeable sand and gravel (Novotny and Olem[30]). Todd[56] indicated that the normal range for groundwater velocities is 1.5 m/yr to 1.5 m/day. However, highly permeable glacial outwash deposits, fractured basalts and granites, and cavernous limestone formations may allow much higher velocities.

Groundwater flow rates depend on aquifer properties such as hydraulic conductivity. Typical hydraulic conductivity values of some formations are (Novotny and Olem[30]): $10^{-6} - 10^{-4}$ cm/sec for clay, sand, and gravel mixes; $10^{-3} - 0.1$ cm/sec for glacial outwash; $10^{-6} - 0.01$ cm/sec for fractured or weathered rock (aquifers); $10^{-6} - 10^{-3}$ cm/sec for sandstone; and $<10^{-8}$ cm/sec for dense solid rock. If the hydraulic conductivity is uniform at all points within the aquifer, the formation is homoge-

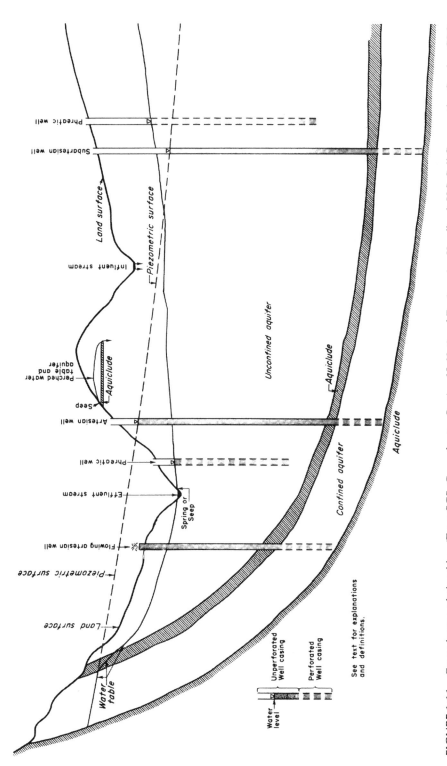

FIGURE 1.11 Groundwater relationships. (From SCS, *Groundwater, Section 18. National Engineering Handbook*, U.S. Soil Conservation Service, GPO, Washington, DC, 1968.)

23

neous. If the hydraulic conductivity varies with location, the formation is heterogeneous. The aquifer is isotropic if the hydraulic conductivity is the same in all
directions. The hydraulic conductivity varies with direction in an anisotropic aquifer.

Groundwater and surface water are interrelated through recharge and discharge.
Groundwater is recharged from movement of soil moisture through the vadose zone
to the saturated zone or through areas where the waterbearing formation is exposed
to the atmosphere. Recharge of groundwater also occurs from surface water bodies.
Recharge rates are highly variable.

Natural discharge from groundwater occurs through springs, spring-fed lakes,
wetlands, stream channels, and oceans. The relatively low flow velocities of groundwater and its long residence time produce a continuous discharge flow rate to streams
and lakes (Serrano[55]). This phenomenon maintains a minimum water level in lakes
and a minimum flow rate called base flow in streams during periods without rainfall.
Base flow can last for several weeks or even months in some cases. Discharge from
groundwater also occurs through pumping for a variety of uses.

1.2.5.1 Groundwater Flow Estimation

The governing equation for groundwater flow is Richards equation, just as it was for
soil moisture movement. When considering soil moisture movement earlier, Richards
equation was written for flow in the vertical direction. The equation can be expanded
to three dimensions and describe flow for an anisotropic aquifer

$$K_x \frac{\delta^2 H}{\delta x^2} + K_y \frac{\delta^2 H}{\delta y^2} + K_z \frac{\delta^2 H}{\delta z^2} = S_s \frac{\delta H}{\delta t} \tag{1.16}$$

where S_s is specific storage (L^{-1}), defined as the volume of water that a unit volume
of porous medium releases from storage per unit change in hydraulic head, and other
variables are as defined previously. For a homogeneous, isotropic material, the
hydraulic conductivities are equal and constant and the equation reduces to

$$K \left(\frac{\delta^2 K}{\delta x^2} + \frac{\delta^2 H}{\delta y^2} + \frac{\delta^2 H}{\delta z^2} \right) = S_s \frac{\delta H}{\delta t} \tag{1.17}$$

For steady flow (i.e., $\delta H / \delta t$ is zero), the equation simplifies to the Laplace equation

$$\frac{\delta^2 H}{\delta x^2} + \frac{\delta^2 H}{\delta y^2} + \frac{\delta^2 H}{\delta z^2} = 0 \tag{1.18}$$

The solution to the Laplace equation gives the hydraulic head in terms of x, y, and z.
The solution of the full equation for transient flow in an anisotropic medium gives
H in terms of t as well as x, y , and z.

In practice, groundwater modeling applications have often used simplified
boundary conditions (Shaw[54]). In addition, assumptions of an isotropic aquifer or
steady-flow conditions or both are often made to facilitate the solution and yet provide acceptable accuracy.

REFERENCES

1. Brooks, K. N., Ffolliott, P. F., Gregersen, H. M., and Thomas, J. L., Hydrology and the Management of Watersheds, First edition, Iowa State University Press, Ames, 1992.
2. Eagleson, P. S., Dynamic Hydrology, McGraw-Hill, New York, 1970.
3. Laws, J. O. and Parsons, D. A., The relation of raindrop-size to intensity, Trans. Am. Geophys. Union, 24, 452, 1943.
4. Laws, J. O., Measurements of the fall-velocity of water-drops and raindrops, *Trans. Am. Geophys. Union,* 22, 709, 1941.
5. Hershfield, D. N., Rainfall Frequency Atlas of the United States, U.S. Weather Bureau Technical Paper 40, May, 1961.
6. Richardson, C. W., A comparison of three distributions for the generation of daily rainfall amounts, in *Statistical Analysis of Rainfall and Runoff,* Singh, V. P., Ed., Water Resources Publications, Littleton, CO, 1981, 67.
7. Mills, W. C., Stochastic modeling of rainfall for deriving distributions of watershed input, in *Statistical Analysis of Rainfall and Runoff,* Singh, V. P., Ed., Water Resources Publications, Littleton, CO, 1981, 103.
8. Jimoh, O. D. and Webster, P., The optimum order of a Markov chain model for daily rainfall in Nigeria, *J. Hydrol.,* 185, 45, 1996.
9. Khaliq, M. N. and Cunnane, C., Modelling point rainfall occurrences with the modified Bartlett-Lewis Rectangular Pulses Model, *J. Hydrol.,* 180, 109, 1996.
10. Hutchinson, M. F., A point rainfall model based on a three-state continuous Markov occurrence process, *J. Hydrol.,* 114, 125, 1990.
11. Rodriguez-Iturbe, I., Cox, D. R. and Isham, V., Some models for rainfall based on stochastic point processes, *Proc. R. Soc. London, A,* 410, 269, 1987.
12. Mellor, D., The modified turning bands (MTB) model for space-time rainfall. I. Model definition and properties, *J. Hydrol.,* 175, 113, 1996.
13. Singh, V. P., Ed., Statistical Analysis of Rainfall and Runoff, Water Resources Publications, Littleton, CO, 1981.
14. O'Connell, P. E. and Todini, D., Eds. Special issue - Modelling of rainfall, flow and mass transport in hydrological systems, *J. Hydrol.,* 175, 1996.
15. Linsley, R. K., Kohler, M. A., and Paulhus. J. L. H., Hydrology for Engineers, McGraw-Hill Book Company, London, 1988.
16. Huggins, L. F. and Burney, J. R., Surface runoff, storage, and routing, in *Hydrologic Modeling of Small Watersheds,* Haan, C. T., Johnson, H. P., and Brakensiek, D. L., Eds., ASAE, St. Joseph, MI, 1982, 167.
17. SCS, Hydrology, Section 4. National Engineering Handbook, U.S. Soil Conservation Service, GPO, Washington, DC, 1972.
18. Rasmussen, T. C. and Evans, D. D., Water infiltration into exposed fractured rock surfaces, *Soil Sci. Soc. Am. J.* 57, 324, 1993.
19. Edwards, D. R. and T. C. Daniel, Abstractions and runoff from fescue plots receiving poultry litter and swine manure, *Trans. ASAE,* 36, 405, 1993.
20. Hauser, V. L. and O. R. Jones, Runoff curve numbers for the southern High Plains, *Trans. ASAE,* 34, 142, 1991.
21. Bondelid, T. R., R. H. McCuen, and T. J. Jackson, Sensitivity of SCS models to curve number variation, *Water Resources Bull.,* 18, 111, 1982.
22. Hawkins, R. H., The importance of accurate curve numbers in the estimation of storm runoff, *Water Resources Bull.,* 11, 887, 1975.
23. Bingner, R. L., Comparison of the components used in several sediment yield models, *Trans. ASAE,* 33, 1229, 1990.

24. Ma, Q. L., Wauchope, R. D., Hook, J. E., Johnson, A. W., Truman, C. C., Dowler, C. C., Gascho, G. J., Davis, J. G., Sumner, H. R., and Chandler L. D., GLEAMS, Opus, and PRZM-2 model predicted versus measured runoff from a coastal plain loamy sand, *Trans. ASAE,* 41, 77, 1998.

25. Chung, S. O., A. D. Ward, and Schalk, C. W., Evaluation of the hydrologic component of the ADAPT water table management model, *Trans. ASAE,* 35, 571, 1992.

26. Strelkoff, T., One-dimensional equations of open channel flow, *Trans. Hyd. Div. ASCE,* 95, 861, 1969.

27. Lighthill, M. J. and Whitman, G. B., On kinematic waves 1. *Proc. Royal Soc., London,* A, 229, 281, 1955.

28. Overton, D. E., Kinematic flow on long impermeable planes, *Water Resources Bull.,* 8, 1198, 1972.

29. Foster, G. R., Huggins, L. F., and Meyer, L. D., Simulation of overland flow on short field plots, *Water Resources Res.,* 4, 1179, 1968.

30. Novotny, V. and Olem, H. Water Quality: Prevention, Identification, and Management of Diffuse Pollution, Van Nostrand Reinhold, New York, 1994.

31. Woolhiser, D. A. and Liggett, J. A., Unsteady one-dimensional flow over a plane—the rising hydrograph, *Water Resources Research,* 3, 753, 1967.

32. Sherman, L. K., Stream flow from rainfall by the unit-graph method, *Eng. New-Rec.,* 108, 501, 1932.

33. Slack, D. C., Modeling infiltration under moving sprinkler irrigation systems, *Trans. ASAE,* 23, 596, 1980.

34. Slack, D. C. and Larson, C. L., Modeling infiltration: the key process in water management, runoff, and erosion, in *Tropical Agricultural Hydrology,* Lal, R. and Russell, E. W., Eds., John Wiley and Sons, Ltd., New York, 1981.

35. Hillel, D., Soil and Water: Physical Principles and Processes, Academic Press, New York, 1971.

36. McWhorter, D. B., Vertical flow of air and water with a flux boundary condition, *Trans. ASAE,* 19, 259, 1976.

37. Brustkern, R. L. and H. J. Morel-Seytoux, Description of water and air movements of soils, *J. Hydrol.,* 24, 21, 1975.

38. Skaggs, R. W. and Khaleel. R., *Infiltration,* in *Hydrologic Modeling of Small Watersheds,* Haan, C. T., Johnson, H. P., and Brakensiek, D. L., Eds., ASAE, St. Joseph, MI, 1982, 119.

39. Richards, L. A. Capillary conduction through porous mediums, *Physics,* 1, 313, 1931.

40. McIntyre, D. S., Permeability measurements of soil crusts formed by raindrop impact, *Soil Sci.,* 85, 185, 1958.

41. Duley, F. L., Surface factors affecting the rate of intake of water by soils, *Soil Sci. Soc. Am. Proc.* 4, 60, 1939.

42. Mannering, J. V., The relationship of some physical and chemical properties of soils to surface sealing, unpublished Ph.D. Thesis, Purdue University, Lafayette, IN, 1967.

43. Edwards, W. M. and Larson, W. E., Infiltration of water into soils as influenced by surface seal development, *Trans. ASAE,* 12, 463, 1969.

44. Holtan, H. N., A concept for infiltration estimates in watershed engineering, *USDA-ARS Bull.* 41–51, 1961.

45. Green, W. H. and Ampt, G. A., Studies on soil physics. 1. The flow of air and water through soils, *J. Agric. Sci.,* 4, 1, 1911.

46. Holtan, H. N. and Lopez, N. C., USDAHL-70 Model of watershed hydrology, Tech. Bull. No. 1435, USDA-ARS, 1971.

47. Holtan, H. N. and N. R. Creitz, Influence of soils, vegetation and geomorphology on elements of the flood hydrograph, in *Proc. Symposium on Floods and Their Computation,* Leningrad, Russia, 1967.

48. Huggins, L. F. and Monke, E. J., The Mathematical Simulation of the Hydrology of Small Watersheds, TR1, Purdue Water Resources Research Center, Lafayette, IN, 1966.

49. Smith, R. E., Approximations for vertical infiltration rate patterns, *Trans. ASAE,* 19, 505, 1976.

50. Mein, R. G. and Larson, C. L., Modeling infiltration during a steady rain, *Water Resources Res.,* 9, 384, 1973.

51. Rawls, W. J., Stone, J. J. and Brakensiek. D. L., Infiltration, in *Water Erosion Prediction Project: Hillslope Profile Version,* Lane, L. J. and Nearing, M. A., Eds., National Soil Erosion Laboratory Report No. 2., USDA-ARS, West Lafayette, IN, 1989.

52. Brakensiek, D. L. and Rawls, W. J., Agricultural management effects on soil water processes Part II. Green-Ampt parameters for crusting soils, in *Proc. Specialty Conf. Adv. Irrig. Drain.,* ASCE, Jackson, WY, 1983.

53. Rawls, W. J. and Brakensiek, D. L., Comparison between Green-Ampt and Curve Number runoff predictions, *Trans. ASAE,* 29, 1597, 1986.

54. Shaw, E. M., Hydrology in Practice, Chapman & Hall, London, 1994.

55. Serrano, S. E., Hydrology for Engineers, Geologists, and Environmental Professionals, HydroScience Inc., Lexington, KY, 1997.

56. Todd, D. K., Groundwater Hydrology, 2d. ed., Wiley, New York, 1980.

2 Soil Erosion and Sedimentation

Mark A. Nearing, L. D. Norton, and Xunchang Zhang

CONTENTS

1-56670-222-4/01/$0.00+$.50
© 2001 by CRC Press LLC

2.1   INTRODUCTION

Soil erosion includes the processes of detachment of soil particles from the soil mass and the subsequent transport and deposition of those sediment particles on land surfaces. Erosion is the source of 99% of the total suspended solid loads in waterways in the United States[1] and undoubtedly around the world. Somewhat over half of the approximately 5 billion tons of soil eroded every year in the United States reaches small streams. This sediment has a tremendous societal cost associated with it in terms of stream degradation, disturbance to wildlife habitat, and direct costs for dredging, levees, and reservoir storage losses. Sediment is also an important vehicle for the transport of soil-bound chemical contaminants from nonpoint source areas to waterways. According to the USDA,[1] soil erosion is the source of 80% of the total phosphorus and 73% of the total Kjeldahl nitrogen in the waterways of the U.S. Sediment also carries agricultural pesticides. Solutions to nonpoint source pollution problems invariably must address the problem of erosion and sediment control. The purpose of this chapter is to discuss the basic processes of soil erosion as it occurs in upland areas. Most of the discussion is focused on rill and interrill erosion. Erosion modeling concepts are presented as a vehicle for discussing our current understanding of soil erosion by water, and some process-based soil erosion models are discussed and contrasted in some detail.

2.1.1   TERMINOLOGY

It is useful here to define some basic terms commonly used in formulating concepts relating to soil erosion. The term soil detachment implies a process description: the removal of one or many soil particles as a function of some driving force (erosivity) such as raindrop impact or shear stresses of flowing water or wind. For purposes of clarity we distinguish between the terms soil and sediment. Soil is considered, for modeling purposes, to be material that is in place at the beginning of an erosion event. If the soil material is detached during an event, it is considered to be sediment. The terms sediment transport and deposition also imply process descriptions. Transport of sediment may be in terms of transport downslope by small-channel flow or it may refer to movement of soil particles across interrill areas via very shallow sheet flow or raindrop splash mechanisms.

The exact meaning of the term deposition has received considerable discussion in erosion literature. In the framework of an empirical erosion model, it is clear that deposition refers to the time-averaged amount of sediment (detached soil) that does not leave the boundaries of the area of interest. We refer to this as total deposition. In process-based models, the use of the term is dependent on how the process of deposition is represented in the source/sink term of the continuity equation and is related to the concept of transport capacity. In certain models, the deposition term represents a net movement of sediment to the bed from the flow, whereas, in other models, deposition is considered to be an instantaneous and continuous process that occurs at all points on the hillslope, including those portions that experience a net flux of sediment to the flow from the bed. This process will be discussed in more detail below.

What is considered to be a sediment source is somewhat dependent on the scale of the process descriptors. Often, in erosion representations, interrill areas are modeled as sediment yield areas that feed sediment to small channels, or rills, for subsequent downslope transport. In this case, the rill flow is considered to be the primary transport mechanism, and interrill sediment movement as a downslope transport mechanism is neglected. It is argued that this approach is justified given the relatively short transport distances of sediment in interrill areas versus the potential longer transport distances of sediment in rills. This argument is probably reasonable if interrill sediment delivery rates to rills, including accurate sediment size distributions, are accurately estimated. Most often, an empirical sediment delivery term and size distribution function are used for estimating sediment delivered to rills from interrill areas. Recently, attempts have been made to model the processes of detachment, transport, and deposition on interrill areas to provide estimates of sediment delivery to rills.[2–4]

Because significant deposition occurs within field boundaries, knowledge of soil loss on the field (and also of soil loss models for erosion) is of limited value in terms of understanding nonpoint source sediment loadings. The sediment delivery ratio is the proportion of sediment that leaves an area relative to the amount of soil eroded on the area. If the interest is in terms of sediment delivery to waterways, then the sediment delivery ratio may represent the amount of sediment that reaches the waterway divided by the total erosion within the watershed. This ratio varies widely and depends on the size and shape of the contributing area; the steepness, length, and shape of contributing surfaces; sediment characteristics; buffer zones; storm characteristics; and land use.

2.1.2 MODELS

Models of soil erosion play critical roles in soil and water resource conservation and nonpoint source assessments, including sediment load assessment and inventory, conservation planning and design of sediment control, and the advancement of scientific understanding.

On-site measurement and monitoring of soil erosion is expensive and time consuming. Erosion events are intermittent, and long-term records would be required to measure the erosion from a specific site. For these reasons, erosion models are, in most cases, the only reasonable tools for making erosion assessment. The USDA Soil Conservation Service, for example, uses the Universal Soil Loss Equation in making periodic resource inventories of soil erosion over large land areas.[1]

Conservation planning is also based on erosion models. Models are helpful when the land use planner must decide whether a specified land management practice will meet soil loss tolerance goals. Design of hydrologic retention ponds, sedimentation ponds, and reservoirs make use of erosion predictions from models for design calculations. For example, an engineer would use an erosion model to assess the expected sediment delivery to a reservoir to estimate expected siltation rates in the reservoir. The designer could use the model to predict the effect of anticipated future land use changes on sediment delivery to the reservoir.

Erosion models play at least two roles with respect to the science of soil erosion. Erosion models are necessarily process integrators. Most often, our knowledge of erosion mechanisms from experimental data is limited in scope and scale. Information may sometimes be misleading in terms of the overall effects on large integrated systems where many processes act interdependently. If individual processes that are well described from erosion experiments are correctly integrated via a process-based model, the result can be used to study model predictions and to assess the behavior of the integrated system. Erosion models also help us to focus our research efforts—to see where gaps in knowledge exist and where to best direct our efforts to increase our overall erosion prediction capabilities.

A goal of most erosion models is to predict or estimate soil loss or sediment yield from specified areas of interest. Soil loss refers to a loss of soil from only the portion of the total area that experiences net loss. It does not integrate, and is not appropriate, to describe areas that contain net depositional regions. The time period considered depends on the objectives of the model, and thus may range from a small portion of a single storm event to a long-term average annual value. The Universal Soil Loss Equation (USLE),[5] for example, is an empirical model that provides estimates of average annual soil loss. The natural runoff plots used to develop the USLE were laid out on essentially uniform slope elements, whereby sediment deposition was considered to be negligible. In other words, the USLE does not address deposition or sediment yield; it is strictly a soil loss model. Other empirical models have been developed that incorporate the USLE for estimating soil loss, but also provide empirically based estimates of sediment yield. Sediment yield refers to the total amount of sediment leaving a delineated area or crossing a specified boundary over a specified time period. Thus, sediment yield is the balance between soil loss and net sediment deposition on the area of interest. The term sediment delivery is equivalent to sediment yield, although sediment delivery is sometimes used also to refer to the delivery of sediment from interrill areas to rills.

The two primary types of erosion models are process-based models and empirically based models. Process-based (physically based) models mathematically describe the erosion processes of detachment, transport, and deposition, and through the solutions of the equations describing those processes provide estimates of soil loss and sediment yields from specified land surface areas. Erosion science is not sufficiently advanced for there to exist completely process-based models that do not include empirical aspects. The primary indicator, perhaps, for differentiating process-based from other types of erosion models is the use of the sediment continuity equation discussed later in this chapter. Empirical models relate management and environmental factors directly to soil loss or sediment yields through statistical relationships. Lane et al.[6] provided a detailed discussion regarding the nature of process-based and empirical erosion models, as well as a discussion of what they termed conceptual models, which lie somewhere between the process-based and purely empirical models. Current research effort involving erosion modeling is weighted toward the development of process-based erosion models. On the other hand, the standard model for most erosion assessment and conservation planning is the empirically based USLE. Active research and development of USLE-based erosion prediction technology continues.

2.2   SOIL EROSION PROCESSES

2.2.1   CONCEPTUALIZATION OF RILL AND INTERRILL EROSION PROCESSES

The concept of differentiating between rill and interrill erosional areas outlines a useful, if somewhat arbitrary, division between dominant processes of erosion on a hillslope surface. In the original description of the processes, Meyer et al.[7] differentiated between areas of the hillslope dominated by shallow sheet flow and raindrop impact and those of small concentrated flow channels, which they termed rills. The concept is useful in terms of mathematical descriptions of erosion and serves as a basis for many process-based erosion simulation models. The concept is also useful in terms of focusing experimental research on the two primary sources of eroded soil. The separation of the two primary sediment sources facilitates the mathematical modeling of nonpoint source pollutants in surface runoff. However, the concept is somewhat arbitrary because it implies a clear delineation between dominant processes on a given area, where, in fact, overlap occurs. Flow depths on a hillslope would be more correctly described in terms of frequency distributions of depth, where processes tend more toward rill or interrill depending on the flow depth.[8] Nevertheless, the introduction of the concept of rill versus interrill sediment source areas is the cornerstone of current erosion research and development of process-based erosion prediction technology. It is the subdivision of the erosion process that opened the "black box" that was employed by earlier, statistically based erosion models such as the USLE[5].

Rills are conceived as being the primary mechanism of sediment transport in the downslope direction. Depths of flow in rills are considered to be relatively large (normally on the order of cm) compared with average broad sheet flow depths (on the order of mm). Detachment of soil in rills is primarily by scour, whereas the principal mechanism of detachment in interrill areas is by raindrop splash. Models of rill and interrill erosion generally treat interrill areas as being sediment feeds for rills. The rills then act to transport the sediment generated in the interrill areas and the soil detached by scour in the rills, down the slope.

2.2.2   RILL EROSION

The hydrodynamics of the surface flow of water is the driving force for detachment of soil in rills. The common parameters used to characterize the capacity of the flow to cause detachment are flow shear stress, τ, and streampower, ω. The flow shear stress is calculated directly from force balance relationships and is given by

$$\tau = \rho \, g \, h \, S \qquad \qquad (2.1)$$

where ρ (kg/m^3) is the density of water, g (m/s^2) is the acceleration of gravity, h (m) is depth of flow, and S is the bed slope. The exact equation for shear stress would include sin θ, where θ is the slope angle, in place of S, which is equal to tan θ; but, at low slopes, the two terms are approximately equal. Units of τ are Pa [kg/(m s^2)].

Streampower, as discussed by Bagnold,[9] is the rate of dissipation of flow energy to the bed per unit area. Calculation of streampower is given by

$$\omega = \tau u = \rho g q S \tag{2.2}$$

where u (m/s) is the average flow velocity, q (m^2/s) is unit discharge of flow, and units of ω are kg/s^3.

Either shear stress or streampower is generally used to characterize the detachment capacity of surface flow. Both terms are borrowed from analogous sediment transport capacity relationships developed for predicting bedload transport of sand in streams. There is no existing evidence that one term more accurately describes detachment capacity, and in fact, there is some evidence that neither accurately reflects detachment capacity under all conditions.[10-11]

Streampower and shear stress are functionally related. For the case of uniform sheet flow, and using the Chezy depth versus discharge relationship,

$$q = C h^{1.5} S^{0.5} \tag{2.3}$$

and steampower can be written as

$$\omega = \rho g C h^{1.5} S^{1.5} \tag{2.4}$$

where C is the Chezy hydraulic roughness coefficient. Thus, assuming the Chezy relationship to be correct, streampower is linearly related to the 3/2 power of shear stress for sheet flow.

The detachment rate of soil in rills by clear water (detachment capacity, D_{rc}) is a function of the driving force described by the hydrodynamics of the flow and resistance forces in the soil. Several types of functions have been used to describe this relationship. A commonly used form of the function for detachment rate capacity that uses flow shear stress is

$$D_{rc} = a(\tau - \tau_c)^b \tag{2.5}$$

where τ_c (Pa) is the critical shear stress of the soil, and "a" (s/m) and "b" (unitless) are coefficients. Both τ_c and "a", and possibly also "b," represent the resistance of the soil to detachment by flow. These are the rill erodibility parameters. It is important to note here that the values for rill erodibility for a given soil and condition will be dependent on the form of the equation describing detachment rate capacity. A linear relationship (b = 1) using stream power instead of shear stress in Equation 2.5 has also been used to describe detachment by flow water.[12]

2.2.3 INTERRILL EROSION

Raindrop impact is the mechanism responsible for detaching soil particles on interrill areas.[13] The physical characteristics of impacting raindrops influence the quantity and nature of detached soil materials. Overland flow, soil characteristics, canopy, and surface cover may also affect raindrop detachment.

Foster[14] developed a conceptual model of the delivery rate of detached particles from interrill areas to rill flow. Interrill sediment delivery may be limited by transport capacity at small slope steepness, especially on relatively rough surfaces. Detachment may be a constraint to sediment delivery on steeper slopes.

Several equations have been proposed for relating soil detachment to raindrop characteristics. Raindrop diameter and velocity were used as variables in empirical detachment formulas developed by Ellison[15] and Bisal.[16] The effect of a rainfall erosivity factor, EI, on soil detachment was evaluated by Free.[17] Park et al.[18] used rainfall momentum to predict splash erosion.

Kinetic energy was used in detachment formulas proposed by many scientists.[19–23] Kinetic energy, kinetic energy per unit of drop area, momentum and momentum per unit of drop area were factors suggested by Meyer[24] to be of potential importance to soil erosion. Kinetic energy and momentum per unit of drop circumference were identified by Al-Durrah and Bradford[25] as rainfall factors of possible significance. Gilley and Finkner[4] found that kinetic energy multiplied by the unit of drop circumference could be used to estimate soil detachment.

Natural rainfall contains drops with a distribution of diameters. Raindrop terminal velocity, in turn, varies with raindrop diameter.[26] The size distribution of raindrops is a function of rainfall intensity.[27] Mathematical models have been developed that predict raindrop size distribution and kinetic energy from rainfall intensity.[28–29] Thus, rainfall intensity must be considered when soil detachment is related to physically based raindrop parameters.

Meyer and Wischmeier[30] proposed an equation of the following form to relate interrill sediment delivery rate, D_i [kg/(m^2 s)] to effective rainfall intensity, I (mm/h)

$$D_i = K_i I^p \tag{2.6}$$

where K_i is an empirical interrill erodibility parameter and p is a regression coefficient. A value of 2 was suggested by Meyer[31] for the regression coefficient p. This suggestion was based on extensive data collection in the field using a rainfall simulator.

An equation with a form similar to Equation 2.6 was proposed by Rose et al,[32] but that equation actually represents a different process. The equation was

$$e = a I^p \tag{2.7}$$

where e is rainfall detachment rate, and a and p are empirical parameters. Equation 2.6 is an interrill sediment yield relationship. It combines processes of detachment, transport, and deposition to describe empirically the delivery of sediment from interrill areas to (presumably) a small concentrated flow area (an incised or nonincised rill) where it might be transported downslope. Rose's equation, on the other hand, was intended to describe only the process of detachment by splash. Deposition in Rose's[32] model was described in a separate term essentially as a product of sediment concentration times settling velocity. In describing the model,[32] Rose indicated that the exponent, p, of Eq. 2.7 was probably close to the value of 2, based on the sediment delivery experiments of Meyer mentioned above. In later model formulations, the difference became apparent.[33] The difference results in lower values for p. Proffitt et al.[34] calculated values of p on the order of 0.7 to 0.9.

Slope has a significant effect on interrill sediment delivery, primarily because it influences the sediment transport capacity of the interrill flow. The general form that includes a slope factor, S_f, is[35]

$$D_i = K_i I^p S_f \qquad (2.8)$$

and Watson and Laflen[36] used a slope factor of

$$S_f = S^z \qquad (2.9)$$

where S is slope steepness (m/m) and z is a regression coefficient. Foster[14] identified the slope factor term as

$$S_f = 2.96 (\sin\theta)^{0.79} + 0.56 \qquad (2.10)$$

where θ is the interrill slope angle. This equation is normalized to a 9% slope (i.e., S is equal to one at $\tan\theta$ equal to 0.09). The slope factor term proposed by Liebenow et al.[37] was

$$S_f = 1.05\ 0.85\ e^{-4\ \sin\theta} \qquad (2.11)$$

This equation is normalized to a 1 to 1 slope, thus S is equal to one at $\tan\theta$ equal to 1.0 ($\theta = 45°$). The slope to which the slope adjustment function is normalized is relatively unimportant, as long as the interrill erodibility term, K_i, is calculated from experimental data in a way that is consistent with the model formulation. The product of rainfall intensity, slope gradient, and runoff rate has also been used in estimating interrill erosion.[38–39] The equations with runoff term is considered to be superior to that without runoff term because two processes (i.e., detachment by raindrop impact and transport by thin overland flow) are represented when runoff term is included. In addition, the inclusion of runoff term indirectly accounts for the effects of infiltration on soil loss rate. In the WEPP model, interrill sediment delivery is calculated as[38]

$$D_i = K_i I I_e S_f \qquad (2.12)$$

where I_e is the interrill runoff rate (m/s), and S_f is from Equation 2.11.

2.2.4 SEDIMENT TRANSPORT

Sediment in water is subjected to several forces, including gravity, buoyancy, and turbulence. Sediment moves downward toward the bed from gravity forces, whereas buoyancy and turbulent forces tend to support and suspend sediment particles. Large amounts of detached sediment can also tend to move by rolling, hopping, or sliding in proximity to the bed. In shallow flows (typical of interrill areas), raindrop impact can greatly enhance the turbulent suspension effect as well as keep greater portions of the bedload materials in motion. As flow depth increases (typical of flow in rills and ephemeral channels), rainfall effects become minimal. The capacity of a flow to transport sediment is conceptualized as being a balance between the rates of sediment falling to the bed and the maximum rate of lifting of sediment from the bed. Thus, for a given sediment type and set of flow characteristics, there will be some finite amount of sediment that the flow can carry. This level of sediment load is referred to as the sediment transport capacity. Sediment transport capacity of flowing water on a hillslope in general is a function of the slope steepness and flow discharge. Thus, transport capacity is higher on longer and steeper slopes and lesser on toeslopes and depressional areas. Transport capacity can also be altered by changes in soil roughness, crop residues, and standing plants, all of which affect overland flow hydraulics.

Sediment transport capacity concepts are used in most erosion models; the major difficulty in application is the selection of an acceptable sediment transport equation. There is a large group of equations for prediction of the sediment transport capacity of river flows; however, no widely accepted equation or set of equations has yet been developed for the shallow flows and nonuniform sediment typical of upland agricultural situations. A wide range of sediment transport relationships have been developed and tested.[40–44]

2.2.5   ERODED SEDIMENT-SIZE FRACTIONS AND SEDIMENT ENRICHMENT

The size distribution and surface area of the eroded sediment and of the sediment yield is important in erosion modeling both in terms of erosion (especially deposition) processes and prediction of the chemical-carrying capacity of the sediment. Fine particles, especially clay and organic matter, which have a large surface area and relatively high electrical surface charge, are the major adsorbents and vehicles for transporting agricultural chemicals of strongly adsorbed inorganic nutrients and organic pesticides. Dispersed clay particles and organic matter can be transported as far as water moves because of their low settling velocities. Thus, predicting the fine fraction of sediment is essential in estimating the chemical-carrying capacity of the sediment. With growing concern over surface water quality and continuing effort in modeling the transport of nonpoint source contaminants in surface water bodies, it becomes increasingly important to be able to estimate the capacity of sediment to carry adsorbed chemicals.

One simple way to estimate the chemical transport in sediment is to multiply chemical concentration of matrix soil by an enrichment ratio, which is considered to be greater than 1. This approach assumes no chemical exchange between adsorbents and runoff water in the course of transport. Enrichment ratio is defined as the ratio of the adsorbed chemical concentration in sediment to that in matrix soil. If the clay fraction is assumed to be the only adsorbents, the enrichment ratio can be calculated as the ratio of clay fraction in sediment to that in matrix soil. Note the enrichment ratio is calculated based on the total clay rather than dispersed clay fraction. Thus, the enrichment ratio does not necessarily reflect the potential of sediment for transporting adsorbed chemicals because the clay fraction that is transported in aggregates is deposited near its source areas.[14] Primarily, it is the clay fraction that is transported as primary clay particles, which poses the potential problem for downstream water body chemical contamination.

Studies have shown that most sediment is eroded and transported in aggregates, especially the clay portion of the sediment.[45–47] However, silt- and clay-sized particles may be enriched during any phase of the erosion process (detachment, transport, and deposition). The detachment process has a relatively smaller impact on the enrichment ratio compared with transport and deposition processes. The enrichment ratio can be understood in terms of the interrill-rill erosion concept. For interrill erosion, raindrop impact is the predominant detachment agent and shallow overland flow is the dominant transport force. Because of the limited transport capacity of thin overland flow, selective removal of fine particles tends to occur rapidly in interrill areas. The degree of enrichment depends on soil particle size distribution and aggregate stability, rainfall intensity, runoff rate, soil surface cover and vegetation, soil roughness, local topography, and water chemistry. The fraction of finer particles increases as rainfall intensity and slope gradient decrease and as surface cover and roughness increase because of a resultant reduction in transport capacity of thin overland flow.[46,48–49] Miller and Baharuddin[50] reported that sandy soils tend to have a greater enrichment ratio compared with clayey soils. This may be because sandy soils tend to be less well aggregated than other soils. High sodium exchange percentage and a low electrolyte concentration in soils also tend to enhance clay particle enrichment.

The size distribution of eroded sediment has been reported to change with time during a storm. In certain studies of interrill erosion, the sediment with diameter of <0.1 mm tended to increase with time, whereas sediment of >0.5 mm tended to decrease; and the sediment between 0.1 and 0.5 mm remained unchanged.[39,46,50] This is caused by continuous breakdown of soil aggregates by raindrop impact during rainfall. In general, fine-particle enrichment of eroded sediment from interrill erosion can take place under certain conditions, but the size distribution of primary particles of eroded sediment resembles those of dispersed surface soil from which sediment eroded. This also indicates that the proportion of particles that made up soil aggregates is similar to that of matrix soil.

Sediment from rill erosion has a greater proportion of larger aggregates than that from interrill erosion because of the massive removal of matrix soil by concentrated flow.[45] Detachment of sediment by rill flow is not selective because of the high erosive and transport power of concentrated flow. However, considerable enrichment can occur through transport and deposition processes. When sediment transport capacity is reduced by the changes in slope steepness or surface roughness, such as on toeslopes or in grass strips, deposition takes place. Because the deposition rate depends on the settling velocity of sediment particles in water, which in turn is dependent on sediment size and density, deposition selectively removes coarse sediment particles, which have higher settling velocities, and enriches the sediment in the finer sediment fraction.

Several approaches have been taken to predict the size distribution of eroded sediment and enrichment ratio based primarily on the size distribution of matrix soil. Foster et al.[51] developed a set of empirical functions that relate soil texture and organic matter content to the size distribution and composition of eroded sediment. They divided the sediment into five size fractions, those being primary clay, primary silt, primary sand, small aggregates, and large aggregates. To each of these size classes they designated a representative particle diameter and density. They further developed a set of equations relating enrichment ratio to sediment delivery ratio with an exponential decay function for each size group. Menzel et al.[52] found that the enrichment ratio decreased exponentially with increasing soil loss rates measured from small watershed and runoff plot data. In newer, process-based erosion models, because the sediment is routed by different size classes, enrichment ratio can be directly computed for any time and at any location based on the sediment composition. This is discussed later in this chapter.

The use of soil amendments such as gypsum and organic polymers and management practices that effect increased soil organic matter at the surface (both of which increase aggregation and reduce clay dispersion) is highly desirable in reducing clay-facilitated chemical transport.

2.3   SOIL EROSION MODELS

2.3.1   Early Attempts to Predict Erosion by Water

In the USA, one of the first attempts to estimate soil loss was an equation relating the loss to slope length and gradient.[53] However, the first major soil erosion model that later received wide use and is still used today in many parts of the world was the Universal Soil Loss Equation.[5,54] The equation has been used, often in modified form to suit the circumstances, in nearly all geographic regions around the world.

2.3.2   The Universal Soil Loss Equation (USLE)

The USLE can be considered a lumped parameter model in that each of the factors of the equation may contain a number of other parameters. As stated before, the USLE was developed from erosion plot and rainfall simulator databases. In certain cases, it is the statistical summarization of those data, making it difficult to extrapolate into other areas. The USLE is composed of six factors to predict the long-term average annual soil loss (A). The equation includes the rainfall erosivity factor (R), the soil erodibility factor (K), the topographic factors (L and S), and the cropping management factors (C and P). The equation takes the simple product form

$$A = R\,K\,L\,S\,C\,P \tag{2.13}$$

The USLE has another concept of experimental importance, which is that of the unit plot. The unit plot is defined as the standard plot condition to determine the erodibility of the soil. These conditions are when the LS factor = 1 (slope = 9% and length

= 22.1 m) where the plot is fallow and tillage is up and down slope and no conservation practices are applied (CP = 1). In this state

$$K = A/R \qquad\qquad (2.14)$$

The parameter estimation equation for K^{55} requires the particle size of the soil, organic matter content, soil structure, and profile permeability. The soil erodibility factor K can be approximated from a nomograph if this information is known. The LS factors can easily be determined from a slope effect chart by knowing the length and gradient of the slope. The cropping management factor (C) and conservation practices factor (P) are more difficult to obtain and must be determined empirically from plot data. The values of C and P are quantitatively expressed as soil loss ratios. In the case of the C factor, this is the ratio of soil loss for the management practice in question to the soil loss for a bare plot. In the case of the P factor, it is the ratio of soil loss with the conservation practice to the soil loss without the practice.

The USLE has been a very successful model for helping to conserve soil around the world. It is quite effective as a tool for choosing best land management practices for controlling erosion. It can also be very effective in making regional or national surveys of erosion to track progress in controlling erosion. The USLE is also a very useful tool when used conceptually for education purposes because the factors that contribute to increased erosion are easily understood.

Problems exist in obtaining accurate parameter values for the USLE, particularly in countries other than the United States. Because the equation was developed for the U.S., the relationships should not be expected to hold up in areas where very dissimilar soils occur, such as the tropics. Likewise, the topographic factors were developed from relatively moderate slope lengths and gradients and may not hold up for steep lands. In many areas, the rainfall erosivity factor is difficult to obtain because of limited data. Cropping and management factors must be determined experimentally, and much effort is needed to obtain data for different systems throughout the world. A major limitation to the USLE is that it explicitly predicts the long-term annual average soil loss, and it estimates spatial averages of erosion on a hillslope. In other words, USLE provides no information on nontemporal and spatial variability of erosion.

The critical deficiency in terms of nonpoint source pollution is that the USLE predicts average soil loss only over the area of net soil loss. It does not predict deposition or sediment delivered from a field or end of slope, nor does it provide any information on the chemical-carrying capacity or enrichment ratio of the sediment generated by erosion.

2.3.3   THE SEDIMENT CONTINUITY EQUATION

Process-based models of erosion have a distinct advantage over current empirical models of erosion for use in nonpoint source pollution applications because they are generally designed to provide estimates of spatial and temporal distributions of both soil loss and net sediment deposition, sediment delivery rates and amounts from field and watershed areas, and the size distribution of the sediment generated and delivered off-site.

Process-based (also termed physically based) erosion models attempt to address soil erosion on a relatively fundamental level using mass balance differential equations for describing sediment continuity on a land surface. The fundamental equation for mass balance of sediment in a single direction on a hillslope profile is given as

$$\partial(cq)/\partial x + \partial(ch)/\partial t + S = 0 \qquad (2.15)$$

where c (kg/m^3) is sediment concentration, q (m^2/s) is unit discharge of runoff, h (m) is depth of flow, x (m) is distance in the direction of flow, t (s) is time, and S [kg/(m^2 s)] is the source/sink term for sediment generation. Equation 2.15 is an exact one-dimensional equation. It is the starting point for development of physically based models. The differences in various erosion models are primarily: a) whether the partial differential with respect to time is included, and b) differing representations of the source/sink term, S. If the partial differential term with respect to time is dropped, the equation is solved for the steady state, whereas the representation of the full partial equation represents a fully dynamic model. The source/sink term for sediment, S, is generally the greatest source of differences in soil erosion models. It is this term that may contain elements for soil detachment, transport capacity terms, and sediment deposition functions. It is through the source/sink term of the equation that empirical relationships and parameters are introduced.

The sediment continuity equation in physically based models is normally written in terms of a single flow direction, x. The equation could be written and solved for the x and y directions to describe sediment continuity on a two-dimensional surface. To date, however, the approach taken to describe sediment continuity on two-dimensional surfaces has been to use the unidirectional equation with the x direction being the direction of water flow at a given point on the landscape surface. Modeling of erosion on watersheds in current process-based erosion models generally involves dividing the watershed area into overland flow elements and channel elements. The overland flow elements are typically either rectangular, representing hillslopes adjacent to channel elements, or they are squares within a pattern of a grid that overlays the watershed. In both cases, rill and interrill erosion processes are described in the overland flow elements, and sediment generated from those overland flow elements is considered to be delivered to the channel elements to be transported through the channel network. In some cases, sediment from an overland flow element may be routed to and potentially through another overland flow element before reaching a channel element.

We first address the application of the sediment continuity equation to the routing of sediment within overland flow elements. In doing so, we focus on three of the many existing models to exemplify the concepts introduced. The Water Erosion Prediction Project Hillslope Profile Model (WEPP)[38,56–57] derives from a family of models developed by Foster,[14] and shares common descriptions of erosion with CREAMS.[52] WEPP is a steady-state model that is intended to be used at the field planning level in much the same way as the USLE is currently used for conservation planning. As such, the model places a strong emphasis on the effects of soil and plant management practices on erosion. It is a continuous simulation model that operates on a daily time-step. The RUNOFF model[58-59] is a single-event, dynamic erosion model. The erosion routines within RUNOFF are driven by the solution of the kinematic wave equation that describes the hydrologic routing of surface runoff. The Hairsine and Rose model[2-3] is also a single-event, dynamic model.

2.3.4 FORMS OF THE SEDIMENT CONTINUITY EQUATION

For the case of steady-state conditions, and using the concepts of rill and interrill erosion, the sediment continuity equation (Equation 2.15) can be rewritten as it is in the WEPP model as

$$dG/dx = D_r + D_i \qquad (2.16)$$

where G [kg/(m s)] is sediment load per unit width in the flow (equal to cq in Equation 2.15), D_r [kg/(m^2 s)] is net rill erosion rate per unit area of rill bottom, and D_i [kg/(m^2 s)] is interrill sediment delivery to the rill (as with rill erosion, expressed on a per unit rill area basis), which was discussed above. For a given set of conditions, the interrill sediment delivery can be calculated and set as a constant in Equation 2.16. For the case of net detachment in a rill, the D_r term will be positive, indicating a net increase in sediment load with downslope distance. For the case of deposition, the D_r term is negative.

In the WEPP model, the sediment continuity equation is applied within the rills, which are described hydraulically as small rectangular channels. This approach contrasts with most other erosion models, such as CREAMS, KINEROS, RUNOFF, and the model of Rose et al.,[32] which use uniform flow hydraulics to describe detachment of soil and transport of sediment by flowing water. The recent model of Hairsine and Rose,[2-3] however, also uses rill hydraulics for describing rill erosion processes.

In formulating Equation 2.16 from Equation 2.15, already several major assumptions and decisions regarding the representation of erosion have been made. In dropping the dynamic term, one must be able to establish a representative steady-state erosion rate and erosion time period that will provide a good estimate of the overall erosion rate for a storm. It has also been decided in formulating Equation 2.16 that the rill and interrill formulation is appropriate and will provide a reliable framework for making erosion predictions. The fact that D_r and D_i represent "net" rather than instantaneous terms is important also. For the interrill case, D_i is an estimate of the amount of sediment delivered to the rill from interrill areas. It does not explicitly account for the individual processes of splash detachment, deposition of splashed materials on interrill areas, and transport of the splashed materials in the shallow interrill flow. For the rill case, D_r represents a net movement of soil to the flow from the bed. This implies physically that detached sediment, once in the flow of the rill, will be transported downslope in the rill flow until an area of net deposition is reached whereby the sediment may fall out and rest on the bed. The net rill detachment rate, D_r, is a function of four primary factors: (1) the amount of sediment in the flow, (2) the hydrodynamics of the flow, (3) the resistance of the soil to detachment by flow, and (4) ground surface cover. The mathematical representations of each of these factors are addressed below.

The sediment continuity equation for overland flow elements used in the RUNOFF model is in dynamic form, and is written as

$$\partial Q_s/\partial x + \partial(CA)/\partial t = g \qquad (2.17)$$

where Q_s (m³/s) is the volumetric sediment discharge, C (m³/m³) is the volumetric concentration of sediment, A (m²) is the cross sectional area of flow, and g [m³/(s m)] is the net volumetric rate of material exchange with the bed per unit length. The RUNOFF model uses uniform flow hydraulics for the sediment continuity relationships, including erosion by flow, thus, Equation 2.17 is expressed on a unit plot or field width basis. The bed exchange rate, g, includes terms for erosion by flow and erosion by raindrop splash, as discussed below, but those terms are not strictly additive.

The Hairsine and Rose erosion model[52,53] describes erosion as a balance of several instantaneous processes rather than net detachment or deposition in rill or interrill areas. In that model, net detachment or deposition rate is conceived as a balance between several processes that occur simultaneously, those being a) the movement of sediment particles that are in the flow to the bed, b) the movement of previously detached sediment into the flow, and c) the detachment of soil particles from the bulk soil mass. The model assigns a separate term for each of these individual processes. The movement of sediment particles that are in the flow to the bed is "deposition," the movement of previously detached sediment into the flow is "re-entrainment," and the detachment of soil particles from the bulk soil mass is "entrainment." Hairsine and Rose introduce entrainment and re-entrainment terms for both rill and interrill erosion. Hairsine and Rose's model uses a sediment continuity equation of the form

$$\partial(c_i\,q)/\partial x + \partial(c_i\,h)/\partial t = e_i + e_{di} + r_i + r_{ri} - d_i \qquad (2.18)$$

where e_i is entrainment by rainfall, e_{di} is re-entrainment by rainfall, r_i is entrainment by surface water flow, r_{ri} is re-entrainment by surface water flow, d_i is the continuous deposition term, and the subscript i indicates the particle settling velocity class of the sediment. Net rates of detachment, deposition, and sediment transport capacity are implicit concepts embodied in this type of representation.

Because the concepts of transport capacity, T_c, and detachment rate capacity, D_{rc}, are not introduced *a priori,* Rose[12] argues that the model of Rose et al.[32] (which is a predecessor to Hairsine and Rose[2-3]) is conceptually simpler than the model of Foster and Meyer[60] (which is a predecessor to WEPP). On the other hand, as noted by Rose, both of the models result in similar patterns of erosion behavior for similar conditions. Furthermore, it should be recognized that each additional source term in the sediment continuity equation requires empirical parameters for driving and resistance functions, and that the terms delineated by Hairsine and Rose[2-3] are inherently difficult to measure in any direct way.

2.3.5   THE SEDIMENT FEEDBACK RELATIONSHIP FOR RILL DETACHMENT

Net detachment rates by flowing water are a function of the amount of sediment in the flow, as was mentioned previously. This is an important factor and should be accounted for in formulating the sediment continuity equation. The flow of water in a rill has, obviously, a finite amount of flow energy at any given time and location. Flow energy is expended both by detachment of soil and by transport of sediment. As the flow picks up increased sediment load from rill and interrill detachment sources, or alternatively, as flow energy decreases along a concave slope, a greater proportion of the flow energy will be expended in transporting the sediment and less of the energy will be available for detaching soil. Detachment rates in the rill will necessarily decrease as a result. The two extreme cases that illustrate the effect of sediment load on rill detachment rates are a) clear water flow ($G/T_c = 0$) and b) when sediment load reaches sediment transport capacity ($G/T_c = 1$) (where T_c is the transporting capacity of the flow expressed in units of mass per unit time per unit width of rill flow, kg/(m s)).

For the case of clear water on bare soil, essentially all of the available flow energy may be expended to detach soil, thus detachment rate will be maximized. The rate of detachment for the clear-water case can be thought of as a detachment potential. Foster and Meyer[60] refer to this potential as the detachment rate capacity, D_{rc}.

The other extreme case is where sediment transport capacity is filled. In this case, all of the flow energy is expended to transport the sediment that is already in the flow and therefore none is available to detach more soil particles. In this case, the net detachment rate, D_r, will necessarily be zero.

Between the two extreme cases the detachment rate, D_r will range between zero and D_{rc}. The functional relationship between these limiting cases is unknown. Foster and Meyer[60] assumed that the relationship was linear; in other words, that the detachment rate, D_r, is proportional to the amount of sediment in the flow up to the point where transport capacity is filled. In that case, the functional form of the detachment rate is given by

$$D_r = D_{rc} \left(1 - G/T_c\right) \qquad (2.19)$$

where T_c [kg/(m s)] is the sediment transport capacity. Equation 2.19 represents the sediment feedback term for rill detachment rates and is used in the WEPP erosion model.

A similar approach to representing rill erosion was taken by Lane et al.[6] for a dynamic model, where net rill detachment was represented as

$$D_r = k_r \left(Tc - G\right) \qquad (2.20)$$

where k_r was an empirical coefficient. Conceptually, the k_r term from Lane et al.[6] would be related to the Foster and Meyer equation as

$$k_r = D_{rc} / T_c. \qquad (2.21)$$

Hairsine and Rose[2-3] take a different approach to describe the sediment feedback relationship. They define a term, H, which is the fractional covering of the soil bed by sediment. They maintain that the entrainment of soil, either by flow or by splash, must be proportional to the fractional exposure of the original bed, (1-H). Because H is dependent on the deposition rate of sediment from the flow, d_i, which, in turn, is dependent on the sediment concentration in the flow, the entrainment rates are also indirectly a function of the sediment concentration of the flow. Thus, there is a similar tendency here, as in the WEPP model as discussed previously, that the greater the sediment concentration, the less the entrainment rates of soil. Also, although the logic is definitely different, the two approaches may not be as diverse as may first appear. WEPP uses an "independent" sediment transport capacity function for estimating T_c (the Yalin equation). The Yalin equation, as with other sediment transport relationships, is based on the concept of balancing the falling-out of particles from the flow (analogous to the continuous deposition term from Hairsine and Rose) with the picking up of previously deposited material (analogous to the re-entrainment terms).

The key difference between the two approaches in terms of the sediment feedback relationship (WEPP and the Hairsine and Rose model) is the concept of shielding by the sediment "layer" in the Hairsine and Rose model as opposed to a reduction of available flow energy in the case of WEPP. Proffitt et al.[34] estimated H visually in experiments on a tilting flume experiment where only interrill processes were active. From those visual estimates of H, they calibrated coefficients of splash entrainment and re-entrainment. From controlled laboratory experiments it is possible, although perhaps difficult, to estimate H.

The RUNOFF model takes into account the sediment in the flow and the sediment layer on the bed in calculating the detachment of soil by flowing water. The model calculates a volumetric potential sediment exchange rate based on the concentration of sediment in the flow that represents the amount of sediment that the flow could take from the bed to fill transport capacity. Any loose sediment on the bed, as well as any interrill sediment contribution, would be taken into the flow toward filling that transport capacity, and any remaining transporting capacity would be available to be filled in part by soil detached directly from the bed. This approach of first allowing the movement of previously detached and deposited sediment from the bed (during the same rain event) to the flow is important in a dynamic model. In a steady-state model the flow depths are representative; they do not change with time. In a dynamic model, variations in flow depth and velocity with time during the erosion event may cause a (net) depositional bed to form during a period of low flow that might then be re-introduced into the flow if the runoff flows later increase.

In RUNOFF, a sediment concentration at sediment transport capacity C_p is computed. Then the potential sediment exchange is assumed to be the difference between the sediment in the flow and that which the flow can carry. Thus, the volumetric potential sediment exchange rate per unit length, g_p (m^2/s), is calculated as

$$g_{p\,i,j} = A/\Delta t_j \, [C_{p\,i,j} \, C_{i1,j1}] \qquad (2.22)$$

where i is the subscript representing a discrete point along the x-axis (downslope distance), j is the subscript representing a discrete point along the time axis, A (m^2) is the cross-sectional area of flow, C_p (m^3/m^3) is the volumetric sediment concentration at potential (capacity) rate, and $C_{i-1,j-1}$ (m^3/m^3) is the volumetric sediment concentration in the flow during the previous time and space increment. The sign of the term g_p serves as an indicator of deposition or erosion mode.

If $g_p > 0$, the transport capacity exceeds the amount of material in transport, and the flow will tend to pick up additional material from the bed. If the detached soil available on the bed is not sufficient to fill the capacity, the flow will erode soil from the parent bed material by expending more energy. Therefore, two erosion cases are considered, depending on the volume of detached soil available on the bed. An available soil volume per unit length is calculated by adding soil detachment from raindrop impact, if any, during Δt_j to the volume of loose sediment left on the bed from interval Δt_{j-1} as

$$v_{i,j} = Pf(e_{i,j1} + E_r\Delta t_j)(1 - \lambda) \tag{2.23}$$

where $v_{i,j}$ (m^3/m) is the volume of detached soil on the bed per unit length, $e_{i,j-1}$ (m^3/m) is the volume of loose sediment per unit length left on the bed from the previous time step, E_r $[m^3/(s\ m)]$ is the raindrop impact erosion rate per unit downslope length, P (m) is the wetted perimeter of flow (unit width for overland elements), f is the fraction of the sediment size group in the distribution, and λ (m^3/m^3) is the porosity of the sediment bed. RUNOFF solves the erosion equations for individual particle-size classes of the sediment distribution, which is discussed in a later section.

If $v_{i,j} \geq g_p\Delta t_j$, then the available detached soil is sufficient to supply sediment to the flow to fill transport capacity. In this case, no additional detachment of original soil occurs, and the rate exchange from the bed, g $[m^3/(s\ m)]$, is computed as

$$g = g_p \tag{2.24}$$

If $v_{i,j} < g_p\Delta t_j$, the available detached soil is not sufficient to fill the available sediment transport capacity, and additional soil is detached from the parent bed material. Erosion from the parent bed material requires additional energy, and a flow detachment coefficient is used to compute the additional erosion from the undetached soil. In this case, the bed exchange rate is computed as

$$g = 1/\Delta t_j\ [v_{i,j} + a_f(g_p\Delta t_j\ v_{i,j})] \tag{2.25}$$

where a_f (dimensionless) is the flow detachment coefficient. Equations 2.24 and 2.25 express the rate of exchange from the bed for the time increment Δt_j and distance increment Δx_i used in the numerical solution of Equation 2.17 for the case of detachment on overland flow elements. The depositional case is discussed below.

2.3.6 DETACHMENT OF SOIL IN RILLS

Foster[14] derived a rill detachment function from the data of Meyer et al.,[61] where the coefficient "b" of Equation 2.5 was assumed equal to 1 and τ_c was nonzero. This relationship was derived from channelized rill erosion data rather than from plot data and uniform flow assumptions. The WEPP model uses a "b" coefficient of 1. The critical shear stress and the coefficient "a" are considered to be properties of the soil and soil surface conditions. This is appropriate because the WEPP erosion model partitions rill flow and calculates rill hydraulics for use in shear stress and transport capacity relationships, rather than using broad sheet flow calculations for rill erosion. Thus, in the WEPP model, the equation for calculating detachment in rills, including the sediment feedback relationship, is

$$D_r = K_r (\tau - \tau_c) (1 - G/T_c) \tag{2.26}$$

where K_r is called the rill erodibility parameter. The units of K_r are mass per unit time per unit shear force [kg/(s N)] or simplified as (s/m).

Detachment of soil by flow in the RUNOFF model is addressed by Equation 2.25. Because g_p is calculated with Equation 2.22, it is a function of the sediment transport capacity of the flow. Thus, the sediment transport relationship describes the driving hydraulic force for rill detachment in RUNOFF.

Rill detachment in the Hairsine and Rose model is a function of streampower. The model considers that flow detachment occurs when streampower exceeds a critical value, ω_c, and that a fraction, 1-F, of the streampower is lost to heat and noise. Thus

$$r_i = (1-H) F (\omega - \omega_c) \; \omega > \omega_c \tag{2.27}$$

and

$$r_i = 0 \; \omega \le \omega_c \tag{2.28}$$

where H is the fraction of the surface shielded by sediment. The Hairsine and Rose model also calculates (as does the RUNOFF model) detachment for individual particle-size classes, which is discussed in a further section. Thus, Equations 2.27 and 2.28 are solved for individual size fractions of material.

2.3.7 MODELING INTERRILL EROSION

Interrill erosion rate in the WEPP model is predicted from Equation 2.12 using the slope adjustment from Equation 2.11. The Hairsine and Rose model uses essentially Equation 2.7 to describe the splash detachment term in Equation 2.18, except that the shielding factor is added, thus

$$e_i = (1-H) \, a \, P^p \tag{2.29}$$

As for the case of entrainment by flow, all of the source terms in Equations 2.18 and 2.29 are actually written for individual settling velocity classes.

Equations 2.12 and 2.29 represent interrill sediment delivery and entrainment by rainfall, respectively. The empirical coefficients, K_i and a, in those equations are assumed to have characteristic values for a given soil. Temporal changes in interrill erosion may be reflected in adjustment terms used to represent canopy cover, ground cover, and potentially soil surface sealing.

The RUNOFF model uses an equation similar to Equation 2.8, also using an exponent (p) value of 2.0, but with a term added to account for the effect of a water layer on splash. The existence of a thin water layer on the soil surface may significantly affect raindrop detachment. A thin water layer may result in greater soil losses than would occur if the water layer were not present. As water depth is increased beyond a critical limit, Palmer[62] found that soil detachment was reduced. Mutchler and Young[63] suggested that a water depth of more than three times the median drop size essentially eliminated detachment by raindrop impact. Moss and Green[64] determined that depth of flow also significantly influenced sediment transport by shallow overland flow. The rainfall detachment equation in RUNOFF basically accounts for a reduction in splash amounts for increasing water depths. Thus, the basic rainfall detachment function in RUNOFF is

$$E_r = a_r \, I^2 \, [1 \; (h + e)/(3d_{50})] \qquad \text{if } (h + e) < 3d_{50} \tag{2.30}$$

and

$$E_r = 0 \qquad \text{if } (h + e) \geq 3d_{50} \tag{2.31}$$

where E_r (m/s) is the rate of soil detachment caused by raindrop impact, a_r is an empirical raindrop detachment coefficient, h (m) is the water depth on the soil surface, e (m) is the thickness of existing detached soil on the bed, and d_{50} is the median raindrop diameter. Equations 2.30 and 2.31 give detachment rate for the entire size distribution used in the simulation. The rate for each size group is calculated by multiplying this rate by the fraction of the corresponding size group in the distribution. Adjustment terms for the effects of canopy and ground surface residue covers are discussed below.

2.3.8 MODELING SEDIMENT TRANSPORT

The WEPP model computes sediment transport capacity, T_c, at points down a hillslope using a simplified form of the Yalin[42] transport equation[65]

$$T_c = k_t \, \tau_s^{1.5} \tag{2.32}$$

where k_t is a sediment transport coefficient and τ_s (Pa) is grain shear stress (see detailed definition below). This coefficient is calibrated by applying the full Yalin equation to compute T_c at the end of an equivalent, uniform hillslope profile. The result is a computationally efficient algorithm that is an extremely close approximation to using the full Yalin equation at all points down the slope.[65]

2.3.9 MODELING SEDIMENT DEPOSITION

If an erosion model makes use of the concept of sediment transport capacity, net deposition is considered to occur when the amount of sediment in the flow exceeds the sediment transport capacity. Often, a first-order decay coefficient, usually being a function of the fall velocity of the sediment, is used to assess the rate of deposition. This concept of deposition represents a net rate of accumulation of sediment on the bed for an instant in time (in the dynamic case) or at steady-state (for the steady-state case).

If the source/sink term in Equation 2.15 includes a term that describes the continual falling-out of sediment particles from the flow to the bed alone, rather than the net balance described above, this process itself is referred to as deposition. In their model, Hairsine and Rose incorporated this factor explicitly in the source/sink term of Equation 2.18. This model also includes, as it must, a term for the simultaneous movement of available sediment from the bed into the flow, which is essentially the other part of the net deposition term discussed above. In other words, the Hairsine and Rose model explicitly includes factors in the source/sink term to describe the balance between falling out and lifting of sediment particles to and from the bed. Thus, in that model, the concept of both net deposition and sediment transport capacity is implicit.

For the WEPP model, when sediment load, G, exceeds the sediment transport capacity, T_c, the net rill erosion rate, D_r, in Equation 2.16 is negative. In that case D_r is calculated as

$$D_r = (\beta v_{ef}/q) [T_c - G] \qquad (2.33)$$

where β is a rainfall-influenced turbulence factor, v_{ef} (m/s) is the effective particle fall velocity of WEPP, and q (m²/s) is unit discharge of flow in the rill. The term $\beta v_f/q$ acts a first-order coefficient in terms of Equation 2.33, which describes how rapidly the sediment load, G, approaches the transport capacity, T_c, in the deposition mode. The WEPP model computes a total deposition rate based on an effective particle fall velocity, v_{ef}, which represents the entire sediment mass, rather than computing deposition rates in each class and summing the result. Deposition for each size class is determined, but only for computing sediment enrichment, as discussed below. As such, the fall velocity term is an effective fall velocity that represents the whole sediment. The β term is empirical, with a value set in the model currently at 0.5 for cases where raindrop impact is active. For snowmelt and furrow irrigation, β is set to 1.0.

RUNOFF works in a similar way to WEPP in that the deposition equations are put into use when sediment concentration exceeds that indicated by transport capacity. Thus, the deposition equation is used if the potential exchange rate with the bed, g_p, is less than zero. The amount of deposition of a particular size class in a given time and space increment depends on the settling velocity of the particle size class. Thus,

$$g = -g_p \quad \text{if } (2v_f \Delta t_j /h) \geq 1 \qquad (2.34)$$

and

$$g = -(2v_f \Delta t_j /h) g_p \text{ if } (2v_f \Delta t_j /h) < 1 \qquad (2.35)$$

where v_f (m/s) is the fall velocity of individual size fractions of sediment and g is the source term for Equation 2.17 as previously defined.

In the Hairsine and Rose model, the continuous deposition term in Equation 2.18 is simply

$$d_i = \alpha_i \, v_i \, c_i \qquad\qquad\qquad (2.36)$$

where $(v_i \, c_i)$ represents the concentration of the settling velocity class i near the bed. Thus the α term is introduced to account for nonuniform distribution of sediment concentration in the flow.

2.3.10 MODELING ERODED SEDIMENT SIZE FFRACTIONS AND SEDIMENT ENRICHMENT

The functions developed by Foster et al.[51] are used in the WEPP model for estimating the size distribution of eroded sediment at the point of detachment. As described earlier, WEPP uses an effective fall velocity term to compute total sediment deposition rates. For computing selective deposition, WEPP solves the sediment continuity equation for each individual sediment-size class at the end points of a depositional area on the hillslope. The total sediment at each such downslope distance is then partitioned proportionally among the five size classes based on the computations for each individual class.

The RUNOFF and the Hairsine and Rose models assume that the particle composition of the eroded sediment is the same as that for the original soil. In that case, either an estimate or a measurement of the particle-size classes, including the aggregate composition, is required to use the models. A difference between RUNOFF and the Hairsine and Rose model is the manner in which the sediment fractions are divided. In RUNOFF, the entire sediment-size distribution is divided into several size groups represented by their median sizes, and the amount of sediment contained in each group is measured and expressed as a fraction of the whole. In the Hairsine and Rose model, however, the sediment is divided not by size but by settling velocity classes, and the detached sediment is divided into classes of equal mass. Then each settling class is assigned a representative settling velocity. This approach makes the solutions of the overall erosion based on the sediment continuity equations for each settling velocity class straightforward and relatively simple. The technique of Lovell and Rose[66] is recommended for measuring the settling velocity distribution of the sediment.

Both the Hairsine and Rose model and RUNOFF obtain the total erosion amounts of net detachment and net deposition in space and time by solving the continuity equations for individual sediment classes and summing the responses. WEPP takes a different approach by computing a single deposition rate based on an effective fall velocity. WEPP then computes the delivered sediment distribution by solving the deposition equation for each sediment size class only at the end of the hillslope or depositional area and fractioning the total sediment load respective to the total calculated yield.

2.4   CROPPING AND MANAGEMENT EFFECTS ON EROSION

2.4.1   EFFECTS OF SURFACE COVER ON RILL EROSION

The effect of ground surface cover on reducing rill detachment rates, as well as sediment transport capacity, is reflected through shear stress or streampower partitioning. Again, as with the effect of sediment load on detachment rates discussed previously, we recognize that the flow has a finite amount of flow energy at any given time and location. When plant residue or rocks are on the soil surface, a portion of the flow energy is dissipated on that cover material and is not available either to detach soil or transport sediment. Therefore, both sediment transport capacity and detachment capacity are reduced.

The relationship used to partition the flow energy between that acting on the soil and that acting on the ground cover is analogous to that used to account for form roughness in streams. The energy is partitioned through the hydraulic roughness coefficients. The basic concepts have been discussed previously.[67-68] Application of the concept to ground surface cover effects on rill erosion was discussed by Foster.[14]

We begin with the assumption that hydraulic friction, as quantified by the Darcy-Weisbach friction factor, is additive, and thus

$$f = f_s + f_r \tag{2.37}$$

where f (unitless) is the total friction factor, f_s is the friction factor for the bare soil, and f_r is the friction factor associated with the surface cover, including rocks and plant residue. Flow velocity, v (m/s) is related to f as

$$v^2 = 8\, g\, R\, S\, /\, f \tag{2.38}$$

where R (m) is the hydraulic radius of the rill. Equation 2.38 is related closely to Equation 2.3 where R = h for the case of uniform sheet flow and $C = (8g/f)^{0.5}$. Using Eqsuations 2.37 and 2.38, hydraulic radius can be written as

$$R = v^2\, (f_s + f_r)\, /\, (8\, g\, S) \tag{2.39}$$

Using this function for R, shear stress for rill flow can be written as

$$\tau = \rho\, g\, R\, S = (\rho\, v^2\, f_s\, /\, 8) + (\rho\, v^2\, f_r\, /\, 8) \tag{2.40}$$

or

$$\tau = \tau_s + \tau_r \tag{2.41}$$

where $\tau_s = (\rho\, v^2\, f_s\, /\, 8)$ is the shear stress acting on the soil bed and $\tau_r = (\rho\, v^2\, f_r\, /\, 8)$ is the shear stress acting on the surface cover. Combining Equations 2.38, 2.39, 2.40, and 2.41 yields

$$\tau_s = \tau\,(f_s/f) = \rho\,g\,R\,S\,(f_s/f) \tag{2.42}$$

The shear stress acting on the surface cover is dissipated and only the fraction of the total shear stress that acts on the soil bed remains available for detachment of soil and transport of sediment. Thus, the rill detachment equation used in WEPP (Equation 2.26) can be rewritten accounting for the effect of surface cover on rill detachment rates as

$$D_r = K_r\,(\tau_s - \tau_c)\,(1 - G/T_c) \tag{2.43}$$

or

$$D_r = K_r\,[\tau(f_s/f) - \tau_c]\,(1 - G/T_c) \tag{2.44}$$

The transport capacity term in WEPP is calculated with the Yalin equation, which also uses the partitioned shear stress term, τ_s, as the driving hydraulic parameter. Thus, in WEPP, both detachment and transport capacity are reduced as a function of ground surface cover roughness using the shear stress partitioning concept.

Conceptually, a similar type of approach of energy partitioning may be taken with respect to streampower and flow detachment. The model of Hairsine and Rose[2–3] assumes that a portion of streampower is lost to heat and noise. The presence of residue on the surface of the soil would increase the portion of streampower lost, thus decreasing the value of F in Equation 2.27 A systematic mechanism for making such adjustments is needed.

2.4.1 EFFECTS OF SOIL CONSOLIDATION AND TILLAGE ON RILL EROSION

It has been recognized that soil erodibility changes with time during the year.[69–71] Existing data indicate that variations in rill erodibility through time are greater than variations in interrill erodibility. Brown et al.[72] studied changes in rill erodibility of a Russell silt loam soil in Indiana as a function of time after tillage. Rill erosion rates were measured at 0, 30, and 60 days after tillage on bare plots. Rill erosion rates were reduced at 60 days to between 12% and 30% (depending on rill flow rates) of the erosion rates measured immediately after tillage.

The principle mechanisms that increase the mechanical stability of a soil (i.e., which cause consolidation) after it has been disturbed are effective stress history[73–75] and time via thixotropic hardening and development of interparticle bonds.[76–77] For erosion, surface sealing and crusting may also cause changes in stability in interrill areas and increased rill erosion because of increased runoff. Primary factors that destabilize the resistance of a soil to erosion are tillage and thawing.

The mechanisms of consolidation, time, and suction, were studied by Nearing et al.[78] for a clay soil and by Nearing and West[79] for fine sand, silt loam, and clay soils. Results of those studies indicated that, although both time and suction influenced soil stability, the soil water suction effect was much more significant than time effects.

The rill erodibility consolidation model of Nearing et al.[80] provides a theoretical framework for accounting for the effects of soil consolidation on rill erosion rates. The model was tested on one site with some success, but model parameters need to be derived and tested for a range of soil types.

Tillage implements have varying effects on mixing the soil and decreasing soil bonding that comes from consolidation processes. One way of characterizing the effect of tillage implements is through a tillage intensity coefficient that ranges in value from 0 to 1. In such a scheme, an implement that causes a large disturbance to the soil, such as a moldboard plow, would have a high intensity coefficient.

2.4.3 BURIED RESIDUE EFFECTS ON RILL EROSION

Buried plant residue may affect rill erosion mechanically and biologically. Mechanically, the plant residue may act to anchor soil as a rill incises the soil and uncovers the buried residue. In that case, one would expect that hydraulic roughness of flow would be affected by the buried residue, and that rill erosion rates would be decreased in a manner analogous to the effect of surface residue discussed above. It can be hypothesized that, as residue decays with time, the microbial degradation products from the residue act as a binding material that increases interaggregate cohesion and hence reduces rill erodibility.

In practice, given the inherent variability associated with even well-controlled field erosion experiments, the mechanical effect of buried residue on hydraulic friction, and hence shear stress, is difficult to document. However, an overall reduction in rill erosion rates as a function of buried residue has been experimentally measured[72,81–82] and should be accounted for in erosion models.

The WEPP model accounts for buried residue by adjusting the rill erodibility factor, K[cf15r, as a function of buried residue mass. The function for K_{rbr}, which accounts for buried residue in the WEPP model, is

$$K_{rbr} = e^{-0.4Mb} \qquad (2.45)$$

where M_b is the mass (kg/m^2) of buried residue in the upper 15 cm of the soil profile. The erodibility term, K_r, is modified by multiplying with K_{rbr}. Similarly, the effects of live and dead roots on K_r were also adjusted by multiplying with a factor that is calculated using an exponential decay type of equation.

2.4.4 CANOPY AND GROUND COVER INFLUENCES ON INTERRILL DETACHMENT

The existence of a crop canopy may reduce raindrop detachment.[83–84] Laflen et al.[85] developed the following equation for estimating C_e, the effect of canopy on interrill erosion

$$C_e = 1 - F_c\, e^{-0.34\, Hc} \qquad (2.46)$$

where F_c is the fraction of the soil protected by canopy cover, and H_c (m) is effective canopy height.

The presence of ground cover decreases the surface area susceptible to raindrop detachment. Surface cover may also reduce interrill sediment delivery. The following equation has been derived for estimating G_e, the effect of ground cover on interrill erosion[57]

$$G_e = e^{-2.5\,gi} \qquad\qquad (2.47)$$

where g_i is the fraction of interrill surface covered by residue. This equation is used in the WEPP model.

The RUNOFF model takes a simpler approach to represent canopy and ground cover effects on splash erosion. The right side of Equation 2.30 is multiplied by the terms $(1-F_c)$ and $(1-g_i)$, thus reducing splash detachment proportionally to the fraction of surface covered by canopy and ground cover, respectively.

REFERENCES

1. USDA-Soil Conservation Service. The Second RCA Appraisal: Analysis of Conditions and Trends. U.S. Govt. Printing Office, Washington, D.C. 1989.
2. Hairsine, P. B. and Rose, C. W. Modeling water erosion due to overland flow using physical principles. 1. Sheet Flow. *Water Resource Res.* 28(1):237–243. 1992.
3. Hairsine, P. B. and Rose, C. W. Modeling water erosion due to overland flow using physical principles. 2. Rill flow. *Water Resource Res.* 28(1):245–250. 1992.
4. Gilley, J. E. and Finkner, S. C. Estimating soil detachment caused by raindrop impact. *Trans. Am. Soc. Agric. Eng.* 28(1):140–146. 1985.
5. Wischmeier, W. H. and Smith, D. D. Predicting Rainfall Erosion Losses. A guide to conservation planning. Agriculture Handbook No. 537. USDA-SEA, U.S. Govt. Printing Office, Washington, DC, 1978.
6. Lane, L. J., Shirley, E. D., and Singh, V. P. Modeling erosion on hillslopes, in *Modeling Geomorphological Systems,* Anderson M.G., Ed. John Wiley, Publ., NY. 1988. p. 287–308.
7. Meyer, L. D., Foster, G. R., and Romkens, M. J. M. Source of soil eroded by water from upland slopes, in *Present and Prospective Technology for Predicting Sediment Yields and Sources, ARS-S-40.* 1975. USDA-Agric. Research Service, pp.177–189.
8. Lewis, S. M., Barfield, B. J., and Storm, D. E. Probability distributions for rill density and flow. Winter meeting ASAE. Chicago, IL Paper No. 90–2558. 1990.
9. Bagnold, R. A. Bedload transport by natural rivers. *Water Resource Res.* 13:303–312. 1977.
10. Nearing, M. A. A probabilistic model of soil detachment by shallow turbulent flow. *Trans. Am. Soc. Agric. Eng.* 34:81–85. 1991.
11. Nearing, M. A., Bradford, J. M., and Parker, S. C. Soil detachment by shallow flow at low slopes. *Soil Sci. Soc. Am. J.* 55:339–344. 1991.
12. Rose, C. W. Developments in soil erosion and deposition models, in *Advances in Soil Science,* vol. 2. Springer-Verlag, New York. 1985.
13. Young, R. A. and Wiersma, J. L. The role of rainfall impact in soil detachment and transport. *Water Resources Res.* 9(6):1629–1639. 1973.
14. Foster, G. R. 1982. Modeling the erosion process, in *Hydrologic Modeling of Small Watersheds.* ASAE Monograph No. 5, Haan, C. T. Ed., Am. Soc. Agric. Eng., St. Joseph, MI. pp. 297–380.

15. Ellison, W. D. Studies of raindrop erosion. *Agric. Eng.* 25(4):131–136. 1944.
16. Bisal, F. The effect of raindrop size and impact velocity on sand splash. *Can. J. Soil Sci.* 40:242–245. 1960.
17. Free, G. R. Erosion characteristics of rainfall. *Agric. Eng.* 41(7):447–449, 455. 1960.
18. Park, S. W., Mitchell, J. K., and Bubenzer, G. D. Rainfall characteristics and their relation to splash erosion. *Trans. Am. Soc. Agric. Eng.* 26(3):795–804. 1983.
19. Ekern, P. E. Rainfall intensity as a measure of storm erosivity. *Soil Sci. Soc. Am. Proc.* 18:212–216. 1954.
20. Rose, C. W. Soil detachment caused by rainfall. *Soil Sci.* 89:28–35. 1960.
21. Bubenzer, G. D. and Jones, B. A. Drop size and impact velocity effects on the detachment of soils under simulated rainfall. *Trans. Am. Soc. Agric. Eng.* 14(4):625–628. 1971.
22. Quansah, C. The effect of soil type, slope, rain intensity and their interactions on splash detachment and transport. *J. Soil Sci.* 32(2)215–224. 1981.
23. Sharma, P. P. and Gupta, S. C. Sand detachment by single raindrops of varying kinetic energy and momentum. *Soil Sci. Soc. Am. J.* 53(4):1005–1010. 1989.
24. Meyer, L. D. Simulation of rainfall for soil erosion research. *Trans. Am. Soc. Agric. Eng.* 8(1):63–65. 1965.
25. Al-Durrah, M. M. and Bradford, J. M. Parameters for describing soil detachment due to single waterdrop impact. *Soil Sci. Soc. Am. J.* 46(4):836–840. 1982.
26. Chow, V. T. and Harbaugh, T. E. Raindrop production for laboratory watershed experimentation. *J. Geophys. Res.* 70(24):6111–6119. 1965.
27. Laws, J. O. and Parsons, D. A. Relation of raindrop size to intensity. *Trans. Am. Geophys. Union* 24:452–460. 1943.
28. Assouline, S. and Mualem, Y. The similarity of regional rainfall: a dimensionless model of drop size distribution. Trans. *Am. Soc. Agric. Eng.* 32(4):1216–1222. 1989.
29. Mualem, Y. and Assouline, S. Mathematical model of raindrop distribution and rainfall kinetic energy. *Trans. Am. Soc. Agric. Eng.* 29(2):494–500. 1986.
30. Meyer, L. D. and Wischmeier, W. H. Mathematical simulation of the process of soil erosion by water. *Trans. Am. Soc. Agric. Eng.* 12(6):754–758, 762. 1969.
31. Meyer, L. D. How rain intensity affects interrill erosion. *Trans. Am. Soc. Agric. Eng.* 24(6):1472–1475. 1981.
32. Rose, C. W., Williams, J. R., Sander, G. C., and Barry, D. A. A mathematical model of soil erosion and deposition processes: I. Theory for a plane land element. *Soil Sci. Soc. Am. J.* 47:991–995. 1983.
33. Hairsine, P. B. and Rose, C. W. Raindrop detachment and deposition: sediment transport in the absence of flowdriven processes. *Soil Sci. Soc. Am. J.* 55(2):320–324. 1991.
34. Proffitt, A. P. B., Rose, C. W., and Hairsine, P. B. Raindrop detachment and deposition: experiments with low slopes and significant water depths. *Soil Sci. Soc. Am. J.* 1991. 55(2):325–332.
35. Neal, J. H. Effect of degree of slope and rainfall characteristics on runoff and soil erosion. *Agric. Eng.* 19(5):231–217. 1938.
36. Watson, D. A. and Laflen, J. M. Soil strength, slope and rainfall intensity effects on interrill erosion. *Trans. Am. Soc. Agric. Eng.* 29(1)98–102. 1986.
37. Liebenow, A. M., Elliot, W. J., Laflen, J. M., and Kohl, K. D. Interrill erodibility: collection and analysis of data from cropland soils. *Trans. Am. Soc. Agric. Eng.* 33(6):1882–1888. 1990.
38. Flanagan, D. C, and Nearing, M. A. USDA-Water Erosion Prediction Project: Hillslope Profile and Watershed Model Documentation. NSERL Report No. 10. USDA-ARS National Soil Erosion Research Laboratory, West Lafayette, IN 47097–1196. 1995.

39. Zhang, X. C. Hydrologic and Sediment Responses to Soil Crust Formation. Ph.D. Diss., Georgia Univ., Athens, GA (Diss. Abstr. 93–20721). 1993.

40. Julien, P. Y. and Simons, D. B. Sediment transport capacity of overland flow. *Trans. Am. Soc. Agric. Eng.,* 28(3):755–761. 1985.

41. Alonso, C. V., Neibling, W. H., and Foster, G. R. Estimating sediment transport capacity in watershed modeling. *Trans. Am. Soc. Agric. Eng.,* 24(5):1211–1220, 1226. 1981.

42. Yalin, M. S. An expression for bedload transportation. *J. Hydr. Div., ASCE,* 89(HY3):221–250. 1963.

43. Lu, J. Y., Cassol, E. A., and Moldenhauer, W. C. Sediment transport relationships for sand and silt loam soils. *Trans. Am Soc. Agric. Eng.* 32(6):1923–1931. 1989.

44. Guy, B. T., and Dickinson, W. T., and Rudra, R. P. The roles of rainfall and runoff in the sediment transport capacity of interrill flow. *Trans. Am. Soc. Agric. Eng.* 30(5):1378–1386. 1987.

45. Alberts, E. E., Moldenhauer, W. C., and Foster, G. R. Soil aggregates and primary particles transported in rill and interrill flow, *Soil Sci. Soc. Am. J.* 44 (3):590–595. 1980.

46. Gabriels, D., and Moldenhauer, W. C. Size distribution of eroded materials from simulated rainfall: effect over a range of texture. *Soil Sci. Soc. Am. J.* 42:954–958. 1978.

47. Meyer, L. D., Harmon, W. C., and McDowell, L. L. Sediment sizes eroded from crop row sideslopes. *Trans. Am. Soc. Agric. Eng.* 23(4):891–898. 1980.

48. Mitchell, J. K., Mostaghimi, S., and Pond, M. C. Primary particle and aggregate size distribution of eroded soil from sequenced rainfall events. *Trans. Am. Soc. Agric. Eng.* 26 (6):1773–1777. 1983.

49. Yong, R. A. Characteristics of eroded sediment. *Trans. Am. Soc. Agric. Eng.* 23(5):1139–1142,1146. 1980.

50. Miller, W. P., and Baharuddin, M. K. Particle size of interrill-eroded sediments from highly weathered soils. *Soil Sci. Soc. Am. J.* 51:1610–1615. 1987.

51. Foster, G. R., Young, R. A., and Niebling, W. H. Sediment composition for nonpoint source pollution analyses. *Trans. Am. Soc. Agric. Eng.* 28 (1):133–139. 1985.

52. Knisel, W. G. (Ed.). CREAMS: A Field Scale Model for Chemicals, Runoff, and Erosion from Agricultural Management Systems. USDA, Conservation Research Report No. 26. USDA-ARS, Washington, DC. 643 pp. 1980.

53. Zingg, R. W. Degree and length of land slope as it affects soil loss in runoff. *Agric. Eng.* 21:59–64. 1940.

54. Wischmeier, W. H., and Smith, D. D. 1960. A universal soil-loss equation to guide conservation farm planning. *Trans. Int. Congr. Soil Sci.,* 7th, p. 418–425.

55. Wischmeier, W. H., Johnson, C. B., and Cross, B. V. A soil erodibility nomograph for farmland and construction sites. *J. Soil Water Conserv.* 26:189–193. 1971.

56. Lane, L. J. and Nearing, M. A. (Ed.). USDA Water Erosion Prediction Project: Hillslope Profile Model Documentation. NSERL Report No. 2, USDA ARS National Soil Erosion Research Laboratory, West Lafayette, IN. 1989.

57. Nearing, M. A., Foster, G. R., Lane, L. J., and Finkner, S. C. A process based soil erosion model for USDA water erosion prediction technology. *Trans. Am. Soc. Agric. Eng.* 32(5):1587–1593. 1989.

58. Borah, D. K. Runoff simulation model for small watershed. *Trans. Am. Soc. Agric. Eng.* 32(3):881–886. 1989.

59. Borah, D. K. Sediment discharge model for small watersheds. *Trans. Am. Soc. Agric. Eng.* 32(3):874–880. 1989.

60. Foster, G. R. and Meyer, L. D. A closed-form soil erosion equation for upland areas. In: *Sedimentation (Einstein),* Shen, H.W. Ed., Colo. State Univ., Ft. Collins, CO 1972.

61. Meyer, L. D., Foster, G. R., and Nikolov, S. Effect of flow rate and canopy on rill erosion. *Trans. Am. Soc. Agric. Eng.* 18:905–911. 1975.

62. Palmer, R. S. The influence of a thin layer on waterdrop impact forces. International Association of Scientific Hydrology, Pub. No. 65, 1964. pp. 141–148.

63. Mutchler, C. K. and Young R. A. Soil detachment by raindrops. *Proc. Sediment Yield Workshop,* Oxford, MS, 1972, USDA ARS S 40, 113–117. 1975.

64. Moss, A. J. and Green, P. Movement of solids in air and water by raindrop impact. Effects of dropsize and water-depth variations. *Aust. J. Soil Res.* 21:257–269. 1983.

65. Finkner, S. C., Nearing, M. A., Foster, G. R., and Gilley, J. E. Calibrating a simplified equation for modeling sediment transport capacity. *Trans. Am. Soc. Agric. Eng.* 32 (5):1545–1550. 1989.

66. Lovell, C. J. and Rose, C. W. Measurement of soil aggregate settling velocities, I, A modified bottom withdrawal tube, *Aust. J. Soil Res.* 26:55–71. 1988.

67. Einstein, H. A. and Banks, R. B. Fluvial resistance of composite roughness. *Trans. Am. Geophys. Union* 31:603–610. 1950.

68. Einstein, H. A. and Barbarossa, N. L. River channel roughness. *Trans. Am. Soc. Civil Eng.* 117:1121–1132. 1952.

69. Pall, R., Dickinson, W. T., Green, D., and McGirr, R. Impacts of soil characteristics on soil erodibility, in *Recent Development in the Explanation and Prediction of Erosion and Sediment Yield.* IAHS Publ. No. 137. 1982. pp. 39–47.

70. Dickinson, W. T., Pall, R., and Wall, G. J. Seasonal variations in soil erodibility. Paper No. 82–2573, ASAE, St. Joseph, MI. 1982.

71. Mutchler, C. K. and Carter, C. E. Soil erodibility variation during the year. *Trans. Am. Soc. Agric. Eng.* 26 (4):1102–1104. 1983.

72. Brown, L. C., Foster, G. R., and Beasley, D. B. Rill erosion as affected by incorporated crop residue and seasonal consolidation. *Trans. Am. Soc. Agric. Eng.* 32 (6):1967–1978. 1989.

73. Holtz, R. D. and Kovacs, W. D. Introduction to Geotechnical Engineering. Prentice-Hall, Inc., Englewood Falls, NJ. 1981.

74. Lambe, T. W. and Whitman, R. V. Soil Mechanics, SI Version, John Wiley & Sons, New York, NY. 1969.

75. Towner, G. D. and Childs, E. C. The mechanical strength of unsaturated porous granular material. *J. Soil Sci.* 23:481–498. 1972.

76. Bjerrum, L. and Lo, K. Y. Effect of aging on the shear-strength properties of a normally consolidated clay. *Geotechnique* 13:147–157. 1963.

77. Mitchell, J. K. Fundamental aspects of thixotropy in soils. *J. Soil Mech. Foundation Div., ASCE* 86(SM3):19–52. 1960.

78. Nearing, M. A., West, L. T., and Bradford, J. M. Consolidation of an unsaturated illitic clay soil. *Soil Sci. Soc. Am. J.* 929–934. 1988.

79. Nearing, M. A. and West, L. T. Soil strength indices as indicators of consolidation. *Trans. Am. Soc. Agric. Eng.* 31 (2):471–476. 1988.

80. Nearing, M. A., West, L. T., and Brown, L. C. A consolidation model for estimating changes in rill erodibility. *Trans. Am. Soc. Agric. Eng.* 31 (3):696–700. 1988.

81. Van Liew, M. W. and Saxton, K. E. Slope steepness and incorporated residue effects of rill erosion. *Trans. Am. Soc. Agric. Eng.* 26 (6):1738–1743. 1983.

82. Dedecek, R. A. Mechanical effects of incorporated residies and mulch on soil erosion by water. Ph.D. Diss. Purdue University, West Lafayette, IN (Diss. Abstr. 84-23352). 1984.

83. Morgan, R. P. C. Splash detachment under plant covers: Results and implications of a field study. *Trans. Am. Soc. Agric. Eng.* 25(4):987–991. 1982.

84. Finney, H. J. The effect of crop cover on rainfall characteristics and splash detachment. *J. Agric. Eng. Res.* 29(4):337–343. 1984.

85. Laflen, J. M., Foster, G. R., and Onstad, C. Simulation of individual storm soil losses for modelling the impact of soil erosion on cropland productivity, in *Soil Erosion and Conservation,* El Swaify, S., Moldenhauer, W., and Lal, R., Eds., SWCS, Anakey, IA, pp. 285–295. 1985.

3 Nitrogen and Water Quality

William F. Ritter and Lars Bergstrom

TABLE OF CONTENTS

1-56670-222-4/01/$0.00+$.50
© 2001 by CRC Press LLC

3.1   THE NITROGEN CYCLE

Nitrogen is one of the major nutrients for all living organisms and one of the most important factors limiting crop yield. Therefore, considerable research efforts have been undertaken over the years, trying to elucidate all the processes controlling N cycling in various ecosystems. The biogeochemical N cycle is very complex because N can occur in many valance states depending on redox potential. Certain processes occur only aerobically and others only anaerobically, regulated to a large extent by microbial processes occurring in a complex soil structure under nonsteady-state conditions.

Because of its importance for crop yields, high amounts of N are usually given to soils in agricultural production systems in North America and western Europe. This has led to considerable environmental problems, such as eutrophication of inland and coastal waters and potential depletion of the ozone layer in the stratosphere. Along with these problems, many diverse agricultural practices have been developed, all with the main goal to reduce harmful emissions of N to a minimum. For such practices to be successful, we need to understand not only the N transformation processes but also the interactions among the various components of the N cycle.

3.1.1   MINERALIZATION AND IMMOBILIZATION

Nitrogen mineralization is the process through which organically bound N, which is the major N constituent in terrestrial systems, is converted to ammonium nitrogen (NH_4-N). This process is mainly carried out by microorganisms. The subsequent fate of NH_4-N in soil depends on several biotic and abiotic factors and processes that compete for available NH_4-N (e.g., nitrification and plant uptake). This ongoing competition usually results in very low NH_4-N levels in cropped agricultural soils. Indeed, in many cases NH_4-N concentrations are below 5 mg N/kg soil, even though mineralization rates are quite high.[1]

The carbon/nitrogen ratio of a substrate added to the soil compared with that of the decomposing microorganisms is determining whether N will be mineralized or immobilized. The switch between net immobilization and mineralization of N is about 15 in well-balanced arable soils.[2] If the substrate has a lower C/N ratio, excess N will be available and NH_4-N will be released. Because of the low N concentration in most undecomposed plant litter, net mineralization (the difference between mineralization and immobilization of N) occurs mainly from soil organic matter. As

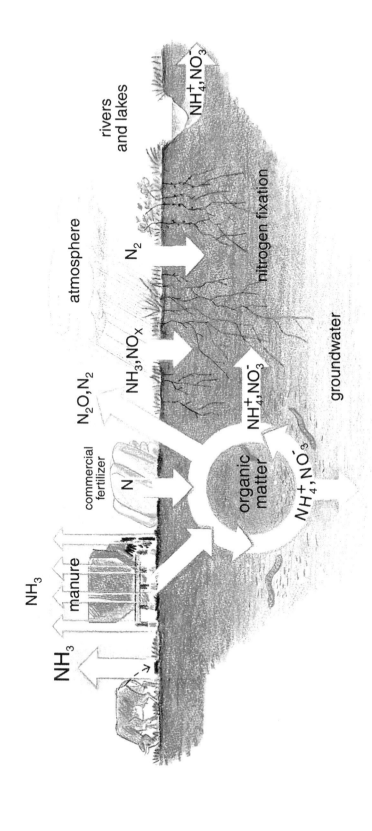

FIGURE 3.1. Nitrogen cycle

61

decomposition of fresh organic material proceeds, N is concentrated into microbial biomass and secondary decomposition products, and carbon is mineralized to CO_2. Release of NH_4-N from microorganisms results from catabolism of nitrogenous substrates such as amino acids when these are assimilated in excess of growth demands.[3]

However, large differences exist in plant litter C/N ratios between different species and also between different parts of the same species. For example, in mixed pastures of grasses and legumes, it is usually the legume leaf litter with a lower C/N ratio than the above-ground grass residues that contributes to net N mineralization during decomposition.[4] Therefore, introduction of N-fixing legumes will not only provide an atmospheric N input to the system but also reduce immobilization of N and hence improve the general soil fertility.[5] On the other hand, although legume leaf litter mostly has a more favorable C/N ratio than leaf litter of grass, their roots have commonly less favorable C/N ratios for mineralization, leading to higher immobilization of N than expected for grass roots. Also, senescent leaves of some grain legumes, such as soybean, can have sufficiently small N contents that N is immobilized when added to soil.[6]

Whether net mineralization will occur or not cannot be judged based only on knowledge about the C/N ratio of a substrate. Indeed, the C/N ratio is merely an approximation of the energy/N ratio, which is important to keep in mind.[2] The assimilation efficiency of the heterotrophic microorganisms responsible for mineralization is also dependent on other quality parameters. Some of the C and N consitutents of the substrate undergoing decomposition, such as nitrogen-free lignins and polyphenols, are not readily available to microorganisms and are therefore not easily mineralized. These microorganisms can also affect immobilization, such that plant materials containing a large proportion of lignins (for example) will not cause any substantial net immobilization of N, even though they have a relatively high C/N ratio.

Soil animals also play a major role in regulating N mineralization and can be of direct importance by excreting NH_4-N.[7] In this respect, microbial feeders protozoa and nematodes have been shown to be especially important.[8] Their relatively low biomass C/N ratio, which is similar to those of microorganisms, results in liberation of NH_4-N as they are grazing on the microbes. This pattern is influenced by the presence of roots because rich root exudates stimulates growth of bacteria that are subsequently consumed by the microbial feeders such as protozoa.[9] When digesting the bacteria, the protozoa release some of the bacterial N as NH_4-N on the root surface, where it can be taken up by the root.[10] Also, nematodes can mineralize substantial amounts of N that can be used by plants. Anderson et al.[7] estimated this mineralization to be 14–124 kg N/ha/yr under field conditions.

3.1.2 PLANT N UPTAKE

Through photosynthesis, green plants convert the energy provided by sunlight into chemical energy. By doing this, plants play a key role in most ecosystems, being the main suppliers of energy to heterotrophic soil organisms. Also, plants and their residues are fundamental sources and sinks of nutrients.[11]

Considering nutrient demands by plants, N is clearly one of the most critical of all essential elements in its effect on growth. Olson & Kurtz[12] summarized the major roles of N in plant growth as follows: (1) component of the chlorophyll molecule; (2) component of amino acids, and therefore essential for protein synthesis; (3) essential for carbohydrate utilization; (4) component of enzymes; (5) stimulative to root development and activity; and (6) supportive to uptake of other nutrients.

Before N can be taken up by plants, it must be transported to the surfaces of roots for absorption. This movement normally occurs by convective flow of water in response to transpiration of a growing crop. When the potential uptake exceeds the N supplied by such mass flow, the N concentration near the root surface drops and movement by diffusion begins. Plants can take up N from the soil solution either in the form of NO_3 or NH_4-N; although, because of chemical and biological processes occurring in the root zone of well-drained agricultural soils and the dominance of mass flow, NO_3 is usually more prevalent and therefore taken up in larger amounts. However, when both ion species are abundantly present in the soil solution, assimilation of NO_3 into organic N is usually retarded and NH_4-N is then preferentially used.[13] Also, early in the growing season, when low soil temperatures limit nitrification rates, it appears as if many crops favor uptake of NH_4-N as an adaptation to the prevailing conditions.[12]

After being taken up by plants, N undergoes certain transformations before it can be used. In terms of NO_3, the initial step is reduction to NO_2, which is subsequently reduced to NH_3. The reductions are catalyzed by NO_3 and NO_2 reductase in the respective transformation, of which the first process (the reduction of NO_3 to NO_2) is the rate limiting step. Accordingly, the activity of NO_3 reductase is often considered as a good indicator of crop growth rates.[14] The level of nitrate reductase in plant tissues shows a considerable variation over time—over the short term as well as over a growing season.[15] Failure to produce NO_3 reductase can be caused by several factors, of which reduced light intensity, soil moisture stress, and other nutrient deficiencies in the plant are some of the most important.[16] The result of such adverse effects can be problems with lodging, winter hardiness, and accumulation of high amounts of NO_3 in leafy parts of plants that potentially could lead to nitrate poisoning of cattle grazing feeds. In contrast, NH_3 seldom accumulates in plants but is readily metabolized and incorporated into amino acids and proteins.[16]

The total amounts of N taken up by plants vary considerably depending on the type of crop and also between different genotypes of the same species. There is also substantial variation in crop N-uptake depending on soil type, climate, and other environmental conditions. Overall, however, there is no doubt that N uptake by plants in most cases represents the largest N sink in croplands, of which a substantial portion is normally exported from the field. For agricultural crops, the harvested portion of the total N uptake is clearly higher than 50%. For some crops (e.g., wheat and soybeans), it may be as high as 75%.[12]

3.1.3 LEACHING AND SURFACE RUNOFF

Leaching and runoff of N to surface waters and groundwaters have gained increasing attention during the last few decades. This is attributed to both the negative effects on rivers, lakes, and coastal waters and to deteriorating drinking water quality. Accordingly, much emphasis has been put on finding counter measures to reduce such losses to acceptable levels.

The overwhelming part of N leaching through agricultural soils occurs as NO_3, whereas NH_4-N, as a cation, is mostly adsorbed to the net negatively charged soil matrix. In clay soils, NH_4-N may also be fixed between the layers of 2:1 type clay minerals,[17] such as the vermiculites, which considerably reduce mobility and availability of NH_4-N to plants. In sandy soils, however, in which adsorption affinity is much less than in clay soils and pH is usually lower (nitrification is thereby reduced), leaching of NH_4-N may constitute a significant part of the total N that is leached.

Two prerequisites have to be met before any notable leaching takes place. First, the NO_3 levels in the soil solution have to be sufficiently high, and second, the downward movement of water has to be enough to displace the available NO_3 below the rooting depth of plants. The first criterion is met in most agricultural soils, except during the growing season when crop uptake of N is high. The second condition is most commonly met in soils of humid and subhumid zones, where precipitation clearly exceeds evopotranspiration. In such areas, considerable amounts of NO_3 may leach through soil after the growing season, depending on soil type, amounts of fertilizer used, hydrogeologic conditions, and management practices.[19] In terms of soil type, sandy soils are usually considered to be more susceptible to NO_3 leaching than clay soils, mainly because of their smaller water-retaining capacity.[20, 21] In some cases, leaching losses in clay soils may certainly also exceed those in sandy soils exposed to similar condition (i.e., if preferential flow processes in the clay rapidly move newly applied NO_3 to deeper soil layers beyond reach of plant roots[22]). In most cases, however, nonequilibrium flow in structured soils tends to reduce NO_3 leaching. This is because NO_3 is mostly mixed with and protected in the smaller pores of the soil matrix, and water flowing through macropores does not interact with the soil matrix.[23] In addition to soil type, hydrogeologic conditions that determine the net vertical pressure gradient in the groundwater flow, and climate are factors that have a major influence on NO_3 leaching and groundwater contamination, although they are more or less impossible to control. In contrast, fertilizer type and intensity and management strategies (e.g., tillage practices and use of cover crops) can be altered or refined, which can reduce leaching of N considerably.[24–26]

In addition to leaching, N can also reach rivers and lakes through surface runoff if precipitation exceeds the infiltration capacity of a soil. Accordingly, this type of loss mechanism is strongly coupled to rainfall intensity and the hydraulic properties of a soil, and certainly also to factors such as topography and degree of soil cover. In total for the U.S., it has been estimated that about 4.5×10^9 kg N is lost yearly by soil erosion,[27] which is compatible with estimates of N leaching. Little of this N is in soluble form. The overwhelming part is in organic form, which is ultimately deposited in freshwater and marine sediments, with small chances of being recycled into agricultural systems.[27]

Because of the great importance of the amount and intensity of rainfall to trigger surface runoff, problems with this loss mechanism are especially widespread in the tropics. However, runoff problems in these regions are associated more with high soil loss rates than losses of N.[28] Also, in cold climates, surface runoff related to snowmelt may cause substantial soil erosion and losses of N. For example, Nicholaichuk & Read[29] estimated runoff losses of N to be about 10 kg N/ha/yr after fallow in Saskatchewan, primarily due to intensive snowmelt.

As for NO_3 leaching, several management practices have been developed with great potential of reducing N losses in surface runoff. The importance of ground cover in N transport by surface runoff was shown by Burwell et al.[30] In a study on a loamy soil in Minnesota, they found that runoff losses could be reduced from 23.8 to 3.3 kg N/ha by switching from continuous corn to hay in rotation. For fields on steep slopes, large runoff reductions could be obtained by tillage practices against the slope (contouring and terracing and combinations thereof).[31] Measures that protect soil against direct raindrop impact, such as cropping systems with multicanopy structure, can also significantly decrease runoff losses of N.[32]

3.1.4 AMMONIA VOLATILIZATION AND DENITRIFICATION

The most important N compounds lost as gases from agricultural cropping systems are ammonia (NH_3), nitrous oxide (N_2O), nitrogen oxide (NO), nitrogen dioxide (NO_2), and diatomic nitrogen (N_2).

Ammonia volatilization to the atmosphere is a complex process controlled by a combination of biological, chemical, and physical factors.[33] Examples of such factors are the balance between NH_4-N and NH_3, which is affected by pH among other things; presence or absence of plants; wind speed; and NH_3 concentration in the air space adjacent to the soil surface. The main source of NH_3 volatilization from agriculture is excreta from animals. Indeed, an average of 50% of the N excreted by farm animals kept in intensive agriculture is released to the atmosphere directly from animal barns during storage, during grazing, and after application of manure to soil.[34] However, substantial amounts of NH_3 emitted to the atmosphere also originate from microbial decomposition of amino acids and proteins in dead plant residues, soil fauna, and microorganisms. It has been estimated that about 90% of all NH_3 volatilization in western Europe originates from agriculture and, therefore, less than 10% from other sources.[34] This corresponds to about 11 and 1 kg N/ha/yr. Near large animal farms, however, considerably larger emissions may occur, reaching toxic levels for the surrounding vegetation. An NH_3 source of increasing importance during recent years is composting of source-separated household wastes. During such composting, 20–70% of the total N initially present in the wastes is typically lost as NH_3.[34]

Because emitted NH_3 is highly water soluble, it will be washed out by clouds and return to the soil surface with precipitation; it will also be deposited as dry deposition near the source. Because NH_3 is a basic compound in the atmosphere, it will form salts with acidic gases that can be transported long distances, especially in the absence of clouds. The most direct environmental consequence of large NH_3 depositions is its contribution to eutrophication of freshwater and marine ecosystems. This eutrophication may lead to decreased biological diversity and also to increased carbon storage in sediments and forest soils, which, over the long term, will likely affect the global carbon budget. Also, NH_3 deposition contributes to acidification of soils if nitrified and leached.

Denitrification, which is the other major source of N loss to the atmosphere, is the process whereby NO_3 and NO_2 are reduced to gaseous forms of N (NO, N_2O, and N_2). Biological denitrification is usually performed under anaerobic conditions by a heterogeneous group of bacteria, including both autotrophs and heterotrophs. The energy generated by using NO_3 as a terminal electron acceptor is almost compatible with that released during aerobic respiration and much more than the regular fermentative pathways. In general, the main end products in the denitrification process are N_2O and N_2, whereas NO is usually quantitatively of less importance. If O_2 concentrations increase, the ratio between N_2O and N_2 also increase, whereas NH_4-N concentrations do not affect production of either of these constituents.

In addition of being responsible for losses of an essential nutrient often limiting plant growth, denitrifying bacteria contribute to regulation of N_2O concentrations in the atmosphere. Nitrous oxide entering the stratosphere is involved in catalytic reactions where ozone is consumer.[35] Several studies have shown that this depletion may have increased during the last decades as a result of elevated atmospheric N_2O levels resulting from enhanced N-fertilization rates.[36] In a recent global assessment, the average yearly N_2O emission from fertilizers was estimated to be 1 kg N/ha $+$ 1.25 $\pm$ 1% of the fertilizer N applied.[37] Still, the atmospheric concentration of N_2O is quite small compared with, for example, CO_2; although its contribution to the "greenhouse effect" is considerable, mainly because of the long residence time and high relative absorption capacity of N_2O per mass unit.

3.2 SOURCES OF GROUNDWATER CONTAMINATION

3.2.1 FERTILIZERS

More intensive farming methods have led to higher rates of fertilization. A rapid increase in N fertilizer use occurred during the 1960s and 1970s. In 1980, U.S. farmers used 11,300,000 mg of N fertilizer, whereas 6,800,000 mg were used in 1970 and only 2,400,000 mg in 1960.[38] In 1997, 13,900,000 mg were used.[39, 40] During the 1980s, groundwater contamination became a national concern. Irrigated area have also increased gradually over the last 25 years. In 1974, the irrigated cropland area in the U.S. was 14,180,000 ha, and in 1998 it was 25,296,000 ha.[41] In the past, the main interest in N management and irrigation was related to agronomic and economic factors, but in the past 15 years, NO_3 leaching under irrigation has become a major environmental concern.

Madison and Brunett[42] did the first comprehensive nationwide mapping of area distribution of NO_3 in groundwater. They used 25 years of records of more than 87,000 wells from the U.S. Geological Survey's Water Storage and Retrieval System (WATSFORE). Nitrate concentration exceeded 3 mg N/L in agricultural areas of Maine, Delaware, Pennsylvania, central Minnesota, Wisconsin, western and northern Iowa, the plains states of Texas, Oklahoma, Kansas, Nebraska, and South Dakota, eastern Colorado, southeastern Washington, Arizona, and central and southern California. Lee and Nielsen[43] used Madison's and Burnett's data together with N fertilizer usage and aquifer vulnerability. They eliminated areas with elevated NO_3 concentrations in northern Maine and added areas in Ohio, Indiana, and Illinois when WATSFORE data were sparse. The studies of Madison and Brunitee[42] and Lee and Nielson[43] indicated there is a higher occurrence and prediction of NO_3 in groundwater in the central and western U.S. than other parts of the country.

There have been a number of comprehensive statewide surveys of NO_3 in groundwater. A study in Texas of 55,495 wells indicated some NO_3 contamination.[44] However, only 8.2% of the wells had NO_3 concentrations above 10 mg N/L. Spalding and Exner[45] concluded, after reviewing the North Carolina survey and other studies in the Southeast, that high temperatures and abundant rainfall and the relatively high-organic-content soils in the Piedmont Plateau and Coastal Plain of the southeastern U.S. promote denitrification below the root zone and therefore, naturally remediate NO_3 loading of the groundwater. Baker et al.[46] found in a statewide survey in Ohio of 14,478 domestic wells that only 2.7% exceeded the EPA drinking water of 10 mg N/L for NO_3 and only 12.7% of the wells exceeded 3.0 mg N/L. The average concentration was 1.3 mg N/L. In their review, Spalding and Exner[45] concluded most leachate was intercepted by tile drainage and never reached the groundwater.

A statistic-based statewide rural well survey in Iowa showed that the regional distribution of NO_3 concentrations above 10 mg N/L was not uniform and skewed.[48] The highest incidents of contamination were in the glaciated areas of southwestern and northwestern Iowa, where 31.4 and 38.2%, respectively, of all the wells were above 10 mg N/L. In northcentral Iowa, only 5.8% of the wells had NO_3 concentrations above 10 mg N/L. The major difference between the high and low contamination areas was related to well construction and well depth.

Halberg[48] has reported decreasing NO_3 concentrations with increasing depth in Iowa aquifers. Intensive irrigation has caused high NO_3 levels in groundwater in certain areas. Exner and Spaling found the NO_3 concentrations exceeded 10 mg N/L in 20% of 5826 sampled between 1984 and 1988 in Nebraska. Slightly more than half of the wells with NO_3 concentrations above 10 mg N/L were in areas highly vulnerable to leaching. These areas are characterized by fence-row-to-fence-row irrigated corn grown on well- to excessively well-drained soils and a vadose zone less than 15 m thick in the Central Platte region.

California has the most irrigated cropland and a history of high NO_3 concentrations in groundwater beneath intensively farmed and irrigated basins in central and southern California.[50] Keeney,[51] in reviewing data from a number of studies in California, concluded that the NO_3 levels in groundwater under normal irrigated cropland in general will range from 25 to 30 mg N/L. Only when N application rates exceed those that are efficiently used by crops does the leaching of N become excessive. For many crops, with good agronomic practices and profitable production, about 20 mg N/L of NO_3 in drainage effluents may be the best achievable.

Devitt et al.[52] measured annual NO_3 losses that ranged from 23 to 155 kg N/ha/yr on six irrigation sites with tile drainage in southern California. On sites where a low leaching fraction was used, NO_3 concentrations in the tile effluent were higher than on sites with a high leaching volume. However, higher mass amounts of NO_3 were lost under irrigation management where a high leaching volume was used.

In the Sand Plain Aquifer region of Minnesota, where 20% of the wells had NO_3 concentrations above 10 mg N/L, nearly 50% of the wells had NO_3 concentrations above 10 mg N/L in the irrigated cropland area.[53] Concentrations averaged 17 mg N/L in the irrigated area and 5.4 mg N/L in the nonirrigated cropland area.

In 1991, the USGS initiated the National Water Quality Assessment (NAWQA) Program in 20 areas and phased in work in more than 30 additional areas in 1997.[54] Results from the first 20 areas have been summarized. Concentrations of NO_3 exceeded 10 mg N/L in 15% of the samples collected in shallow groundwater beneath agricultural and urban areas. Concentrations of NO_3 in 33 major drinking water aquifers were generally lower than those in the shallow groundwater. Four of the 33 major drinking water aquifers had NO_3 concentrations above 10 mg N/L in 15% or more of the samples. All four of the aquifers were relatively shallow in agricultural areas, and were composed of sand and gravel that is vulnerable to contamination by application of fertilizers. Nitrate concentrations in the shallow groundwaters in the Central Columbia Plateau study area of Washington were among the highest of the 20 study areas. The highest NO_3 concentrations occurred where fertilizer use and irrigation were greatest.

3.2.2   LIVESTOCK WASTES

Nitrate contamination of groundwater can occur as a result of seepage from manure storage basins and lagoons, dead animal disposal pits, stockpiled manure, and livestock feedlots. Reese and Louden[55] conducted a literature review on seepage from earthen livestock waste storage basins and lagoons on data from 1970 to 1982. They concluded natural sealing takes place that results in very low seepage rates occurring in earthen manure storage basins and lagoons. Initially, this seal takes time to develop, which could result in a shock load of pollutants moving down and reaching groundwater. There is also the possibility of initial seal breakage because of drying, and the potential for another shock load upon refilling a manure basin after cleanout. Ritter and Chirnside[56] concluded that seals may break and cause serious groundwater contamination. They found a swine waste lagoon with a clay liner that was pumped dry twice a year and had NH_4-N concentrations above 1,000 mg N/L in the shallow monitoring wells around the lagoon.

Westerman et al.[57] found that seepage from old unlined lagoons in North Carolina was much higher than previously believed. Two swine lagoons that received swine waste for 3.5 to 5 years had high NO_3 and NH_4-N concentrations in the shallow groundwater. In a follow-up study, Huffman[58] evaluated 34 swine lagoons for impacts for seepage. About two-thirds of the sites had NO_3 concentrations above 10 mg N/L at 38 m down gradient in the shallow groundwater.

Several researchers have found that livestock feedlot soil profiles develop a biological seal similar to earthen manure storage basins and lagoons.[59, 60] The feedlot usually contains a compacted interfacial layer of manure and soil that provides a biological seal that reduces water infiltrations to less than 0.05 mm/hr. Norstadt and Duke[59] measured soil NO_3 levels that decreased from 80 mg N/kg at the top of feedlot soil profiles to less than 10 mg N/kg at the 1.0 to 1.5 m depth.

On the Delmarva Peninsula and in the southeastern U.S., where broiler production is concentrated, dead birds are most often disposed of on the farm. In the past, many farms used disposal pits that could be a source of groundwater contamination. Today, many farms use composting for dead bird disposal but some still use disposal pits. Most of the disposal pits do not have lined floors. Hatzell[61] found the median NO_3 concentration increased by 2.0 mg N/L in the vicinity of a dead bird disposal pit relative to two wells upgradient of the pit in northcentral Florida. Ritter et al.[62] measured NO_3 and NH_4-N concentrations in groundwater around six disposal pits on Coastal Plain soils in Delaware. Elevated NH_4-N concentrations were detected in the groundwater at three of the six disposal pits. Ammonium nitrogen concentrations as high as 366 mg N/L were measured. Average NO_3 concentrations ranged from 0.46 to 18.3 mg N/L, with three of the disposal pits having NO_3 concentrations above 20 mg N/L. Disposal pits used on the Delmarva Peninsula are old metal feed bins with the bottoms cut out. Many of these pits are partially in the groundwater because of the high groundwater table in many parts of the Delmarva Peninsula.

Recent research has shown that old poultry houses themselves may be causing groundwater contamination. Ritter et al.[63] investigated N movement under 12 poultry houses constructed from 1959 to 1985. Total mass of NH_4-N in the top 150 cm of the soil profile varied from 3420 to 12,580 kg N/ha, and the NO_3 concentrations in the groundwater around a set of two poultry houses was 45.5 mg N/L in northcentral Florida. Lomax et al.[64] sampled 30 broiler houses with house floor types, caterorized as loose soil, compacted (hard) soil, and concrete. They took borings to a depth of 150 cm in the spring and fall in each house. The loose, compacted, and concrete floor types had average total kjeldahl nitrogen (TKN) concentrations of 1063, 1077, and 213 mg N/kg, NH_4-N concentrations of 404, 460, and 24 mg N/kg and NO_3 concentrations of 245, 263, and 14 mg/N/kg, respectively. These studies clearly indicate that poultry houses with dirt floors may be a source of groundwater contamination.

3.2.3 LAND APPLICATION OF MANURES, SLUDGES, AND WASTEWATER

Excessive applications of manures or sludges may cause NO_3 contamination of groundwater. Land application of wastewater has been used in the food processing industry for years, and over the past 20 years has become a more popular method of disposal of municipal wastewater. Nitrogen is the limiting design parameter in many cases. Today approximately 55% of the sludge generated in the U.S. is applied to land or used as a soil amendment.[65] Other forms of solid waste are also used as soil amendments.

Overapplication of poultry litter has been shown to cause elevated levels of NO_3 in soil solution and groundwater.[66] Adams et al.[67] evaluated NO_3 leaching in soils fertilized with both poultry litter and hen manure at 0, 10, and 20 mg/ha. They found that the amount of NO_3 leaching into the groundwater was a function of litter application rate.

Liquid swine and dairy manure are commonly applied to forage crops in the southeastern U.S. Vellidis et al.[68] evaluated the environmentally and economically sustainable liquid dairy manure application rates on a year-round forage production system on a loamy sand Coastal Plain soil in Georgia. Nitrate concentrations increased in the soil solution at 1.5 and 2.0 m depths at application rates of 600 and 800 kg N/ha, remained relatively unchanged under the 400 kg N/ha rate, and decreased at an application rate of 250 kg N/ha.

Westerman et al.[69] found applying swine lagoon effluent at a rate of 450 kg N/ha of available N was too high for coastal bermuda grass in a sandy soil with a high water table and caused increased NO_3 concentrations in the groundwater. Hubbard et al.[70] found NO_3 concentrations exceeded drinking water standards on a Georgia Coastal Plain plinthic soil when dairy manure was applied to coastal bermuda grass at rates of 44 and 91 kg N/ha per month. It appears an annual application rate of 400 kg N/ha to coastal bermuda grass is the maximum application rate that should be used. Other forage systems would probably use less N and should have lower manure application rates. Stone et al.[71] indicated, from reviewing a number of research studies that groundwater in swine manure spray fields often has NO_3 concentrations above 20 mg N/L, whereas most row crop studies have NO_3 concentrations below 20 mg N/L and pastures have NO_3 concentrations below 5 mg N/L.

Nitrate contamination of groundwater from land application of municipal effluents and sewage sludges can be controlled to a great extent because their application is regulated in the U.S. Irrigation of crops with treated effluent has not been regulated at the federal level but is regulated at the state level. Land application of sludges was not regulated at the federal level until the EPA promulgated "Standards of the Use and Disposal of Sewage Sludge" in 1993.[72] Regulations do not allow wastewater or sludges to be applied at greater than agronomic N rates of the crop.

Research has shown that excessive sludge application rates will contaminate groundwater. Higgins[73] found that the upper rate of sludge application to corn on a Sassafras sandy loam soil to protect groundwater was 22.4 mg of dry solids/ha. Chang et al.[74] found that large concentrations of NO_3 accumulated in profiles of sludge-treated soils when rates of sludge application exceed crop requirements for N. Greater than optimum rates of sludge addition increased NO_3 leaching from course and fine loamy soils as linear functions of increasing total N inputs. Other researchers have also found NO_3 leaching is a linear function of sludge application rates above crop N requirements on sandy soils, but occurred only above a certain threshold on clay soils.[75]

One of the problems in estimating sludge application rates is in determining the mineralization rates of organic N. Typically treated sludges contain from 1 to 6% N on a dry-weight basis, with a large portion being in the organic form in some sludges. The rate of mineralization of sludge-borne organic N in soil ranges from a high of essentially 100% per year to a low of a few percent during the initial year of application. Nitrogen not mineralized the first cropping year is mineralized in subsequent years but usually at diminishing rates. Laboratory incubation studies of the N release characteristics of sewage sludge mixed with soil have proven useful in developing sludge application rates. For example, the mineralizable N content of anaerobically digested sludge during the year of application has been estimated at 15% of the organic N fraction by this approach.[76] Based upon reported N availabilities, the decreasing potential risk of NO_3 leaching from various types of sewage sludge the first year after application is liquid, digested > dewatered, digested > liquid, undigested > dewatered, undigested.

Applying liquid manure to fields with tile drainage may have an increased impact on tile effluent water quality. Dean and Foran[77] found high concentrations of bacteria and N and P in tile drainage discharge when rainfall occurred shortly before or shortly after manure spreading. In a study in southwestern Ontario on a Brookston clay loam soil, McLellan et al.[78] found tile discharge NH_4-N concentrations increased from 0.2 to 0.3 mg N/L before spreading to a peak of 53 mg N/L shortly after manure was spread. Land application of liquid manure did not increase NO_3 concentrations in the tile effluent. Blocking the drains to simulate controlled drainage decreased NH_4–N and bacteria concentrations.

3.3   SOURCES OF SURFACE WATER CONTAMINATION

3.3.1   FERTILIZERS

In the U.S. Geological Survey NAWQA study, the estimated background total N concentrations in streams from 28 watersheds in 20 study units was 1.0 mg N/L.[54] Average annual concentrations of total N in about 50% of agricultural streams ranks among the highest of all streams sampled in the 20 study units. In these streams, total N was about 2.9 mg N/L. Total N input from fertilizer, manure, and atmospheric sources was generally above 56 kg N/ha for the county.

One of the major sources of N input to surface water in the Corn Belt is through subsurface drainage discharge. Zucker and Brown[79] reviewed water quality impacts and subsurface drainage in the Midwest. Water quality and agricultural drainage are discussed in detail in Chapter 8.

Field studies have shown that N losses in surface runoff are correlated with fertilizer rates. In Georgia, TKN concentrations in surface runoff from watersheds cropped were related to N application rate.[80] Fields fertilized at the recommended rate did not contribute large quantities of N in runoff. Corn Belt research indicated N application rates greater than 168–196 kg N/ha for corn increased N runoff losses but did not significantly improve yields.[81] In another study, N fertilizer applied at a rate of 448 kg/ha/yr had annual losses of 50.2 kg/ha total N in surface runoff and an application rate of 174 kg/ha/yr had annual losses of 28.1 kg/ha.[82]

Methods of fertilizer application and farm management practices can significantly affect N losses in surface runoff. Research in the Corn Belt demonstrate conclusively that most of the total N lost in surface runoff is associated with sediment losses.[83] Therefore, sediment control practices should effectively reduce total N losses in surface runoff. Kissel et al.[84] also concluded that controlling sediment losses and following soil test results for proper fertilizer application rates can reduce N losses in the southwestern prairies.

In simulated rainfall studies in Minnesota,[85] it was shown that fertilization methods can be varied to control N losses in surface runoff. Greatest N losses came from plots upon which fertilizer was broadcast on a disked surface and the lowest losses were with fertilizer broadcast onto a plowed surface. Corn, forage, small grain, and soybean growers in New York are advised to band and sidedress fertilizer to best meet economic and water quality objectives.[86]

3.3.2   ANIMAL WASTES

The major potential pollution source of N from animal wastes in addition to land application of manure is from feedlot runoff. Both the volume and pollution concentration of feedlot runoff are highly variable. Precipitation is more important than either slope or stocking rate in determining feedlot runoff rates. In a study in South Dakota at six locations, the annual N loss varied from 0.1 to 6.6% of the total N in the manure.[87] More N is usually lost through N volitalization than runoff. Bierman[88] estimated 53 to 63% of the N voided was lost by volitalization, whereas runoff loss of N was only 5% for normal fiber diets, 7% for high-fiber diets, and 21% for low-fiber diets.

Westerman and Overcash[89] measured N concentrations from an open dairy lot in North Carolina. Nitrogen concentrations were lower than data reported for beef feedlots. In the same study, runoff N concentrations from the lot were 4 to 6 times higher than from a pasture that received 1 cm of dairy lagoon water every other week by irrigation.

3.3.3   LAND APPLICATION OF MANURES AND SLUDGES

In the U.S., sludge application sites require permits, so runoff on these sites should be controlled through regulations. Many swine, dairy, and poultry layer operations in the southeastern U.S. use liquid waste management systems that include a lagoon and land application of the lagoon water by irrigation. Bermuda grass and tall fescue are used on many of the land application sites. Nitrogen application rates of 400 to 600 kg/ha may be used.[69] With such high N application rates, there is the potential for surface runoff losses of N. Westerman et al.[90] applied swine manure slurry at a rate of 670 kg N/ha/yr and swine lagoon effluent to supply 600 and 1200 kg N/ha/yr for four years to tall fescue on a Cecil silt loan soil and compared these rates with 201 kg/ N/ha/hr of commercial fertilizer. They concluded both surface-water and groundwater contamination can occur by applying manure and effluent at these high rates. Pollution by runoff was more likely when rainfall occurred soon after manure application.

Many dairy farmers with small herds do not have manure storage systems and spread it daily in states like New York, Wisconsin, and Pennsylvania. By spreading manure on frozen or snow-covered ground, there is an increased potential for surface runoff. Hensler et al.[91] reported that up to 20% of the N was lost from manure applied to frozen, tilled soil. Young and Mutchler[92] found that soil cover influenced runoff and nutrient losses. Up to 20% of the N was carried away in the spring runoff from manured alfalfa plots, whereas no more than 3% of the N was lost from manure spread on fall plowed soils. Klausner et al.[93] found little difference in nutrient losses between different manure application rates when the soil was not frozen, but nutrient losses rose with increasing rates of application when the soil was frozen. Steenhuis et al.[94] found the fate of the first melt water after spreading manure on frozen soils largely determined the fate of the total N application. If this water infiltrates, the N losses will be small. If, however, the water runs off, the losses will be high. Thus, if manure is spread on frozen soil covered with an ice layer or on melting snow, high N losses can be expected.

There are concerns regarding surface water quality impacts of using poultry litter as a nutrient source. Nitrogen losses in surface runoff from litter and poultry manure from numerous studies are summarized in Table 3.1. The interval between manure application and rainfall affect the quality of runoff water. Westerman and Overcash[99] found that concentrations of TKN decreased by approximately 90% following a 3-day delay between application of poultry manure to fescue plots and simulated rainfall.

McLeod and Hegg[95] compared water quality impacts of commercial fertilizer, municipal sludge, dairy manure, and poultry manure applied to all fescue plots. One day after application runoff from the plots treated with poultry manure had 40 mg N/L TKN, 16 mg N/L NH_4, and 2.5 mg N/L NO_3. Simulated rainfall was used to produce runoff events at weekly intervals and after that, N concentrations decreased by 80% with increasing number of runoff events. Edwards and Daniels[100] also found highest N concentrations occurred in the first runoff event from tall fescue plots receiving poultry litter and inorganic fertilizer and that background concentrations (control) were approached after 2 to 5 runoff events.

Several authors have studied the effect of sludge application on the quality of runoff water from agricultural lands. Kelling et al.[101] found significant reductions in runoff and sediment losses from sludge treated areas compared with commercial

TABLE 3.1
Nitrogen Concentrations In Runoff From Areas Receiving Poultry Waste

Location	Soil	Waste	Loading Rate (Mg/ha)	Total N	NH_4 (mg N/L)	NO_3 (mg N/L	Reference
South Carolina	Clay	Litter	2.8–8.9	6–40	1–15	2–2.5	McLeod et al.[95]
Maryland	Silt loam	Litter	6.4	10–35	—	0.5–1.4	Magette et al.[96]
Maryland	Silt loam	Litter	4.7–6.7	3–7	0–0.5	—	Magette et al.[97]
North Carolina	Clay, sandy loam	Litter	4.1–8.2	129–165	19–39	1.3–2.1	Westermann et al.[98]
North Carolina	Clay, sandy loam	Manure	3.0–6.0	106–230	15–38	0.2–0.4	Westermann et al.[98]
North Carolina	Sandy loam	Manure	3.3	8–132	—	—	Westerman and Overcash[99]

fertilized plots. However, NO$_3$ losses in runoff water from sludge-treated plots increased compared with the control plots. Dunnigan and Dick[103] found that surface application of sewage sludge resulted in increased N losses relative to incorporated sludge. Bruggeman and Mostaghimi[103] found that surface application of sludge at a rate of 75 kg N/ha reduced runoff, sediment, and N losses compared with plots where no sludge was applied. Sludge applications of 150 kg N/ha increased the infiltration capacity of the soil, thereby reducing runoff but greatly increasing N yields. A sludge application of 75 kg N/ha on no-till plots seemed to be the best alternative for sludge disposal from a surface water quality standpoint.

3.4   GROUNDWATER–SURFACE WATER INTERACTIONS

There are three ways that groundwater interacts with streams. Streams may gain water from inflow of groundwater through the streambed, they lose water to groundwater by outflow through the streambed, or they do both, gaining in some reach areas, losing in other reaches.[104] In general, this shallow groundwater that interacts with streams is more susceptible to contamination because changing meteorological conditions strongly affect surface water and groundwater patterns. Precipitation, rapid snowmelt, or release of water from a reservoir upstream may cause a rapid rise in stream stage that causes water to move from the stream into the streambanks by a process known as bank storage (Figure 3.2). As long as the rise in stage does not overtop the strea banks, most of the volume of stream water that enters the streambanks returns to the stream within a few days or, in some cases, weeks. If large areas of the land surface are flooded, widespread recharge to the water table can take place in the flood area. In this case, the time it takes the recharged flood water to return to the stream by groundwater flow may take weeks, months, or years. Depending upon the frequency, magnitude,

BANK STORAGE

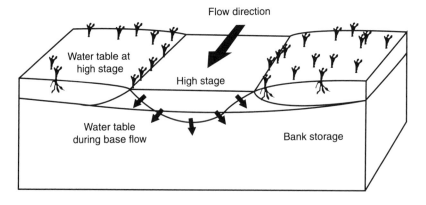

FIGURE 3.2   Bank storage in streams[104]

and intensity of storms and on the related magnitude of increases in stream stage, some streams and adjacent shallow aquifers may be in continuous readjustment from interactions related to bank storage and overbank flooding.

Other processes may also affect the exchange of water between streams and adjacent shallow aquifers. Pumping can cause changes in stream flow between gaining and losing conditions. In headwater areas, changes in stream flow between gaining and losing conditions may be extremely variable. The headwater segments of streams may be completely dry except during storms or during certain seasons when snowmelt or precipitation is sufficient to maintain continuing flow for days or weeks. During dry periods, the stream loses water to the unsaturated zone beneath its bed. However, as the water table rises through recharge in the headwater area, the losing reach may become a gaining reach as the water table rises above the level of the stream.

Significant denitrification has been found to take place at locations where oxygen is absent or present at very low concentrations and where suitable electron donor compounds, such as organic carbon, are available. Such locations include the interface of aquifers with silt- and clay-confining beds and along riparian areas adjacent to streams. McMahon and Bohlke[105] examined the effects of denitrification and mining on NO_3 loadings to surface water in Nebraska's South Platte River alluvial aquifer, which is affected by irrigation. Denitrification and mixing between river water and groundwater on the floodplain deposits and riverbed sediments substantially reduced NO_3 concentrations between recharge area and discharge area groundwater. Denitrification accounted for about 15–30% of the apparent decrease in NO_3 concentrations. Mass balance measurements indicated that discharging groundwater accounted for about 18% of the NO_3 load in the river. However, the NO_3 load in discharging groundwater was about 70% less than the load that would have resulted from the discharge of unaltered groundwater from the recharge area.

Several studies have shown that riparian zones can lower groundwater NO_3 concentrations to below 2 mg/L. Martin et al.[106] found that two riparian headwater stream zones in southern Ontario removed nearly 100% of the NO_3 from subsurface waters. Magette et al.[107] concluded NO_3 concentrations will be diluted in the groundwater by buffer areas of native riparian vegetation in the Chesapeake Bay watershed.

In studying surface-water and groundwater quality in a mixed land use watershed, Shirmohammadi et al.[108] concluded that lateral groundwater flow plays a major role in NO_3 loadings to streams in the Piedmont physiographic region. Nutrient management becomes an important priority in upland agricultural fields to reduce these loads. Ritter[109] concluded that groundwater discharge contributed 75% of the N load to the Delaware Inland Bays from nonpoint sources.

3.5 RIPARIAN ZONE PROCESSES

In the Atlantic coastal plain, broad coastal plains are transected by streams, scarps, and terraces. The gentle relief and sandy well-drained soils of the coastal plain make it ideal for agriculture. In many areas, cropland is separated from streams by riparian forests and wetlands. Evapotranspiration directly from groundwater is widespread in the coastal terrain.[104] The land surface is flat and the water table is generally close to the land surface; therefore, many plants have root systems deep enough to transpire groundwater at nearly the maximum potential rate. The result is that the evapotranspiration causes a significant water loss, which affects the configuration of groundwater flow systems.

Movement of nutrients from agricultural fields has been documented for the Rhodes River watershed in Maryland.[110] Application of fertilizer accounted for 69% of the N input to the watershed and 31% from precipitation. Forty-six percent of the N was taken up by harvested crops. Almost all of the rest of the N is transported in groundwater and is taken up by trees in riparian forests and wetlands or is denitrified to N gas before it reaches the stream. It was determined that less than 1% of the N reached the stream.

Martin et al.[106] found riparian zones of two streams in southern Ontario removed almost 100% of the NO_3 from subsurface waters. Attenuation was concentrated in the leading 20–30 m of the riparian zone. Forested riparian zones depleted NO_3 over a shorter distance than grassy riparian zones. Other studies have also shown that riparian zones can lower groundwater NO_3 levels below 2 mg N/L.[110, 111]

Nitrogen in surface runoff is removed in the riparian zones by plant uptake, denitrification, and sediment trapping.[112] Plant uptake alone may not be a permanent removal of required N unless the plants are harvested. Annual plants will die and release the N following decomposition. The relative importance of plant uptake and denitrification is site-specific for a given site and season of the year. Clausen et al.[113] found that neither of the two processes was important pathways for NO_3 removal in a 35-m riparian area of a field planted in corn.

3.6   EFFECT OF TILLAGE ON FATE AND TRANSPORT OF NITROGEN

3.6.1   SURFACE WATER

Conservation tillage will reduce erosion from 50 to 90% and the amount of particulate nutrients in runoff but can increase soluble nutrient concentrations in runoff.[114] The increase in soluble nutrient losses is attributed to the increase in the amount of surface residue and decrease in fertilizer incorporation. Baker and Laflen[115] showed that surface fertilizer significantly increased NH_4–N concentrations in runoff, as high as 5% of the NH_4–N applied was lost in runoff. In another study, Mickelson et al.[116] found surface-applied N losses with no-tillage were 14 times higher than with incorporated fertilizer N treatment

Some studies have shown that most N losses are associated with the sediment fraction. In evaluating six-tillage practices, Barisas et al.[117] found that the sediment fraction was the major carrier of N. In the highly erodible loessial soils in northern Mississippi, N losses from conventional tillage soybeans were 46.4 kg N/ha and 4.7 kg N/ha from no-tillage soybeans.[118] Staver et al.[119] found that the greatest potential for N transport in surface runoff from a coastal plain watershed in Maryland occurs during extreme precipitation events soon after N application. They observed very little annual difference of N surface runoff losses between conventional tillage and no-tillage.

In a comprehensive literature review, Baker[120] concluded that, in general, conservation tillage reduces runoff and losses of N via this route. The reduction in runoff volume has been variable between locations and years, but the average reduction with conservation tillage is probably 20–25%. The reduction in the amount of N in surface runoff as a result of conservation tillage has not been as great as the reduction in the amount of sediments. There is generally higher concentrations of dissolved N in the surface water and higher total N in the sediment. The higher average concentrations of dissolved N is a result of most fertilizer N being applied on the surface.

3.6.2 GROUNDWATER

Many studies have shown that conservation tillage decreases runoff and increases infiltration. Surface residues provide protection against surface sealing that results in increased infiltration before runoff occurs on well-structured soils. Because of the initial higher infiltration, NO_3 losses in surface runoff will be low, and with increased infiltration with conservation tillage, there is the potential for increased NO_3 leaching.

A number of studies have been conducted under different climate and soil conditions to study leaching of NO_3 under different tillage systems. Kitur et al.[121] found equal N fertilizer losses under no-till and conventional tillage systems. Kanwar et al.[122] found higher NO_3 leaching losses under conventional tillage systems in a rainfall simulation study. The results from that study indicated that most of the previously applied NO_3 present in the soil was bypassed by the applied water later, as it infiltrated through the macropores under no-till systems. In another study, Kanwar et al.[123] studied the effects of no-till and conventional tillage and simple and split N applications on the leaching of NO_3 with subsurface drainage of continuous corn. No significant effect of tillage or N management was observed during the first year of the experiments. However, in the third year, a significant reduction of NO_3 in subsurface drainage water with no-till relative to conventional tillage was observed.

An 11-year study in Minnesota showed there was very little difference in NO_3 losses between conventional tillage and no-tillage in subsurface drainage.[79] Nitrate concentrations were lower in the no-till plots, but the amount of subsurface drainage flow was higher, so NO_3 losses were approximately the same.

In Georgia, McCracken et al.[124] found no consistent differences between no-tillage and conventional tillage in their effect on NO_3 leaching and concluded the choice of tillage method will have minor impact on groundwater quality. In another study in western Tennessee and Kentucky, Wilson et al.[125] found there was little difference in NO_3 leaching rates between conventional annual tillage and no-tillage, but cropping systems and rainfall timing had pronounced effects. Cotton was the most susceptible crop to NO_3 losses. Research by Tyler and Thomas[126] in Kentucky demonstrated greater NO_3 leaching with no-tillage than conventional tillage. They concluded no-tillage enhanced the preferential leaching of NO_3 through macropores.

3.7   WHOLE-FARM NITROGEN BUDGETS

One method of predicting NO_3 leaching potential to groundwater is by calculation of N budgets for individual farms. The N budget can be formulated so that a positive balance would indicate the amount of N potentially available for leaching. The average amount of groundwater recharge could then be estimated to predict the mean maximum amount of NO_3 leached to the groundwater. The N budget can be simplified by assuming that soil organic matter, and consequently soil N content, remain constant on a yearly basis on monoculture systems or on a rotation basis for crop rotation systems. Farm N inputs need to be calculated for feed, fertilizer, and seed; nitrogen fixation; and atmospheric deposition. Outputs need to be estimated for animal and grain products leaving the farm along with atmospheric losses through N volatilization and denitrification. The simplified N balance approach for predicting the long-term effect of farming practices on groundwater quality has been described in detail by Fried et al.[127]

Sims and Vadas[128] estimated the N surplus for a poultry farm in Delaware with three poultry houses and 75 ha of cropland was 210 kg N/ha/yr. Klausner[129] estimated the N surplus for a typical New York dairy farm with 120 cows and 100 ha of cropland was 202 kg N/ha/hr. Poultry and livestock farms have much larger N surpluses than grain farms. In applying the N budget approach to farms in Ontario, Barry et al.[130] concluded that denitrification losses were a significant component of the N budget for grain corn and silage corn grown in southwestern Ontario. Neither Sims and Vadas[128] nor Klausner[129] considered denitrification or atmospheric N inputs in their N budget calculations. Barry et al. estimated a groundwater NO_3 concentration of 6.7 mg N/L for a cash grain farm in Ontario and 58.4 mg N/L for a dairy farm.

3.8   NITROGEN AND WATER MANAGEMENT PRACTICES TO REDUCE NONPONT SOURCE POLLUTION

3.8.1   NITROGEN MANAGEMENT PRACTICES

3.8.1.1   Accounting For All Sources

When multiple sources of N are used, it is important to account for all sources of N. Nitrogen available from manure applications, legumes, soil organic matter, and other sources should be accounted for before supplementary applications of N are made. The importance of accounting for all sources of N varies greatly from farm to farm and region to region, depending on the relative contributions of various sources of N to the soil-crop system.

3.8.1.2   Realistic Yield Goals

One of the important facets in determining N requirements for crops is yield. It is important to set realistic yield goals when deciding how much N to apply. Climate, crop genetics, crop management, and the physical and chemical properties of the soil have a significant effect on crop yield. The primary reason for using realistic yield goals is economic. Methods to set realistic yield goals include using farm averages, using a rolling 7- to 10-year field average or adjusting the past average and increase it by a chosen percentage (usually less than 5%) to take advantage of higher-yielding varieties.[131]

3.8.1.3   Amounts of Nitrogen To Apply

Applying only enough N to supply crop requirements should be used. Nitrogen needs can be supplied by commercial fertilizer or manure. When deciding how much manure to apply, it is important to know how much N is in the manure. The manure application method will determine how much NH_3 is lost.

3.8.1.4   Timing of Application

The most efficient method of using N fertilizer and minimizing its loss is to supply it as the crop needs it. Maximum N use occurs near the time of maximum vegetative growth. If irrigation is used, N may be applied through the irrigation system in four or five applications. For nonirrigated crops, split applications or side-dressing are two effective methods for controlling the timing of application. Manure should be applied as soon as possible after planting except when used as a N source to top-dress small grains.

3.8.1.5   Calibration of Equipment

It is important to calibrate manure and fertilizer applicator equipment. The task is simple and easy. Nitrogen in manure can be used more efficiently when a farmer knows how much manure the spreader is applying per unit area. Details on calibrating manure spreaders can be found in the Pennsylvania manure management manual.[132]

3.8.1.6   Early Season Soil And Plant Nitrate Tests

Early-season soil (preside-dress soil NO_3 test) and plant NO_3 tests have been developed for estimating available N contributions from soil organic matter, previous legumes, and manure under the soil and climatic conditions that prevail at specific production locations.[133, 134] These tests are performed 4 to 6 weeks after the corn is planted. Early-season soil NO_3 tests involve taking soil samples in the top 30 cm of the soil profile. Early-season plant NO_3 testing involves determining the NO_3 concentration in the basal stem of young plants 30 days after emergence. One disadvantage of the early season soil and plant NO_3 testing is that there must be a rapid turnaround between sample submitted and fertilizer recommendations from the soil testing laboratory. If side-dress N fertilizer is being used in conjunction with manure, the early-season NO_3 test should help reduce the potential for overfertilization.

3.8.1.7   Nitrification Inhibitors

Nitrification inhibitors are available to stabilize N in the NH_4 form. Stabilizing the N in manure by inhibiting nitrification should increase its availability for crop uptake later in the season, reduce its mobility in soil, and reduce its pollution potential under both conventional and conservation tillage.[135] Sutton et al.[136] found that stabilized swine manure had a similar efficiency for crop production as anhydrous NH_3. Nitropyin will temporarily slow nitrification in the soil.

3.8.1.8   Leaf Chlorophyll Meters

The use of leaf chlorophyll meters is a relatively new method to measure N in corn. Girardin et al.[137] demonstrated a strong relationship between N crop deficiency, photosynthetic activity, and leaf chlorophyll content. Lohry[138] was one of the first researchers to use leaf chlorophyll content to monitor the N status of corn. In recent years, chlorophyll meters have been used to schedule fertigation and side-dress N for corn.[139]

3.8.1.9   Cover Crops

Cover crops are used to prevent the buildup of residual N during the dormant season and prevent N leaching to groundwater in North America and Europe. In the U.S., cover crops are more widely used in the southeastern and Mid-Atlantic regions than other parts of the country. Some of the concerns that have limited their use are depletion of soil water by the cover crop, slow release of nutrients contained in the cover crop and difficulty in establishing and killing cover crops.[140] Nonlegume cover crops are much more efficient than legumes at reducing N leaching.

3.8.2   WATER MANAGEMENT

3.8.2.1   Irrigation Method

The irrigation method, insofar as it determines the uniformity, amount, and application efficiency, plays an important role in determining the irrigation management for obtaining the greatest N use efficiency. The coefficient of uniformity determines how efficiently water is applied to a field. By increasing the coefficient of uniformity, the application efficiency increases and N leaching losses are reduced.[141]

Wendt et al.[142] found that on a loamy, fine sand soil in Texas, less NO_3 was leached using subirrigation systems than with furrow or sprinkler systems. Furrow irrigation had the highest water requirements, whereas automatic subirrigation had the lowest. Water requirements for sprinkler irrigation and manual subirrigation were approximately the same. McNeal and Carlile[143] concluded that the typical furrow irrigation system for potatoes on sandy soils of the Columbia Basin area in Washington used much larger quantities of water than efficient sprinkler irrigation and produced extensive NO_3 leaching. Alternative furrow irrigation (where two adjacent irrigation furrows are never wet concurrently) produced considerably less NO_3 leaching than regular furrow irrigation. Surge-flow furrow irrigation offers improved opportunities for N management with fertigation.[139]

3.8.2.2.   Drainage Volume

Irrigation water management resulting in high leaching volume of 25–50% or more of the water applied will cause considerable leaching of N. Nitrate leaching is significantly reduced by water management techniques that result in very low drainage volumes and contribute relatively low mass emission of NO_3 in the drainage waters.[144] Letey et al.,[145] in studying the amounts of leached NO_3 on various commercial farming sites in California and on a controlled experimental plot, found using multiple regression analysis that the highest correlation was obtained from the amount of leached NO_3 vs. the product of the drainage volume and N fertilizer application. The second highest correlation was for amount leached vs. drainage volume. Smika et al.,[146] in a three-year study in Colorado on a sandy soil, found that for three center-pivot irrigation systems, average annual deep percolation losses were 16, 29, and 73 mm. The corresponding average annual NO_3 losses were 19.0, 30.4, and 59.7 kg N/ha, respectively.

3.8.2.3   Irrigation Scheduling

Irrigation scheduling based on soil moisture measurements or evapotranspiration (ET) requirements is the most practical water management method for controlling NO_3 leaching. With good irrigation scheduling, the required amount of water can be applied at the right time. Duke et al.[147] were able to successfully use the USDA irrigation computer scheduling program to determine the proper timing for irrigation and the amount of water necessary to maintain high crop yields and minimize leaching losses on sandy soils in Colorado. Wendt et al.[142] were able to maintain the N in the root zone for furrow, sprinkler, and subirrigation systems by irrigating on the basis of potential ET. When water applied was greater than the 2–2.5 times potential ET and NO_3 in the soil profile were greater than 200 kg/ha, the leachate concentrations were greater than 20 mg/L on a fine sand/loam soil.

Cassel et al.,[148] in developing a sprinkler irrigation schedule for soybeans on sandy loam soil in North Dakota, examined NO_3 leaching differences occurring with four water levels (dryland, under-irrigation, optimum irrigation, and over-irrigation). They found that NO_3 moved below the crop rooting zone with both heavy fertilizer N applications and water in excess of ET. Agronomists and engineers in the Hall County, Nebraska, Irrigation Management Quality Project[149] demonstrated that, with irrigation scheduling based on soil moisture measurements, reasonable corn yield goals are attainable with less irrigation water and supplemental N than is commonly used.

3.9   SUMMARY

The biogeochemical N cycle is very complex because N occurs in many valence states depending upon redox potential. Important N cycle processes include mineralization and immobilization, plant uptake, leaching, runoff, NH_3 volitalization, and denitrification. Sources of groundwater contamination include fertilizers, manures, and sludges. Shallow groundwater NO_3 concentrations in some parts of the U.S. may be high. The USGS NAWQA study found that 15% of the samples collected in shallow groundwater beneath agricultural and urban areas had NO_3 concentrations above 10 mg N/L. The lowest NO_3 groundwater concentrations are found in the southeastern U.S.

Surface water N concentrations are highest in agricultural areas. One of the major sources of N input to surface waters in the Corn Belt is through subsurface discharge. Field studies have shown that N losses in surface runoff are correlated with fertilization rates.

The best management practices to control N leaching can be classified as N management practices or water management practices. Accounting for all N sources is important before supplemental N applications of manure or fertilizer are made. Other N management practices include setting realistic yield goals, timing of N application, calibration of equipment, and use of cover crops. Newer N management practices being used today include early-season soil and plant NO_3 tests and leaf chlorophyll meters. Water management practices include irrigation application method, reducing drainage volumes, and irrigation scheduling.

REFERENCES

1. Bowen, G. D. and Smith, S. E., The effects of mycorrhizae on nitrogen uptake by plants, in *Terrestrial Nitrogen Cycles, Processes, Ecosystem Strategies and Management,* Clark, F. E. & Rosswall, T. Eds. Ecol. Bull. 33, 237, 1981.
2. Jansson, S. L. and Persson, J., Mineralization and immobilization of soil nitrogen, in *Nitrogen in Agricultural Soils.* Stevenson, F. J., Ed., Agronomy Monograph 22, ASA, Madison WI., 1982, 229.
3. Alexander, M., *Introduction to Soil Microbiology,* John Wiley & Sons, New York, NY., 1977.
4. Palm, C. A. and Sanchez, P. A., Nitrogen release from leaves of some tropical legumes as affected by their lignin and polyphenolic contents *Soil Biol. Biochem.,* 23:83, 1991.
5. Urquiaga, S., Giller, K. E. and Cadisch, G., Tracing mechanisms of nitrogen transfer from legume to grass in tropical pastures, in *Soil Management in Sustainable Agriculture,* H. Lee & H. Cook, Eds. Wye College Press, Wye, Ashford, UK., 1993, 104.
6. Toomsan, B., McDonagh, J. F., Limpinuntana, V. and Giller, K. E., Nitrogen fixation by groundnut and soybean and residual nitrogen benefits to rice farmers' fields in Northeast Thailand, *Plant and Soil,* 175, 45, 1995.
7. Anderson, R. V., Coleman, D. S. and Cole, C. V., Effects of saprotrophic grazing on net mineralization, in *Terrestrial Nitrogen Cycles. Processes, Ecosystem Strategies and Management Impacts,* Clark, F. E. and Rosswall, T., Eds., *Ecol. Bull.* 33, 201, 1981.
8. Rosswall, T., Microbiological regulation of the biogeochemical nitrogen cycle. *Plant and Soil,* 67, 15, 1982.
9. Clarholm, M., Protozoan grazing of bacteria in soil—impact and importance, *Microbiol. Ecol.* 7, 343, 1981.
10. Clarholm, M., Possible rules for roots, bacteria, protozoa and fungi in supplying nitrogen to the plants, Fitter, A. H., Atkinson, D., Read, D. J. & Usher, M. B., Eds. in *Ecological Interactions in Soil:Plants, Microbes and Animal,* Brit. Ecol. Soc. Spec. Publ. Vol. A, Blackwell Sci. Publ., Oxford, 1985, 355.
11. Hansson, A. C., Roots of arable crops: production, growth dynamics and nitrogen content, Swedish Univ of Agric. Sci., Dept of Ecology and Environmental Research, Report 28, 1987.
12. Olson, R. A. and Kurtz, L. T., Crop nitrogen requirements, utilization and fertilization, in *Nitrogen in Agricultural Soils,* Stevenson, F. J., Ed., Agronomy Monograph 22, Madison, WI, 1982, 567.

13. Schrader, L. C., Domska, D., Jung, P. U. and Peterson, A., Uptake and assimilation of ammonium-N and nitrate-N and their influence on the growth of corn (*Zea mays*), *Agron. J.* 64, 690, 1972.

14. Vietz, F. G. Jr, and Hageman, R. H., Factors affecting the accumulation of nitrate in soil, water and plants, Agricultural Handbook No. 413, U.S. Department of Agriculture, Washington, D.C., 1971.

15. Stevenson, F. J., *Cycles of Soil—Carbon Nitrogen, Phosphorus, Sulfur, Micronutrients,* John Wiley & Sons, New York, NY, 1986.

16. Newbould, P., The use of nitrogen fertilizer in agriculture, Where do we go practically and ecologically?, in *Ecology of Arable Land—Perspectives and Challenges, Developments in Plant and Soil Sciences,* Clarholm, M. & Bergstrom, L., Eds. Vol. 39, Kluwer Academic Publ., 1989, 281.

17. Nommik, H. & Vahtras, K., Retention and fixation of ammonium and ammonia in soils, in *Nitrogen in Agricultural Soils,* Stevenson, J. F., Ed., Agronomy Monograph 22, ASA, Madison, WI, 1982, 123.

18. Allison, F. E., Doetsch, J. H. and Roller, E. M., Availability of fixed ammonium in soils containing different clay minerals, *Soil Sci.,* 75, 373, 1953.

19. Gustafson, A., Leaching of nitrogen from arable land into groundwater in Sweden, *Environ. Geol.* 5, 65, 1983.

20. Kissel, D. E., Bidwell, O. W. and Kientz, J. F., Leaching classes of Kansas soils, Kansas State Univ., Agric. Exp. Sta. Bull. 64, 1982.

21. Bergstrom, L. and Johansson, R., Leaching of nitrate from monoligh lysimeters of different types of agricultural soils, *J. Environ. Qual.,* 20, 801, 1991.

22. Priebe, D. L. and Blackmer, A. M., Recovery of urea-derived 15N in calcareous soil following surface applications under wet and dry conditions, in *Agronomy Abstracts,* ASA, Madison, WI, 1985.

23. Bergstrom, L., Leaching of dichlorprop and nitrate in structured soils, *Environ. Poll.,* 87, 189, 1995.

24. Kanwar, R. S., Baker, J. L. and Laflen, J. M., Nitrate movement through the soil profile in relation to tillage system and fertilizer application method, *Trans., ASAE,* 28, 1802, 1985.

25. Bergstrom, L. and Brink, N., Effects of differentiated applications of fertilizer N on leaching losses and distribution of inorganic N in soil, *Plant and Soils,* 93, 333, 1986.

26. Meisinger, J. J., Hargrove, W. L., Mikkelsen, R. L., Williams, J. R. and Benson, V. W., Effects of cover crops on groundwater quality, in *Proc. of Int. Conf. on Cover Crops for Clean Water,* Hargrove, W. L., Ed., Soil and Water Conserv. Soc., Antieny, IA, 1991, 57.

27. Legg, J. O. and Meisinger, J. J. Soil nitrogen budgets, in *Nitrogen in Agricultural Soils,* Stevenson, F. J. Ed., Agronomy Monographic 22, 503, 1982.

28. Kussow, W., El-Swaify, A. A. and Mannering, J. *Soil Erosion and Conservation in the Tropics,* ASA Special Publication No. 43, ASA, Madison, WI, 1982.

29. Nicholaichuk, W. and Read, W. L., Nutrient runoff from fertilized and unfertilized fields in western Canada, *J. Environ. Qual.,* 7:542–544, 1978.

30. Burwell, R. E., Timmons, D. R., and Holt, R. F., Nutrient transport in surface runoff as influenced by soil cover and seasonal periods, *Soil Sci. Soc. Am. Proc.,* 39, 523, 1975.

31. Schuman, G. E., Burwell, R. E., Piest, R. F., and Spomer, R. G., Nitrogen losses in surface runoff from agricultural watersheds on Missouri Valley loess. *J. Environ. Qual.* 2, 299, 1973.

32. Lal, R., Effective conservation farming systems for the humid tropics, Soil Erosion and Conservation in the Tropics, Kussow, W. El-Swaify, S. A. and Mannering, J., Eds., *ASA Special Publication No. 43,* ASA, Madison, WI, 1982, 57.

33. Freney, J. R., Simpson, J. R., and Denmead, O. T., Ammonia volatilization, Clark, F. E. and Rosswall, T., Eds., Terrestrial Nitrogen Cycles, Processes, Ecosystems Strategies and Management Impacts, *Ecol. Bull.,* 33, 291–302, 1981.

34. Kirchmann, H., Esala, M., Morken, J., Ferm, M., Bussink, W., Gustavsson, J. and Jakobsson, C., Ammonia emissions from agriculture–summary of the Nordic seminar on ammonia emission, science and policy, *J. Nutrient Cycling in Agroecosystems,* 51, 84, 1998.

35. Crutzen, P. J., SST's a threat to the earth's ozone shield, *Ambio,* 3:201–210, 1972.

36. Ryden, J. C., N_2O exchange between grassland soil and the atmosphere, *Nature,* 292, 235, 1981.

37. Bouwman, A. F., Report No. 773004004, Natl. Inst. of Public Health and Environmental Protection, Bilthoven, The Netherlands, 1994.

38. Vroomen, H., Fertilizer use and price statistics, 1960–1988, Stat. Bul. 780, USDA, ERS, Washington, DC, 1989.

39. National Agricultural Statistics Service, 1998, Agricultural chemical use estimates for vegetable crops, USDA, NASS, ERS, Washington, DC, 1999.

40. National Agricultural Statistics Service, 1998 agricultural chemical use estimates for field crops, USDA, NASS, ERS, Washington, DC.

41. U.S. Department of Agriculture, 1998 farm and ranch irrigation survey, 1997 census of agriculture, Vol. 3, USDA, NASS, Washington, DC, 1999.

42. Madison, R. J. and Brunett, J. O., Overview of the occurrence of nitrate in groundwater in the United States, Water Supply Paper 2275, USGS, 1985.

43. Lee, J. K. and Nielsen, E. G., Farm chemicals and groundwater contamination, in *Agricultural and Groundwater Quality-Examining the Issues,* J. R. Nelson and E. M. McTernan, Eds., Univ. Center for Water Res., Oklahoma State Univ., Stillwater, OK, 1989. 2.

44. Texas State Soil and Water Conservation Board, a comprehensive study of Texas watersheds and their impact on water quality and water quantity, ISSWCB, Tempe, TX, 1991.

45. Spaulding, R. F. and Exner, M. E., Occurrence of nitrate in groundwater—a review, *J. Environ. Qual.,* 22, 392, 1993.

46. Baker, D. B., Wallrabenstein, L. K., Richards, R. P., and Creamer, N. L., Nitrates and pesticides in private wells of Ohio: a state atlas, Water Quality Lab., Heidelberg College, Tiffin, OH, 1989.

47. Kross, B. C., Halberg, G. R., Bruner, D. R., and Libra, R. D., The Iowa statewide rural well water survey. Water quality data: initial analysis, Iowa Dept. of Nat. Resour. Tech Inf. Ser. 19, Des Moines, IA, 1990.

48. Halberg, G. R. Nitrate in groundwater in the United States, in *Nitrogen Management and Groundwater Protection,* R. F. Follet, Ed., Elsevier, Amsterdam, Netherlands, 1989, 35.

49. Exner, M. E. and Spalding, R. F., Occurrence of pesticides and nitrate in Nebraska's groundwater, Water Center Publ. 1, Inst. of Agric. and Nat. Resour., University of Nebraska, Lincoln, NB, 1990.

50. Ward, P. C., Existing levels of nitrates in waters—the California situation, in *Nitrates and Water Supply: Source and Control,* 12 Sanit. Eng. Conf. Proc., Univ. of Illinois, Urbana, IL, 14, 1970.

51. Keeney, D. R., Nitrogen management for maximum efficiency and minimum pollution, in *Nitrogen in Agricultural Soils,* F. J. Stevenson, Ed., Monograph No. 22, ASA, Madison, WI, 1982, 605.

52. Devitt, D., Letey, J., Lund, J., and Blair, J. W., Nitrate-nitrogen movement through soil as affected by soil profile characteristics, *J. Environ. Qual.,* 5, 283, 1976.

53. Ruhl, J. F., Hydrologic and water quality characteristics of glacial-drift acquifers in Minnesota, Water Resources Invest. Report 87-4224, USGS, Minneapolis, MN, 1987.

54. U.S. Geological Survey, The quality of our nation's waters, nutrients and pesticides, Circular 1225, USGS, Reston, VA, 1999.

55. Reese, L. E. and Louden, T. L., Seepage from earthen manure storages and lagoons, a literature review, Paper No. 83-4569, ASAE, St. Joseph, MI, 1983.

56. Ritter, W. F. and Chirnside, A. E. M., Impact of animal waste lagoons on groundwater quality, *Biol. Wastes,* 34, 39, 1990.

57. Westerman, P. W., Huffman, R. L., and Feng, J. S., Swine lagoon seepage in sandy soil, *Trans. ASAE,* 38, 1749, 1995.

58. Huffman, R. L., Evaluating the impacts of older swine lagoons on shallow groundwater, in *1999 Animal Waste Management Symposium,* Havenstein, G. B., Ed., North Carolina State University, Raleigh, NC, 92, 1999.

59. Norstadt, F. A., and Duke, H. R., Stratified profiles: characteristics of simulated soils in a beef cattle feedlot, *Soil Sci. Soc. Am. J.,* 45, 827, 1982.

60. Schuman, G. F. and McCalla, T. M., Chemical characteristics of a feedlot soil profile, *Soil Sci.,* 119, 113, 1975.

61. Hatzell, H. H., Effects of waste-disposal practices on groundwater quality at five poultry (broiler) farms in north-central Florida, 1992–93, Water Resources Invest., Report 95-4064, USGS, Tallahassee, FL, 1995.

62. Ritter, W. F., and Chirnside, A. E. M., Impact of dead bird disposal pits on groundwater quality on the Delmarva Peninsula, *Bioresources Tech.,* 53, 105, 1995.

63. Ritter, W. F., Chirnside, A. E. M., and Scarborough, R. W., Nitrogen movement in poultry houses and under stockpiled manure, Paper No. 94, 4057, ASAE, St. Joseph, MI.

64. Lomax, K. M., Malone, G. W., Gedamu, N., and Chirnside, A., Soil nitrogen concentrations under broiler houses, *Applied Eng. Agric.,* 13, 773, 1997.

65. U.S. Environmental Protection Agency, Biosolids generation, use, and disposal in the United State, EPA 530, R-99-009, EPA Municipal and Industrial Solid Waste Division, Washington, DC, 1999, Chap. 1.

66. Moore, P. A., Best management practices for poultry manure utilization that enhance agricultural productivity and reduce pollution, in *Animal Waste Utilization: Effective Use of Manure as a Soil Resource,* Hatfield, J. A. and Stewart, B. A., Eds., Ann Arbor Press, Chelsea, MI, 1998, 89.

67. Adams, P. L., Danield, T. C., Edwards, D. R., Nichols, D. J., Pote, D. H., and Scott, H.D., Poultry litter and manure contributions to nitrate leaching through the vadose zone, *Soil Sci. Soc. Am J.,* 58, 1206, 1994.

68. Vellidis, G., Hubbard, R. K., Davis, J. G., Lawarence, R., Williams, R. G., Johnson, J. C., and Newton, G. L., Nutrient concentrations in the soil solution and shallow groundwater of liquid dairy manure land application site, *Trans. ASAE,* 39, 1357, 1996.

69. Westerman, P. W., Huffman, R. L., and Barker, J. C., Environmental and agronomic evaluation of applying swine lagoon effluent to coastal bermuda grass for intensive grazing, in *Proceedings 7th International Symposium on Agricultural and Food Processing Wastes,* Ross, C. C., Ed., ASAE, St. Joseph, MI, 1995, 162.

70. Hubbard, R. K., Thomas, D. L., Leonard, R. A., and Butler, J. L., Surface runoff and shallow groundwater quality as affected by center pivot applied dairy cattle wastes, *Trans. ASAE,* 30, 430, 1987.

71. Stone, K. C., Hunt, P. G., Humenik, F. J., and Johnson, M. H., Impact of swine waste application on ground and stream wate quality in an eastern coastal plain watershed, *Trans. ASAE,* 41, 1665, 1998.

72. U.S. Environmental Protection Agency, Standards for the use and disposal of sewage sludge, final rules, 40CFR parts 247, 405 and 503, *Federal Register,* 58, 9248, 1993.

73. Higgins, A. J., Land Application of sewage sludge with regard to cropping systems and pollution potential, *J. Environ. Qual.,* 13, 441, 1984.

74. Chang, A. C., Page, A. L., Pratt, P. F., and Warneke, J. E., Leaching of nitrate from freely drained-irrigated fields treated with municipal sludges, in *Planning Now For Irrigation and Drainage in the 21ˢᵗ Century, Proc. ASCE Irrigating and Drainage Division Conference,* Hays, E., Ed., Lincoln, NE, 1988, 455.

75. Jansson, R. E., Antel, R. S., and Borg, G. C. H., Simulation of nitrate leaching from arable soils treated with manure, in *Nitrogen in Organic Wastes Applied to Soils,* Hansen, J.A. and Kemiksen, K., Eds., Academic Press, London, England, 1989, 150.

76. Parker, C. F., and Sommers, L. E., Mineralization of nitrogen in sewage sludges, *J. Environ. Qual.,* 12, 150, 1983.

77. Dean, D. M., and Foran, M. E., The effect of farm liquid waste application on tile drainage, *J. Soil Water Conserv.,* 47, 368, 1992.

78. McLellan, J. E., Fleming, R. J., and Bradshaw, S. H., Reducing manure output to streams from subsurface drainage systems, Paper No. 93-2010, ASAE, St. Joseph, MI, 1993.

79. Zucker, L. A., and Brown, L. C., Agricultural drainage, water quality impacts and subsurface drainage studies in the Midwest, Bull. 871, Ohio State University, Columbus, OH, 1998.

80. Langdale, G. W., Leonard, R. A., Fleming, W. G., and Jackson, W. A., Nitrogen and chloride movement in small upland Pierdmont watersheds: II Nitrogen and chloride transport in runoff, *J. Environ. Qual.,* 8, 57, 1979.

81. Whitaker, F. D., Heinemann, H. G., and Burwell, R. E., Fertilizing corn adequately with less nitrogen, *J. Soil Water Conserv.,* 33, 28, 1978.

82. Burwell, R. E., Schuman, G. G., Heinemann, H. G., and Spomer R. G., Nitrogen and phosphorus movement from agricultural watersheds, *J. Soil Water Conserv.,* 32, 266, 1977.

83. Burisas, S. G., Baker, J. L., Johnson, H. P., and Loflen, J. M., Effect of tillage systems on runoff losses of nutrients, a rainfall simulation study, *Trans. ASAE,* 21, 893, 1978.

84. Kissel, D. E., Richardson, C. W., and Burnett, E., Losses of nitrogen in surface runoff in the Blackland prairie of Texas *J. Environ. Qual.,* 5, 288, 1976.

85. Timmons, D. R., Burwell, R. E., and Holt, R. F., Nitrogen and phosphorus losses in surface runoff from agricultural land as influenced by placement of broadcast fertilizer, *Water Res. Res.,* 9, 658, 1973.

86. Cornell Cooperative Extension, 1982. Cornell recommendations for field crops, Cornell University, College of Agriculture and Life Science, Ithaca, NY, 1981.

87. Madden, J. M. and Dornbush, J. N., Measurement of runoff and runoff carried waste from commercial feedlots, in *Proc. Int. Symp. on Livestock Wastes,* ASAE, St. Joseph, Mi, 1971, 44.

88. Baerman, S. J., Nutritional effects on waste management, MS Thesis, University of Nebraska, Lincoln, NE, 1995.

89. Westerman, P. W. and Overcash, M. R., Dairy open lot and lagoon irrigated pasture runoff quantity and quality, *Trans. ASAE,* 23, 1157, 1980.

90. Westerman, P. W., King, L. D., Burns, J. C., Cummings, G. A., and Overcash, M. R., Swine manure lagoon effluent applied to a temperate forage mixture: II Rainfall runoff and soil chemical properties, *J. Environ. Qual.,* 16, 106, 1987.

91. Hensler, R. F., Olsen, R. J., Witzel, S. A., Attoe, O. J., Paulson, W. H., and Johannes, R. F., Effects of methods of manure handling on crop yields nutrient recovery and runoff losses, *Trans. ASAE,* 13, 736, 1970.

92. Young, R. A., and Mutchler, C. K., Pollution potential of manure spread on frozen ground, *J. Environ. Qual.,* 5, 174, 1976.

93. Klausner, S. D., Twerman, P. J., and Coote, D. R., Design parameters for the land application of dairy manure, Report No. 600/2-76-187, EPA, Washington, DC, 1976.

94. Steenhuis, T. S., Burbenzer, G. D., Converse, J. C., and Walter, M. F., Winter spread manure nitrogen loss, *Trans. ASAE,* 24, 436, 1981.

95. McLeod, R. V. and Hegg, R. O., Pasture runoff water quality from application of inorganic and organic nitrogen sources, *J. Environ. Qual.,* 13, 122, 1984.

96. Magette, W. L., Brinsfield, R. B., Palmer, R. E., Wood, J. D., Dillaha, J. A., and Reneau, R. B., Vegetative filter strips for agricultural runoff, Report CBS/TRS 2187, EPA, Chesapeake Bay Office, Annapolis, MD, 1987.

97. Magette, W. L., Brinsfield, R. B., and Hrebenach, D. A., Water quality impacts of land applied broiler litter, Paper No. 88-2050, ASAE, St. Joseph, MI, 1988.

98. Westerman, P. W., Donnelly, T. L., and Overcash, M. R., Erosion of soil and poultry manure—a laboratory study, *Trans. ASAE,* 26, 1070, 1983.

99. Westeman, P. W. and Overcash, M. R., Short term attenuation of runoff pollution potential for land-applied swine and poultry manure, in *Livestock Waste: A Renewable Resource, Proc. 4ᵗʰ Int. Symp. on Livestock Wastes,* ASAE, St. Joseph, MI, 1980.

100. Edwards, D. R. and Daniel, J. C., Quality of runoff from fescue grass plots treated with poultry litter and inorganic fertilizer, *J. Environ. Qual.,* 23, 579, 1994.

101. Kelling, K. A., Walsh, L. M., Keeney, D. R., Ryan, J. A., and Peterson, A. E., A field study of the agricultural use of sewage sludge: II Effects in soil N and P, *J. Environ. Qual.,* 6, 345, 1977.

102. Dunnigan, E. P., and Dick, R. P., Nutrient and coliform losses in runoff from fertilized and sewage sludge-treated soil, *J. Environ. Qual.,* 9, 243, 1980.

103. Bruggeman, A. C. and Mostaghimi, S., Sludge application effects on runoff, infiltration and water quality, Paper No. 89-2623, ASAE, St. Joseph, MI, 1989.

104. Winter, T. C., Harvey, J. W., Franke, O. L., and Allez, W. M., Groundwater and surface water: a single resource, Circular 1139, USGS, Denver, CO, 1998.

105. McMahon, P. B. and Bohlke, J. K., Denitrification and mixing in a stream-aquifer system: effects of nitrate loading to surface water, *J. Hydrol.,* 186, 105, 1996.

106. Martin, T. L., Kaushik, N. K., Whiteley, H. R., Cook, S., and Nduhiu, J. W., Groundwater nitrate concentrations in the riparian zones of two southern Ontario streams, *Can. Water Res.,* 24, 125, 1999.

107. Magette, W. L., Wood, J. D., and Ifft, T. H., Nitrate in shallow groundwater, Paper No. 9-1502, ASAE, St. Joseph, MI, 1990.

108. Shirmohammadi, A., Yoon, K. S., and Magette, W. L., Water quality in mixed land-use watershed-Piedmont region in Maryland, *Trans. ASAE,* 40, 1563, 1997.

109. Ritter, W. F., Nutrient budgets for the Inland Bays, Tech Report, Agri. Eng. Dept., University of Delaware, Newark, DE 1986.

110. Correll, D L., Jordan, T. E., and Weller, D. E., Nutrient flux in a landscape—effects on coastal land use and terrestrial community mosaic on nutrient transport to coastal waters, *Estuaries,* 15, 431, 1992.

111. Osborn, L. L., and Kovaci, D. A., Riparian vegetated buffer strips in water quality restoration and stream management, *Freshwater Biol.,* 29, 243, 1993.

112. Mikkelsen, R. L., and Gilliam, J. W., Transport and losses of animal wastes in runoff from agricultural fields, in *Proc. of 7ᵗʰ Int. Symp. on Agricultural and Food Processing Wastes,* Ross, C. C., Ed., ASAE, St. Joseph, MI, 1995, 185.

113. Clausen, J. C., Wayland, K. G., Saldi, J. A., and Guillard, J., Movement of nitrogen through an agricultural riparian zone, I. Field studies, *Water Sci. Technol.,* 28, 605, 1993.

114. Mannering, J. V., Schertz, D. L. and Julian, B. A., Overview of conservation tillage, in *Effects of Conservation Tillage on Groundwater Quality,* Logan, T. J., Davidson, J. M., Baker, J. L., and Overcash, M. R., Eds., Lewis Publishers, Chelsea, MI, 1987, 3.

115. Baker, J. L. and Laflen, J. M., Effects of corn residue and fertilizer management on soluble nutrient runoff losses, *Trans. ASAE,* 251, 344, 1982.

116. Mickelson, S. K., Baker, J. L., and Laflen, J. M., Managing corn residue to control soil and nutrient losses, Paper No. 83-2161, ASAE, St. Joseph, MI, 1983.

117. Barisas, S. G., Baker, J. L., Johnson, H. P. and Laflen, J. M., Effect of tillage systems on nutrient loss: a rainfall simulation study, *Trans. ASAE,* 21, 893, 1978.

118. McDowell, L. L., and McGregor, K. C., Nitrogen and phosphorus losses in runoff from no-till soybeans, *Trans. ASAE,* 23, 643, 1980.

119. Staver, K., Brinsfield, R., and Magette, W., Nitrogen export from Atlantic coastal plain soils, Paper No. 88-2040, ASAE, St. Joseph, MI, 1988.

120. Baker, J. L., Agricultural areas as nonpoint sources of pollution, in *Environmental Impacts of Nonpoint Source Pollution,* Overcash, M. R., and Davidson, J. M., Eds., Ann Arbor Science Publishers, Ann Arbor, MI, 1983, 90.

121. Kitur, B. K., Smith, M. S., Blevins, R. L., and Frye, W. W., Fate of depleted ammonia nitrate applied to no-tillage and conventional tillage corn, *Agron. J.,* 76, 240, 1984.

122. Kanwar, R. S., Baker, J. L., and Laflen, J. M., Effect of tillage systems and methods of fertilizer application on nitrate movement through the soil profile, *Trans. ASAE,* 28, 1802, 1985.

123. Kanwar, R S., Baker, J. L., and Baker, D. G., Tillage and split N-fertilization effects on subsurface drainage water quality and crop yields, *Trans. ASAE,* 31, 453, 1988.

124. McCracken, B., Boy, J. E. Hargrave, W. L., Cabrera, M. L., Johnson, J. W., Raymer, R. L., Johnson, A D., and Harbers, G. W., Tillage and cover crop effects on nitrate leaching in the southern Piedmont, in *Clean Water Clean Environment–21st Century, Vol. II Nutrients,* ASAE, St. Joseph, MI, 1995, 135.

125. Wilson, G. V., Tyler, D. D., Logan, J., Thomas, G. W., Blevins, R. L., Dravillas, M. C., and Caldwell, W. E., Tillage and cover crop effects on nitrate leaching, in *Clean Water—Clean Environment–21st Century, Vol. II: Nutrients,* ASAE, St. Joseph, MI, 1995, 251.

126. Tyler, D. D., and Thomas, G. W., Lysimeter measurement of nitrate and chloride losses from conventional and no-tillage corn, *J. Environ. Qual.,* 6, 63, 1979.

127. Fried, M., Tanji, K. K., and Van de Pol, R. M., Simplified long term concept for evaluating leaching of nitrogen from agricultural land, *J. Environ. Qual.,* 5, 197, 1976.

128. Sims, T. J. and Vadas, P. A., Nutrient management planning for poultry grain agriculture, Report ST-11, Delaware Cooperative Extension, Univ. of Delaware, Newark, DE 1997.

129. Klausner, S. D., Managing nutrients responsibly, in *1993 Cornell Dairy Nutrition Conf. Proc.,* Dept. of Animal Sci., Cornell Univ., Ithaca, NY, 1993.

130. Barry, D. A., Goorahoo, D., and Gross, M. J., Estimation of nitrate concentrations in groundwater using a whole farm nitrogen budget, *J. Environ. Qual.,* 22, 767, 1993.

131. Taylor, R. W., Realistic yield goals for crops, Agron. Facts AF-3, Delaware Cooperative Extension, University of Delaware, Newark, DE, 1993.

132. Pennsylvania Department of Environmental Resources, Manure management for environmental protection, Graves, R. E., Ed., Commonwealth of Pennsylvania, Harrisburg, PA, 1986.

133. Magdoff, F. R., Ross, D., and Amadon, J., A soil test for nitrogen availability to corn, *Soil Sci. Soc. Am. J.,* 48, 1301.

134. Iversen, J. V., Fox, R. H., and Piekielek, W. P., The relationship of nitrate in young corn stalks to nitrogen availability, *Agron. J.,* 77, 927, 1985.

135. Sutton, A. K., Huber, D. M., Jones, D. D., and Kelly, D. J., Use of nitrification inhibitors with summer application of swine manure, *J. Appl. Eng. Agric.,* 6, 296, 1990.

136. Sutton, A. K., Huber, D. M., Jones, B. D., and Jones, D. D., Management of nitrogen in swine manure to enhance crop production and minimize pollution in *Proc. 7th Int. Symp. on Agricultural and Food Processing Wastes,* Ross, C. C., Ed., ASAE, St. Joseph, MI, 1995, 532.

137. Girardin, P., Tollenoor, M., and Muldon, J. F., The effect of temporary N starvation on leaf photosynthesis rate and chlorophyll content in maize, *Can. J. Plant Sci.,* 65, 491, 1985.

138. Lohry, R. D., Effect of nitrogen fertilizer rate and nitrapyrin on leaf chlorophyll, leaf nitrogen concentration, and yield on three irrigated maize hybrids in Nebraska, Ph.D. dissertation, Univ. of Nebraska, Lincoln, NE, 1989.

139. Schepers, J. S., Varvel, G. E., and Watts, D. G., Nitrogen and water management strategies to reduce nitrate leaching under irrigated maize, *J. Contam. Hydrol.,* 20, 227, 1995.

140. Ritter, W. F., Scarborough, R. W., and Chirnside, A. E. M., Winter cover crops as a best management practice for reducing nitrogen leaching, *J. Contam. Hydrol.,* 34, 1, 1998.

141. Rauschkolb, R. S., and Hornsby, A. G., *Nitrogen Management in Irrigated Agriculture,* Oxford University Press, New York, NY, 1994, 198.

142. Wendt, C.W., Onken, A.B., and Wilke, O.C., Effects of irrigation methods on groundwater pollution by nitrates and other solutes. Report No. EPA-600/2-76-291, EPA, Washington, DC, 1976.

143. McNeal, B. L. and Carlile, B. L., Nitrogen and irrigation management to reduce returnflow pollution in the Columbia Basin. Report No. EPA-600/12-76-158, EPA, Washington, DC, 1976.

144. Ritter, W. F., Nitrate leaching under irrigation in the United States—a review, *J. Environ. Sci Health, Part A., Environ. Sci. Eng.,* A24, 349, 1989.

145. Letey, J. J., Blair, J. W., Devitt, D., Lund, L. J. and Nash, P., Nitrate-nitrogen in effluent from agricultural tile drains in California, *Hilgardia,* 49, 289, 1977.

146. Smika, D. E., Heermann, D. F., Duke, H. R., and Batcheldet, A. R., Nitrate-N percolation through irrigated sandy soil as affected by water management, *Agron. J.,* 69, 623, 1977.

147. Duke, H. R. D., Smika, D. E., Heermann, D. F., Groundwater contamination by fertilizer nitrogen. *J. Irrig. Drain., Eng.,* 140, 283, 1979.

148. Cassel, D. K., Bauer, A., and Whited, D. A., Management of irrigated soybeans on moderately coarse-textured soil in the upper Midwest. *Agron. J.,* 70, 100, 1978.

149. University of Nebraska, Irrigation management demonstration program, Hall County, Water quality project, Cooperative Extension Service, Univ. of Nebraska, Lincoln, NE, 1982.

4 Phosphorus and Water Quality Impacts

Kenneth L. Campbell and Dwayne R. Edwards

TABLE OF CONTENTS

4.1 INTRODUCTION

Phosphorus (P) is a major nutrient that has many important roles and influences in production agriculture and natural ecosystems. It is essential to all forms of life and does not have toxic effects. Phosphorus is an essential element for plant growth, and its input has long been recognized as necessary to maintain profitable crop production. As one of the major plant nutrients, it is required by all plants, in varying amounts, for optimum growth and production. Phosphorus also is an important nutrient in the diet of animals and contributes to animal growth, maintenance, and production. For these reasons, it is often necessary to supplement the native P in the soil and in animals' diets with additional P. Even under good management practices, this can result in excess P available to move from agricultural production areas, especially in areas where animal wastes are being used as fertilizers.[1] In addition to these important roles in production agriculture, P has an important influence on the growth and makeup of both upland and wetland natural ecosystems. Different plants need P in different amounts, so the P concentration in an ecosystem affects the makeup of the ecosystem in both uplands and wetlands. This is especially true in ecosystems that have developed under conditions where P was the limiting nutrient for plant growth. Over many thousands of years, the natural ecosystems developed and were populated with different species of plants and animals, partially based upon their requirements for water, phosphorus, and other nutrients. As a result, the makeup of natural ecosystems where P is a limiting nutrient is very sensitive to the amount of P available in the system. When these natural ecosystems are adjacent to agricultural production systems or other sources of P, the potential exists for over-enrichment of these natural systems.

The presence of P in surface water bodies is recognized as a significant water quality problem in many parts of the world. Some forms of P are readily available to plants. If these forms are released into surface waters, eutrophic conditions that severely impair water quality may result. Phosphorus inputs can increase the biological productivity of aquatic ecosystems, changing their plant species or limiting their use for fisheries, recreation, industry, or drinking. The physical and chemical changes caused by advanced eutrophication (pH variations, oxygen fluctuations or lack in lower zones, organic substance accumulation) may interfere with recreational and aesthetic uses of water. In addition, possible taste and odor problems caused by algae can make water less suitable or desirable for water supply and human consumption.

The fate of P and P cycling in the environment are important factors in understanding the potential for, and impacts of, P transport through watersheds in agricultural and native landscapes. The fate and transport of P depend to a great degree on the behavior of the hydrologic system. Accumulation of P in soils, plants, plant detritus, sediments, and water can result in its movement within the system in ways, and to locations, that are not wanted. In addition, P transformations may occur that affect its characteristics and movement. These transformations are complex processes that are influenced by many characteristics of the soil, water, plant, and atmospheric environment. All these factors combine to make it very difficult to predict the water quality and associated environmental impacts of P in a specific situation. Following sections of this chapter will hopefully shed some light on at least a portion of these complex interrelationships and their expected impacts.

The impacts of P on aquatic systems depend on many factors and relationships among the plants, water, soil, and P. Most commonly, P is the nutrient that limits growth in freshwater aquatic systems.[2] The availability of P to vegetation, depending on its form and other factors, greatly influences the response of the aquatic system to its presence. Lake bottom sediments may be enriched with P from long-term accumulation with minimal adverse impacts on the system until some event occurs to disturb the system (e.g., a strong wind event on a very shallow lake that stirs up the bottom sediments, making large amounts of P available for algae growth and resulting in oxygen depletion and a fish kill).

Effective P control strategies depend on an understanding of the fate and transport of P in the watershed. Effective management of P for improved water quality involves two fundamental approaches: (1) limiting P inputs to the system through more efficient use, and (2) minimizing the transport of P offsite by use of improved management techniques, often called best management practices (BMPs), to reduce the amount of P carried by water. Unlike the case of nitrogen, P losses in the gaseous form do not occur naturally. Some P does become airborne in dust, but most P either remains in the soil or is removed by plants and water. These approaches to P management are addressed later in this chapter.

4.2 PHOSPHORUS SOURCES, SINKS, AND CHARACTERIZATION

Phosphorus is a naturally occurring element in soils. It is present in numerous different forms in the soil, many of which are not available to plants. These P forms can be broadly classified as particulate and dissolved. Phosphorus in the soil originates from the weathering of soil minerals and other more stable geologic materials. At any given time, most of the P in soils is normally in relatively stable forms that are not readily available to plants or dissolved in water. This generally results in low concentrations of dissolved P in the soil solution. Exceptions to this may occur in organic soils, where organic matter may accelerate the downward movement of P, and in sandy soils, where low P sorption capacities result in P being more susceptible to movement. Also, P may be more susceptible to movement in soils that have become anaerobic through waterlogging, where a decrease in soluble iron content and organic P mineralization occurs.[3]

Rainfall, plant residues, commercial fertilizers, animal manures, and municipal, agricultural, and industrial wastes or by-products are the major sources of P that may be introduced into the ecosystem, in addition to the natural weathering process of soil minerals. Land use and management determine which of these P sources are most important in any given location. As P is solubilized by the physical and chemical weathering processes, or added by input from any of the above major sources, it is accumulated by plants and animals, reverts to stable forms in the ecosystem, or is transported by water or erosion into aquatic systems where it is available to aquatic plants and animals or deposited in sediments.

The P cycle includes interactions and transformations occurring through a variety of physical, chemical, and microbiological processes that determine the forms of P, its availability to plants, and its transport in runoff or leaching. These processes and mass pools of P that together make up the P cycle are illustrated in Figure 4.1. Soil P exists in inorganic and organic forms. Fractionation of these P forms describes their relative availability to plants and for water transport in the soil solution. Organic P forms mineralize and replenish the inorganic P pool through microbial activity. Through the immobilization process, inorganic P may be converted to organic P under some conditions. Inorganic P is converted from mineral forms to bioavailable and soluble forms by dissolution through the weathering process. Through a variety

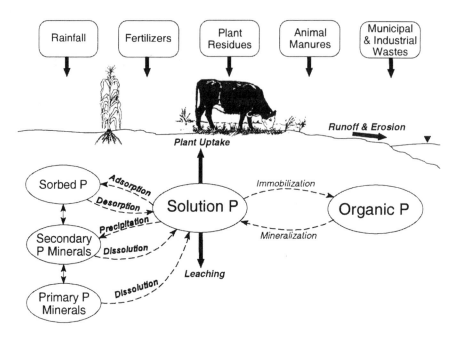

FIGURE 4.1 A representation of the phosphorus cycle in the soil–water–animal–plant system.

of chemical reactions collectively referred to as P fixation or precipitation, soluble and bioavailable P forms may be held in place in the soil. The presence of clays, Al, Fe, organic C, and $CaCO_3$ in soil greatly affects the portion of bioavailable and soluble P through adsorption and desorption relationships. Soil solution P is readily available for uptake by plants and transport by water in leaching or surface runoff. Part of the plant uptake P is removed in harvest of crops, part may be recycled on the soil surface as animal waste from grazing animals, and part may return to the soil in plant residues remaining on the surface and as decaying root mass. Additional sources of P are introduced to the soil system as discussed in the previous paragraph. Phosphorus transported from the soil system in soluble form or adsorbed to eroded sediments may be trapped temporarily or permanently in any of several sinks or transported into streams, wetlands, lakes, or estuaries. A more extensive discussion of the P transformations and processes can be found in Sharpley.[3]

Potential sinks for P include fixation in the soil, deposition with sediment in low areas of the landscape; and deposition or plant uptake in field buffer strips, treatment wetlands, and riparian zones. All of these potential sinks have upper limits to the quantity of P that can be retained and may be more or less effective depending on a range of conditions. Phosphorus that is transported through all of these potential sinks into streams or lakes may be adsorbed by bottom sediments, stored there for significant periods of time, and later released back into the water. Some of this P reaching streams or lakes may remain in the bottom sediments as a long-term sink.

The effectiveness and dynamics of the above-described P sources and sinks in an individual watershed are primary determining factors in the potential of the occurrence of adverse environmental impacts at that location. The potential for transport of P from sources to sinks or aquatic systems is another primary determining factor of offsite impacts. A primary goal of P management is to identify areas with high potential P sources and transport, then implement practices to minimize adverse impacts. Methods to accomplish this are discussed in a later section of this chapter.

4.3   INTRODUCTION OF PHOSPHORUS INTO THE ENVIRONMENT

Because P is a major nutrient required for plant growth, it is frequently applied to meet crop needs. A portion of this fertilizer P often becomes unavailable to the crop because of reactions with soil minerals, so more P may need to be added than will be used by the crop. If this process continues annually, it results in a continuing accumulation of P in the soil and a new equilibrium level of dissolved P in the soil solution. This solution concentration is referred to as the equilibrium phosphorus concentration (EPC). Because the P concentration in solution is particularly important to potential water quality effects, an increase in EPC because of the increasing P content of the soil is an undesirable situation with regard to water quality.[4]

Phosphorus may also be introduced into the environment as a by-product of animal production. In pasture production systems, animal wastes usually occur in quantities that are not a problem in affecting water quality unless the animals spend excessive time in or very near water bodies that may flow off-site. However, many animal production systems are managed in highly concentrated numbers in restricted areas or under confinement where accumulated wastes must be disposed of in some manner. Often these wastes are applied to land, either at agronomic rates for crop production, or in a disposal mode of operation. In either case, an accumulation of P in the soil may occur just as in the application of commercial fertilizer discussed above. This results in an increase in solution P concentration as the EPC increases. Animal waste applications may increase the EPC more than equivalent additions of commercial fertilizer in some cases.[4] The increased residual P levels in the soil from all application sources lead to increased P loadings to surface water, both in solution and attached to soil particles.

The major significance of P as a water pollutant is its role as the limiting nutrient in eutrophication. Eutrophication is the process by which a body of water becomes enriched in dissolved nutrients and, often, seasonally deficient in dissolved oxygen. This is a naturally occurring process characterized by excessive biological activity, but it is often accelerated by pollution from human activities. When P enters surface waters, it often becomes a pollutant that contributes to the excessive growth of algae and other aquatic vegetation and may cause a change in the dominance of aquatic plant species in wetlands. Other nutrients essential for plant growth generally occur naturally in the environment in sufficient quantities to support plant and algae growth in water bodies. Amounts of P in the water exceeding the minimum required for algae growth can lead to accelerated eutrophication.

Consequences of this accelerated eutrophication include reduced aquatic life and species diversity because of the lowered dissolved oxygen levels and increased biological oxygen demand (BOD). It also usually results in degradation of recreational benefits and drinking water quality with associated increased treatment costs. Unlike pathogenic bacteria and nitrates from agricultural sources, eutrophication from excessive P has not been considered a public health issue. However, some toxic algae may flourish in the presence of excessive nutrients, causing a public health concern.

4.4 PHOSPHORUS DYNAMICS IN CROP/SOIL/WATER SYSTEMS

4.4.1 FACTORS INFLUENCING P TRANSFORMATIONS AND PROCESSES

As described earlier, P in soil and water can experience adsorption/desorption, precipitation/dissolution, immobilization/mineralization, and plant uptake/plant decomposition as its characteristics are chemically and biologically altered. The rates at which these opposing processes occur, the relative proportions of P present in a given physical or chemical state, and even which of the opposing processes dominates at a particular time are complex functions of soil, weather, and crop variables.

4.4.1.1 Adsorption/Desorption

Adsorption and desorption are opposing processes that affect the degree to which P is held by chemical bonds to reactive soil constituents and, conversely, the degree to which it exists in solution. The proportions of P presented in adsorbed and solution forms are quite important in the context of pollution by P, because the mechanisms by which pollution occurs differ between the forms. Adsorbed P can cause pollution when transported along with eroded soil, whereas solution P is transported in the runoff itself independently of eroded soil.

Relationships between adsorbed and desorbed (or solution) P concentrations are commonly specified in the form of isotherms, which relate adsorbed P concentration to equilibrium solution P concentration. Figure 4.2 contains examples of isotherms for two hypothetical soils. The isotherm for Soil A is seen to lie above that for Soil B at all points, indicating that more P must be adsorbed by Soil A to achieve the same solution P concentration as Soil B at equilibrium. Another way of viewing the isotherm is that Soil B reaches a given equilibrium solution P concentration with less additional P adsorption than Soil A. The x-intercept of the isotherm is referred to as the equilibrium P concentration at zero sorption, or EPC_0. The EPC is thus the equilibrium concentration of solution P for a given soil in the absence of P addition or extraction. The EPC_0 of Soil B is higher than that of Soil A, indicating that in their original states, the soil solution P concentration is greater in Soil B than in Soil A.

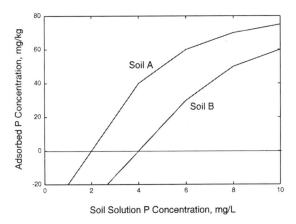

FIGURE 4.2   Example phosphorus isotherms for two hypothetical soils.

Standard equations have been used to describe the relationships between adsorbed and solution P, referred to as the Langmuir and Freundlich isotherm equations. The Langmuir equation is given by

$$C_A = \frac{Q^o b C_S}{1 + C_S} \qquad (4.1)$$

where C_A is adsorbed P concentration, C_S is solution P concentration, Q^o is maximum adsorption at the given temperature, and b is a parameter related to adsorption energy. The Langmuir equation thus considers adsorbed P concentration as an approximately linear function of solution P concentration. If the adsorption energy parameter is not constant, then the isotherm might be better described by the Freundlich isotherm equation, given by

$$C_A = KC_S^{1/n} \qquad (4.2)$$

where K and n are constants. As opposed to the Langmuir isotherm, the relationship between adsorbed and solution P concentrations is nonlinear for the Freundlich isotherm equation. Isotherm equation parameters can be determined empirically or, in the case of Langmuir isotherm parameters, estimated from equations such as those developed by Novotny et al.[5]

A key point about the curves in Figure 4.2 is that these curves demonstrate how amounts of adsorbed and solution P would change as a result of P addition. If solution P were extracted from the soil (e.g., from plant uptake or leaching of solution P) so that the soil solution P concentration fell below its equilibrium value, then P desorption would occur until a new equilibrium was established. This process would not, however, follow the same isotherm that describes adsorption. Desorption does not occur as readily as adsorption. Although a portion of adsorbed P is readily available for desorption (and thus for plant uptake, runoff transport, and leaching), a

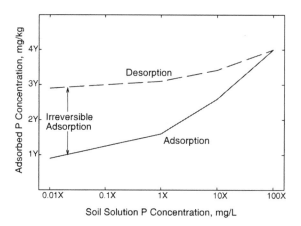

FIGURE 4.3   Illustration of typical hysteresis in the relationship between soil solution and adsorbed phosphorus concentrations.

significant amount of P is desorbed relatively slowly, if at all. This process is illustrated in Figure 4.3, which demonstrates the typical hysteresis in the relationship between soil solution and adsorbed P concentrations. The practical implication is that significant, soil-specific laboratory analyses must be performed before isotherms can be used to reliably predict adsorption/desorption dynamics, and this creates a practical challenge to their use.

The specifics of the chemical bonding that occurs between P and reactive soil constituents during adsorption are not well understood. As a result, much of the evidence regarding how various factors affect adsorption/desorption is empirical. However, published research studies have been very valuable in identifying the variables that influence adsorption/desorption and assessing their general effects. The primary variables controlling P adsorption include soil clay, Fe and Al, $CaCO_3$, and particulate organic matter contents. An increase in any of these variables generally favors P adsorption. In acidic soils, the first three variables are dominant in governing P adsorption, whereas Fe and $CaCO_3$ contents control adsorption in calcareous soils (soil pH is therefore also influential in P adsorption). Irrespective of pH, adsorption is favored at low soil P contents because of relatively low competition for adsorption sites. It follows that the adsorbed proportion of P in weathered soils can be high, because these soils typically contain relatively high clay, Al, and Fe contents. Sandy soils, in contrast, contain relatively low amounts of reactive constituents and thus promote P occurrence in solution. Organic matter can enhance P adsorption, but only within limits; very high organic matter (for example, peat and heavily manured soils) can favor occurrence of solution P, perhaps because the organic matter interferes with P adsorption sites.[6]

4.4.1.2   Precipitation/Dissolution

Precipitation is a P fixation process that denotes the formation of discrete, solid materials. Phosphorus that has been precipitated is generally considered not susceptible to transport by runoff alone and is less susceptible than P associated with fine soil particles. Similar to adsorption/desorption dynamics, precipitation/dissolution dynamics can thus be of considerable importance in the context of pollution by P.

The controlling mineral(s) in precipitation reactions is highly pH dependent. In calcareous soils, P combines with $CaCO_3$ to form apatites. At lower pH, P combines instead with Fe and Al. The amount of P potentially precipitated depends on the presence of Ca or Fe/Al, depending on pH. Dissolution, the opposite of precipitation, is also very pH dependent, with maximum dissolution occurring at pHs of 6–6.5 (which is one reason why most soils used for agricultural production are managed to have slightly acidic pH). In some texts, precipitation/dissolution is not treated as a separate set of opposing processes, but is instead considered part of the adsorption/desorption processes (e.g., Novotny et al.[5]).

4.4.1.3   Mineralization/Immobilization

Mineralization (biological conversion of organic P to mineral P) and immobilization (conversion of mineral P to microbial biomass) are opposing processes that occur continuously and simultaneously. In comparison with processes described earlier, mineralization/immobilization is of low direct importance in the context of P pollution, because the physical form of soil P (adsorbed/precipitated versus solution) is of more importance than the chemical form (inorganic versus organic). Mineralization and immobilization dynamics are of indirect importance, however, in the sense that they influence plant uptake of P, which does have a relatively direct impact on pollution by P.

The term *net mineralization* is often used to denote the difference between amounts of P mineralized and immobilized. Relative to net N mineralization, equations that relate net P mineralization to influential factors are underdeveloped. Rather than equations such as those developed by Reddy et al.[7] for N, the tools most commonly used to estimate P mineralization are empirical rate coefficients for a presumed first-order mineralization model. Sharpley[8] and Stewart and Sharpley[9], for example, reported that from 2% (temperate climate) to 15% (tropical climate) of soil organic P was mineralized annually. Data of a similar nature are most often used to estimate the amount of P mineralized from animal manures; SCS[10] estimates that from 75 to 80% of manure P is mineralized in the first year following land application, with an additional 5–10% per year mineralized in the next two years. One of the most notable exceptions to the simplified methods of describing P mineralization/immobilization is the model developed by Jones et al.[11] and included for use in the Erosion/Productivity Impact Calculator (EPIC) model.[12]

The numbers quoted in the preceding paragraph demonstrate that there can be large differences in P mineralization rates depending on the degree to which organic P is resistant to mineralization. A relatively high proportion of the organic P in animal manures is readily mineralizable to plant-available, or labile, forms with relatively high mineralization rate coefficients. The remaining organic P is more resistant to mineralization and has lower mineralization rate coefficients.

The factors that govern P mineralization are similar to, and in many cases the same as, those that are important in N mineralization. For example, P mineralization occurs at optimum rates during warm, moist conditions. Assuming that no other nutrients (e.g., N) limit microbial biomass production, net P mineralization depends on the C:P ratio. Ratios less than 200 favor net mineralization, whereas ratios of greater than 300 favor net immobilization. Mineralization and immobilization are in approximate balance for C:P ratios of 200:300. Farming techniques that maintain high C:P residues (e.g., no-till) appear to have mixed effects on P mineralization. Data reported by Sharpley and Smith[13] suggest that the tendency of residues to create conditions favorable for mineralization (i.e., maintenance of soil moisture and warm temperatures) might offset the tendency of residues having high C:P ratios (e.g., corn and wheat) to promote immobilization, even when those residues were incorporated.

The balance between mineralization and immobilization can obviously be influenced by addition of P forms to the soil. Treating soil with mineral fertilizer will (at least initially) result in an increased proportion of inorganic P, just as treatment with manures will increase the organic P proportion. As implied in our earlier discussion, other soil amendments can influence the balance between inorganic and organic P forms. Addition of N to soils having high C:N ratios can promote N mineralization and thus P mineralization because the two nutrients are used simultaneously by the mineralizing microbes. Conversely, treatment with materials having high C content (e.g., straw or stalks), especially if incorporated, can favor immobilization and shift the balance in favor of organic P.

4.4.1.4 Plant Uptake

Crops affect the fate and transformations of P through uptake and conversion to plant material. Crop uptake affects pollution by P, but not as clearly or immediately as adsorption/desorption and precipitation/dissolution. Phosphorus that has been extracted by plants is generally considered unavailable for loss in leachate, runoff, or eroded soil. Sharpley[14], however, has shown that P leached from a cotton, sorghum, or soybean canopy can constitute as much as 60% of runoff P. Since plant extraction of P occurs in the root zone, which is some depth beneath the soil surface, any effect of reducing P near the soil surface is not immediate. In fact, it might be possible in some cases for the presence of plants to increase pollution by P. If the distribution of soil P is such that the soil surface is relatively deficient in P, then the contribution of P leached from the crop canopy might cause a net increase in P runoff relative to what would have occurred with no crop present.

The P content of grain crops typically ranges from 0.2 to 0.6% of dry matter harvested; the average P content is similar for forage crops but with a wider range, from approximately 0.1 to 0.9%.[10] A substantial amount of P can thus be tied up in organic form as plant material. For example, corn yielding 11,300 kg/ha can uptake 50 kg P/ha. Typical forage crop uptake of P can range from 25 kg P/ha for fescue (7000 kg/ha) to 50 kg P/ha for clover/grass mixtures (14,000 kg/ha). Examples of typical annual P uptake for selected crops are given in Table 4.1.

As indicated in Table 4.1, the amount of P that can be converted into plant material depends strongly on the crop. Comparing typical annual uptakes of oats and corn, for example, it can be seen that corn takes up more than 2.5 times the uptake of oats

TABLE 4.1
Typical P Uptake of Common Crops

Crop	Yield (kg/ha)	P uptake (kg/ha)
Corn	11,300	50
Soybeans	3,460	26
Grain sorghum	8,400	39
Wheat	9,500	25
Oats	3,600	20
Barley	6,500	32
Tall fescue	13,500	55
Clover	13,500	44
Bermudagrass	18,000	47
Alfalfa	18,000	59

at typical yields. Phosphorus uptake also depends on all other factors that influence crop growth, such as temperature, soil moisture, soil pH, and availability of other nutrients. Conditions that favor plant growth will promote P uptake and thus maximize potential P removal, in turn maximizing the transformation of adsorbed P to solution P.

Plants are considered to use primarily inorganic P extracted from the soil solution. Plant uptake thus decreases soil solution P concentration, in turn promoting desorption of adsorbed P, as described earlier. If the crop is harvested and removed, then, the net effect is one of "mining" adsorbed P. On the other hand, if the crop is not removed but is recycled, as through grazing, the crop production basically has no net effect on quantity of P present.

It should be recognized that P uptake by plants integrates the processes of adsorption/desorption, precipitation/dissolution, and mineralization/immobilization. Each of these pairs of processes impacts on the physical and chemical forms of P present in the soil and is therefore capable of limiting plant uptake of P.

4.5   PHOSPHORUS LOADINGS TO AQUATIC SYSTEMS

4.5.1   FACTORS INFLUENCING P TRANSPORT PROCESSES

Three elements must be present for P from nonpoint sources to enter aquatic systems: P must be available in a transportable form (i.e., in solution or adsorbed to soil particles) at or near the soil surface, there must be an agent to achieve movement of soil P to "edge-of-field," and there must be an agent capable of continuing the transport of P from edge-of-field to the aquatic system. Except where P leaching is significant, the edge-of-field transport agent is runoff, and the continuing transport agent is stream flow. Under conditions favorable for P leaching (e.g., sandy soils, organic soils, high soil P content, low soil Al and Fe contents), however, subsurface water can be thought of as an edge-of-field transport agent. Wind can also be considered a transport agent because of its ability to transport P associated with soil particles. Phosphorus transport can then be thought of as governed by three sets of factors: availability factors, edge-of-field transport factors, and in-stream transport factors.

The Phosphorus Index is a concept currently being considered in many states as a tool to assess the potential risk of P loss from agricultural land to nearby water bodies. Several variations of the P Index are being developed in different regions to best adapt to the concerns and needs related to P sources, transport, and management factors in those regions. The ranking of the P Index identifies sites where the risk of P movement may be relatively higher than that of other sites. Review of the individual parameters making up the index rating may indicate particular factors that are causing a high risk rating and, therefore, may become the basis for planning corrective soil and water conservation practices and management techniques.

4.5.1.1   Surface Transport

Phosphorus in runoff is transported in either soluble form or particulate form. The particulate form is also called "sediment P," denoting its association with eroded soil and other solid materials. The availability factors in the context of surface water are those that govern the amount and physical form (i.e., adsorbed or solution) of P near the soil surface (1–2 cm). The transport availability factors, therefore, include all variables that affect P transformations (e.g., soil pH, cover crop, clay content, and presence of residue) as well as management practices that affect P transport availability. For example, the method of P application (surface versus incorporated) and addition of other soil amendments (e.g., lime) have direct effects on the amount and form of P present near the soil surface. Cultivation can affect P transport availability, particularly when P is surface-applied. Because of the relatively low mobility of P, surface application tends to produce relatively high P concentrations at or near the soil surface, with concentrations decreasing with increasing soil depth. Cultivation can decrease P availability for transport by turning under a high P content soil surface layer and exposing in its stead a layer of relatively P-deficient soil.

As noted earlier, the prime edge-of-field transport mechanism for surface water is runoff. Water erosion can be thought of as another edge-of-field transport mechanism for P, but it is probably more properly considered a subset of runoff because it occurs only in conjunction with runoff and is dependent on runoff amount and rate. The single most important runoff factor is precipitation, particularly in the form of rainfall, and specifically rainfall parameters such as total depth and duration. The next most important transport factor is soil texture, because of its joint role with precipitation parameters in determining the occurrence and amount of runoff. For a given rainfall event, coarse soil textures (for example, high sand content) favor infiltration, whereas fine-textured soils (e.g., high clay content) favor runoff. For a given soil texture, intense precipitation events (relatively large depths and short durations) will favor runoff, while more infiltration occurs during less-intense storms. Soil cover is closely rated to texture, in that low cover promotes high runoff. Soils with a good cover or residue will have relatively low runoff. High soil moisture at the time of the rainfall event diminishes the amount of water that can be stored before runoff occurs and thus favors the occurrence of runoff.

The amount of P experiencing edge-of-field transport is directly related to runoff amount, as is discussed further. To predict P transport or estimate it when data are unavailable, then, it is necessary to be able to predict runoff as a function of the influential factors. The SCS[15] curve number model is a widely used runoff estimation method which can be easily applied to estimate runoff as a function of soil texture, cover, antecedent moisture, and rainfall. The hydraulic properties of a particular soil for given cover and soil moisture are summarized in a single parameter known as the curve number which, taken together with total rainfall, is used directly to calculate the associated runoff.

In some cases, the rates of runoff, in addition to runoff amounts, are important. Detachment and transport of soil particles, for example, increases with runoff rate. The unit hydrograph method is a popular means of estimating runoff rates as a function of physical characteristics such as slope, flow length, and surface roughness. There are abbreviated methods available for estimating only peak flows, if it is not necessary to know flow rates throughout the duration of runoff.

There are many other models and equations that can be used similarly to characterize transport agents, many of which are more physically based. Haan et al.[16] and Chow et al.[17] provide excellent descriptions of runoff estimation procedures that cover a wide range in physical basis and ease of application.

Soil erosion is the pathway by which P associated with soil particles is transported from its origin to the edge-of-field. Similar to runoff estimation, there are a variety of methods available for estimating soil erosion on an annual or event basis. The Modified Universal Soil Loss Equation (MUSLE)[18] is oriented toward event sediment yield estimation based on field properties and runoff characteristics and is one of the simplest erosion prediction methods in general use. Toward the opposite end of the complexity spectrum is the soil detachment and transport algorithm developed by Foster et al.[19], that is included in the Water Erosion Prediction Project (WEPP) model.[20] The Revised Universal Soil Loss Equation (RUSLE)[21] can be used to estimate gross erosion on either an annual or event basis. The RUSLE exists in software form and is relatively easy to implement.

Estimation of P transport from source areas often takes the form of relatively simple empirical equations. Soluble P can be estimated, for example, from the relationship:[22]

$$P_S = \frac{K P_A B D I^\alpha W^\beta}{V} \tag{4.3}$$

where P_S is event average concentration (mg/L) of soluble P in runoff, P_A is soil test (Bray 1) P concentration (mg/kg) in the top 50 mm of soil, B is bulk density (mg/m^3), D is the effective depth (mm) of interaction between runoff and soil, t is the duration of runoff (min), W is the ratio of runoff to suspended sediment volumes, and V is event runoff (mm). The parameters K, α, and β are soil-specific constants that have been determined and reported (e.g., Sharpley[23]) for selected soils. A simpler equation, having the form

$$P_S = CKP_A V \qquad (4.4)$$

is used in the EPIC model[12], where C is a unit conversion coefficient; K is the ratio of runoff to soil P concentrations; and P_s, P_A, and V are as previously defined.

Transport of particulate P is often estimated using the enrichment ratio (ratio of sediment P content to parent soil P content) concept. The first step in this approach is to estimate sediment yield from the field of interest, using methods described earlier or others. It is known that the nutrient content of eroded soil is generally significantly higher than that of the parent soil because of selective transport of finer particles and the association of nutrients with finer particles. Novotny and Olem[24], for example, report that the total P content of eroded soil is approximately twice that in the original soil, resulting in an enrichment ratio of 2.0. Sharpley[8] reported enrichment ratios of approximately 1.5 for six western soils and related enrichment ratio to sediment yield as

$$\ln(R_E) = 1.21 - 0.16\ln(Y) \qquad (4.5)$$

where R_E is the enrichment ratio and Y is the sediment yield (kg/ha). Thus, particulate P content, P_P, can be estimated from

$$P_P = R_E P_A Y \qquad (4.6)$$

where all terms are as defined earlier. Storm et al.[25] and Novotny et al.[5] developed models of P transport that are considerably advanced in terms of their physical basis.

The in-stream transport factors are those related to stream velocity, travel time to the water body of interest, and quality of in-transit inflows. Conditions that promote high stream velocities (e.g., smooth beds and steep slopes) tend to prevent settling of P-bearing soil particles and thus favor high delivery ratios (proportion of P entering the stream that reaches the water body of interest). Since adsorption and desorption can occur during stream flow, the original balance between sediment P and solution P can be altered during transit, and longer travel times favor establishment of a new equilibrium. The quality of downstream inflows can influence in-stream adsorption/desorption dynamics by establishing a new equilibrium between sediment and solution P. If, for example, edge-of-field P loss is primarily as sediment P, a subsequent stream inflow of P-deficient runoff would encourage desorption of the sediment P.

Quantifying how in-stream transport factors influence P delivery to water bodies is a relatively underdeveloped area in the field of nonpoint source pollution analysis, undoubtedly because of the complexity of mathematically describing the numerous processes that are involved. As a compromise, the effects of the in-stream factors on P delivery are often integrated into a single, first-order relationship of the form

$$R_D = e^{-kL} \tag{4.7}$$

where R_D is the delivery ratio, L is the distance from the field to the water body, and k is an empirical constant. The delivery ratio relationship can also be refined so that k is not a constant, but varies with stream flow.

4.5.1.2 Subsurface Transport

Under soil conditions favorable for P leaching, significant amounts of soluble P are present in the soil solution. Many very sandy soils have an extremely low P adsorption capacity so P added to these soils often moves readily in water.[26] Although these conditions do not occur in most soils, in regions where these conditions are present, P transport by subsurface lateral flow may be the primary means of P delivery at the edge-of-field depending upon the hydrologic conditions of the area.[27] The EPC concept discussed in an earlier section indicates that the addition of large amounts of P can result in similar conditions on other soils. No soil has an infinite capacity to adsorb P, and as larger amounts of P are added, the potential for P loss to drainage water is increased accordingly. The current patterns of concentrating animal production and the corresponding large amounts of animal waste being applied to many soils will result in more regions experiencing conditions favorable to P leaching.[3,28] On soils approaching this condition, annual P applications from waste or fertilizer should be limited to the amount of P expected to be removed in the crop in order to prevent excessive P loss to the aquatic systems. In some states it is being proposed that sites assessed as very high risk for P loss by the P Index should have no animal wastes applied.

4.6 IMPACTS OF P LOADINGS TO AQUATIC SYSTEMS

The most commonly discussed impact of P entry into aquatic ecosystems is the tendency to accelerate eutrophication, which is the natural aging process experienced by water bodies. Water bodies generally progress through a series of trophic stages in the order oligotrophic, mesotrophic, eutrophic and hypereutrophic, in order of increasing content of nutrients. The Rocky Mountains contain many examples of oligotrophic lakes having very low nutrient concentrations and low productivity of aquatic flora and fauna. At the opposing end of the spectrum are the eutrophic water bodies, which have sufficient nutrient content to support relatively profuse growth of aquatic vegetation and algal growth. These advancing trophic stages can ultimately lead to depressed dissolved oxygen from decomposition of the increased biomass, diminished biological diversity, and a different aquatic food web involving relatively undesirable species of fish. Eutrophic conditions can also make water treatment for

drinking purposes more difficult and expensive. The surface water impacts of P loadings have relatively little to do with human health concerns and relate instead to aesthetic and economic concerns. Since there are not human health concerns, leaching
of soluble P through the soil is considered to be a problem primarily when, or if, it
emerges into the surface waters as may occur in sandy, high-water-table regions, or
karst regions with springs that discharge into surface waters, for example.

Lake production can be limited by inputs of N, P, light, or other factors. The limiting factor can change with time of year, from light during the warm months (if
shaded by leaves) to N during the cool months. However, a number of studies indicate that eutrophication of inland water bodies is generally limited by P inputs. The
direct result is that decreases in P loadings will lead directly to decreases in lake productivity until another factor becomes limiting. In other locations, P might not be the
limiting factor, in which case there is no reason for any initial focus on P input reduction. It is also possible that lakes that were once P limited might have become, over
time, N limited because of excessive P inputs.

4.7   MANAGING PHOSPHORUS FOR WATER QUALITY

As noted above, the presence of P in soil does not constitute any environmental concern unless it is present in forms that are available for transport and there are transport agents to move the P from its origin to the edge-of-field and onward toward the
water body of concern. Conversely, soil P can be a concern to the degree that it is
available and transport agents exist. This implies two avenues of P management for
water quality: approaches based on availability and those based on transport.

4.7.1   AVAILABILITY-BASED APPROACHES

Availability-based approaches are management options that attempt to limit soil P
content or to limit its susceptibility to transport in either particulate or soluble form.
One of the easier examples of availability-based approaches is to manage the soil P
concentrations so that the soil contains only sufficient P to produce the desired yield
of the crop. In other words, P additions should be based on the needs of the crop and
the amount of residual, plant-available P in the soil. This requires knowledge of plant
P uptake, soil P content, fertilizer P content, and the relationship between gross P
addition and net plant P availability. Management is simplified when inorganic P is
applied. In such cases, routine soil testing can determine current P availability. Many
soil testing laboratories are also equipped to generate fertilizer recommendations,
ultimately in the form of a gross P application to meet a specific yield target for a specific crop. Phosphorus application management is considerably more difficult for
organic sources because of variability in P content and in mineralization rates.
Indeed, organic sources have a high potential for ultimately causing or exacerbating
P transport problems unless the application rates are selected to meet plant P needs.
If organic application rates are selected on the basis of meeting plant N requirements,
then there will almost always be excess P which tends to accumulate and promote

leaching, runoff, etc. Chemical amendments are a recent, novel method of managing P availability. The principle is to alter soil chemical characteristics so that there is less soluble soil P. Alum addition, for example, can cause P to precipitate with Fe and has been successfully applied to organic P to reduce runoff P concentrations.[29,30] This principle also is being used on an experimental basis in treatment wetlands of the Everglades Nutrient Removal Project.[31] Initial results of these studies appear to be positive.

4.7.2 TRANSPORT-BASED APPROACHES

This class of management approaches focuses on reducing the occurrence or magnitude of transport agents, primarily runoff and erosion. Reductions in either runoff or erosion will reduce P transport. Fortunately, there are accepted standard practices for reducing runoff and erosion. Runoff can be reduced, for example, by the presence of cover, terracing, furrow-diking, contour tillage, reduced/minimum tillage, and related practices. These practices are described in detail in Chapter 10. Each of these practices can also reduce erosion and hence transport of particulate P. It should be noted, though, that particulate P can be a small proportion of total P for grassed source areas (e.g., pasture or meadow), because of very low erosion. Erosion can thus be virtually eliminated in such cases with no impact on soluble P concentrations. Also, reduction of runoff will reduce soluble P lost in runoff, but edge-of-field transport of soluble P may still occur in sandy soils with low P adsorption capacity when there are significant amounts of lateral subsurface flow.

REFERENCES

1. Sharpley, A., T. C. Daniel, J. T. Sims, and D. H. Pote. 1996. Determining environmentally sound soil phosphorus levels. *Journal of Soil and Water Conservation* 51(2):160–166.
2. Daniel, T. C., A. N. Sharpley, D. R. Edwards, R. Wedepohl, and J. L. Lemunyon. 1994. Minimizing surface water eutrophication from agriculture by phosphorous management. *Journal of Soil and Water Conservation, Nutrient Management,* Supplement to 49(2):30–38.
3. Sharpley, A. N. 1995. Soil phosphorus dynamics: agronomic and environmental impacts. *Ecological Engineering* 5:261–279.
4. National Research Council. 1993. Soil and water quality: an agenda for agriculture. National Academy Press, Washington, D.C., 516 p.
5. Novotny, V., H. Tran, and G. V. Simsiman. 1978. Mathematical modeling of land runoff contaminated by phosphorus. *Journal of Water Pollution Control Federation* 50:101–112.
6. Pierzynski, G. M., J. T. Sims, and G. F. Vance. 1994. Soils and environmental quality. Lewis Publishers, Boca Raton, Florida, USA.
7. Reddy, K. R., R. Khaleel, M. R. Overcash, and P. W. Westerman. 1979. A nonpoint source model for land areas receiving animal wastes: I. Mineralization of organic nitrogen. *Transactions of the ASAE* 22:863–872.
8. Sharpley, A. N. 1985. The selective erosion of plant nutrients in runoff. *Soil Science*

Society of America Journal 49:1527–1534.

9. Stewart, J. W. B. and A. N. Sharpley. 1987. Controls on dynamics of soil and fertilizer phosphorus and sulfur. SSSA Special Publication Series 19:101–121. Soil Science Society of America, Madison, Wisconsin, USA.

10. Soil Conservation Service. 1992. Agricultural waste management field handbook. U.S. Department of Agriculture, Washington, D.C.

11. Jones, C. A., C. V. Cole, A. N. Sharpley, and J. R. Williams. 1984. A simplified soil and plant phosphorus model: I. Documentation. *Soil Science Society of America Journal* 48:800–805.

12. Williams, J. R., P. T. Dyke, W. W. Fuchs, V. W. Benson, O. W. Rice, and E. D. Taylor. 1990. EPIC—erosion/productivity impact calculator. 2. User manual. Tech. Bull. 1768. USDA-ARS, Washington, D.C.

13. Sharpley, A. N. and S. J. Smith. 1989. Mineralization and leaching of phosphorus from soil incubated with surface-applied and incorporated crop residue. *Journal of Environmental Quality* 18:101–105.

14. Sharpley, A. N. 1981. The contribution of phosphorus leached from crop canopy to losses in surface runoff. *Journal of Environmental Quality* 10:160–165.

15. Soil Conservation Service. 1985. Hydrology. Section 4. Soil Conservation Service National Engineering Handbook. U.S. Department of Agriculture, Washington, D.C.

16. Haan, C. T., B. J. Barfield, and J. C. Hayes. 1994. Design hydrology and sedimentology for small catchments. Academic Press, Inc., San Diego, CA.

17. Chow, V. T., D. R. Maidment, and L. W. Mays. 1988. Applied hydrology. McGraw-Hill Book Company, New York, New York.

18. Williams, J. R. 1975. Sediment-yield prediction with the universal equation using a runoff energy factor. *In: Present and prospective technology for predicting sediment yields and sources.* ARS-S-40. Agricultural Research Service, U.S. Department of Agriculture, Washington, D.C., pp. 244–252.

19. Foster, G. R., D. C. Flanagan, M. A. Nearing, L. J. Lane, L. M. Risse, and S. C. Finkner. 1995. Chapter 11. Hillslope erosion component. *In*: Flanagan, D. C. and M. A. Nearing (editors). Technical documentation. USDA – Water Erosion Prediction Project (WEPP). NSERL Report No. 10. National Soil Erosion Research Laboratory, West Lafayette, Indiana, USA.

20. Flanagan, D. C. and M. A. Nearing (editors). 1995. Technical documentation. USDA—Water Erosion Prediction Project (WEPP). NSERL Report No. 10. National Soil Erosion Research Laboratory, West Lafayette, Indiana, USA.

21. Renard, K. G., G. R. Foster, G. A. Weesies, and J. P. Porter. 1991. RUSLE revised universal soil loss equation. *Journal of Soil and Water Conservation* 46:30–33.

22. Sharpley, A. N., L. R. Ahuja, M. Yamamoto, and R. G. Menzel. 1981. The kinetics of phosphorus desorption from soil. *Soil Science Society of America Journal* 45:493–496.

23. Sharpley, A. N. 1983. Effect of soil properties on the kinetics of phosphorus desorption. *Soil Science Society of America Journal* 47:462–467.

24. Novotny, V. and H. Olem. 1994. Water quality: prevention, identification and management of diffuse pollution. Van Nostrand Reinhold, New York, New York.

25. Storm, D. E., T. A. Dillaha III, S. Mostaghimi, and V. O. Shanholtz. 1988. Modeling phosphorus transport in surface runoff. *Transactions of the ASAE* 31:117–126.

26. Graetz, D. A. and V. D. Nair. 1995. Fate of phosphorus in Florida Spodosols contaminated with cattle manure. *Ecological Engineering* 5:163–181.

27. Campbell, K. L., J. C. Capece, and T. K. Tremwel. 1995. Surface/subsurface hydrology and phosphorus transport in the Kissimmee River Basin, Florida. *Ecological Engineering* 5:301–330.
28. Gilliam, J. W. 1995. Phosphorus control strategies. *Ecological Engineering* 5:405–414.
29. Shreve, B. R., P. A. Moore, Jr., T. C. Daniel, and D. R. Edwards. 1995. Reduction of phosphorus in runoff from field-applied poultry litter using chemical amendments. *Journal of Environmental Quality* 24:106–111.
30. Moore, P. A. 1998. Reducing ammonia volatilization and decreasing phosphorus runoff from poultry litter with alum, in: *Proceedings of 1998 national poultry waste management symposium,* J. P. Blake and P. H. Patterson, editors, pp. 117–124.
31. Bachand, P. A. M., P. Vaithiyanathan, and C. J. Richardson. 1999. Using alum and ferric chloride dosing to enhance phosphorus removal capabilities of treatment wetlands. Presented at the 1999 ASAE/CSAE-SCGR Annual International Meeting, Paper No. 992061. ASAE, 2950 Niles Road, St. Joseph, Michigan.

5   Pesticides and Water Quality Impacts

William F. Ritter

CONTENTS

5.1   INTRODUCTION

Before the 1940s, pesticides consisted of products from natural sources such as nicotine, pyrethrum, petroleum and oils, rotenone, and inorganic chemicals such as sulfur, arsenic, lead, copper, and lime. During and after World War II, phenoxy herbicides and organochlorine insecticides were widely used with the discovery of 2,4 dichlorophenoxyacetic acid (2-4-D) and dichlorodiphenyltrichloroethane (DDT). In the mid-1960s, the use of these classes of pesticides declined; they were replaced by amide and triazine herbicides and carbonate and organophosphate insecticides. Some pesticides have been banned from use mainly because of toxicities. In the past 10 years, the use of triazine herbicides and organophosphate and carbamate insecticide has declined. These groups of pesticides have been replaced by other classes of pesticides that have shorter half-lives and are applied in smaller amounts. Some of the older pesticides such as cyanazine have been banned and the use of others has been

TABLE 5.1
Classes of pesticides

Herbicides	Insecticides	Fungicides
Arylanilines	Carbamates	Azoles
Benzoic Acids	Organochlorines	Benzimidazoles
Bipyridyliums	Organophosphates	Carboxamides
alpha-Chloroacetamides	Organotins	Dithiocarbamates
Cyclohexadione Oximes	Oximinocarbamate	Morpholines
Dinitroanilines	Pyrethroids	Organophosphates
Diphenyl Ethers & Esters		Phenylamides
Hydroxybenzonitriles		Strobilurine Analogs
Imidazolinones		
Organophosphates		
Phenoxyacetic Acids		
Sulfonylureas		
Thiocarbamates		
sym-Triazines		
unsym-triazinones		
Uracil		
Ureas		

restricted. Today there are more than 30 classes of chemicals with pesticidal properties that are registered for weed, insect, and fungal control.[1] These classes are summarized in Table 5.1.

On-farm pesticide use increased from about 182 million kg in the mid-1960s to nearly 386 million kg by 1980. Since the mid-1980s, total pesticide consumption has increased only modestly to 411 million kg in 1996.[1] Atrazine and alachlor are the two most widely used pesticides.[2]

Pesticide formulations include emulsifiable concentrates, wettable powders, granules, and flowables. Emulsifiable concentrates are the bulwark product for pesticide sprays.

5.2   FATE AND TRANSPORT PROCESSES

The environmental fates of pesticides applied to cropland are summarized in Figure 5.1. Pesticides applied to cropland can be degraded by microbial action and chemical reactions in the soil. Pesticides are also immobilized through sorption onto soil organic matter and clay minerals. Pesticides that are taken up by pests or plants either can be transformed to degradation products or, in some cases, can accumulate in plant or animal tissue. A certain amount of pesticides applied are also removed when the crop is harvested. Pesticides not degraded, immobilized, or taken up by the crop or insects are lost to the environment. The major losses of pesticides to the environment are through volatilization into the atmosphere and aerial drift, runoff to surface water bodies in dissolved and particulate forms, and leaching to groundwater.

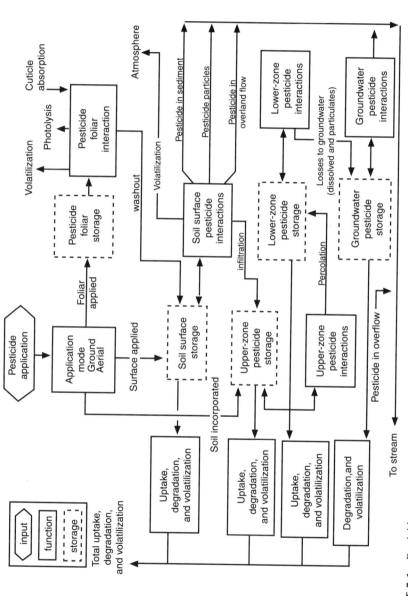

FIGURE 5.1 Pesticide transport and transformation in the soil-plant environment and the vadose zone.[3] (Reprinted with permission of American Society for Agronomy, Crop Science Society of America, and Soil Science Society of America.)

5.2.1   PESTICIDE PROPERTIES.

Chemical characteristics of pesticides that influence transport include strength (cationic, anionic basic or acidic), water solubility, vapor pressure, hydrophobic/hydrophilic characters, partition coefficient, and chemical photochemical and biological reactivity. Pesticides that dissolve readily in water are considered highly soluble. These chemicals have a tendency to be leached through the soil to groundwater and to be lost as surface water runoff from rainfall events or irrigation practices.

Pesticide vapor pressures are extremely low in comparison with other organic chemicals such as alcohols or ethers. Taylor and Spencer[4] cited values ranging over about six orders of magnitude from 2800 m Pr for EPTC to 0.00074 m Pr for picloram. Pesticides with high vapor pressures are easily lost to the atmosphere by volitalization. Some highly volatile pesticides, however, may also move downward into the groundwater.

Pesticides may be sorbed to soil particles, particularly the clays and soil organic matter. The linear and Freundlich isotherm equations have been most often used to describe pesticide adsorption on soils. These equations are given by

$$C_s = k_d C_L \qquad (5.1)$$

and

$$C_S = k_f C_L^N \qquad (N < 1) \qquad (5.2)$$

where k_d and k_f are the sorption coefficients, C is the sorbed-phase concentration (g/g), C_L is the total solute concentration (mg/L), and N is an empirical constant. Green and Karickoff and Koskinen and Harper discuss the pesticide sorption process in detail. Sorption coefficient data has been published for many pesticides.[7,8] The value of k_d or k_f is a measure of the extent of pesticide sorption by the soil. The soil organic C (OC) content is the single best predictor of the sorption coefficient for monionic hydrophobic pesticides. When the pesticide sorption coefficient is normalized with respect to soil OC, it is essentially independent of soil type. This has led to the OC-normalized sorption coefficient, K_{oc} as

$$k_{oc} = \frac{(k_d \text{ or } k_p)}{\% \text{ OC}} \times 100 \qquad (5.3)$$

Pesticides may be degraded by chemical and biological processes. Chemical degradation processes include photolysis (photochemical degradation), hydrolysis, oxidation, and reduction. The degradation of pesticides through microbial metabolic processes is considered to be the primary mechanism of biological degradation.[9]

Rao and Hornsby[8] have summarized pesticide sorption coefficients and half-lives (Table 5.2). They classify pesticides as nonpersistent if they have half-lives of 30 days or less, moderately persistent if they have half-lives longer than 30 days but less than 100 days, and persistent if their half-lives are more than 100 days. Published half-lives are generally based upon laboratory data; it is difficult to predict the half-life of a chemical in the field because of dependent variables such as soil

TABLE 5.2
Sorption Coefficients and Half-Lives of Pesticides Used In Florida

Pesticide (common name)	Sorption Coefficient (ml/g of organic chemical)	Half-Life (days)
Nonpersistent		
Dalapon	1	30
Dicamba	2	14
Chloramben	15	15
Metalaxyl	16	21
Aldicarb	20	30
Oxamyl	25	4
Propham	60	10
2,4,5-T	80	24
Captan	100	3
Fluometuron	100	11
Alachlor	170	15
Cyanazine	190	14
Carbaryl	200	10
Iprodione	1,000	14
Malathion	1,800	1
Methyl parathion	5,100	5
Chlorpyrifos	6,070	30
Parathion	7,161	14
Fluvalinate	100,000	30
Moderately Persistent		
Picloram	16	90
Chlormuron-ethyl	20	40
Carbofuran	22	50
Bromacil	32	60
Diphenamid	67	32
Ethoprop	70	50
Fensulfothion	89	33
Atrazine	100	60
Simazine	138	75
Dichlorbenil	224	60
Linuron	370	60
Ametryne	388	60
Diuron	480	90
Diazinon	500	40
Prometryn	500	60
Fonofos	532	45

(continued)

TABLE 5.2 (continued)

Pesticide (common name)	Sorption Coefficient (ml/g of organic chemical)	Half-Life (days)
	Moderately Persistent	
Chlorbromuron	996	45
Azinphos-methyl	1,000	40
Cacodylic acid	1,000	50
Chlorpropham	1,150	35
Phorate	2,000	90
Ethalfluralin	4,000	60
Chloroxuron	4,343	60
Fenvalerate	5,300	35
Esfenvalerate	5,300	35
Trifluralin	7,000	60
Glyphosphate	24,000	47
	Persistent	
Fomesafen	50	180
Terbacil	55	120
Metsulfuron-methyl	61	120
Propazine	154	135
Benomyl	190	240
Monolinuron	284	321
Prometon	300	120
Isofenphos	408	150
Fluridone	450	350
Lindane	1,100	400
Cyhexatin	1,380	180
Procymidone	1,650	120
Chloroneb	1,653	180
Endosulfan	2,040	120
Ethion	8,890	350
Metolachlor	85,000	120

temperature, moisture, microbial populations, and soil type. Pesticides most likely to contaminate groundwater are those with low sorption coefficients, long half-lives, and a high water solubility.[10]

5.2.2 SOIL PROPERTIES

Soil properties have significant influences on the fate and transport on pesticides. Soil organic matter is the most important soil property in the sorption process of most pesticides. Fine-textured soils have a higher sorptive capacity than coarse-textured soils because of the high clay content. Soil water has an important role in the retention of pesticides by soil in that it is both a solvent for the pesticide and a solute that can compete for adsorption sites. It also plays a direct role in many of the adsorption mechanisms such as water bridging and liquid exchange.

Infiltration rate and hydraulic conductivity influence pesticide transport. Soils with higher infiltration rates will generally have lower surface runoff rates, so a pesticide that readily infiltrates into the soil is more likely to be leached to groundwater than lost in surface runoff. Soil water will also move through soils more rapidly with greater hydraulic conductivity rates, so pesticides will be leached to the groundwater more rapidly and have less time to degrade. In general, coarse-textured soils have greater infiltration rates and hydraulic conductivity rates than fine-textured soils.

Soil pH is an important property for those pesticides degrading by hydrolysis. The hydrolysis or dehalogenation of DBCP occurs in the soil at a faster rate under alkaline conditions.

Soil structure, which reflects the manner in which soil particles are aggregated and cemented, influences erosion and infiltration rates. A soil with a weak structure will likely be eroded and have lower infiltration rates, which will result in sorbed pesticides being lost in runoff. Macropores and cracks can have a major effect on pesticide transport. Under particular water application rate conditions, pesticides will move through the macropores and cracks and reach the water table in a shorter period of time.

5.2.3   SITE CONDITIONS

A shallow depth of groundwater offers less opportunity for pesticide sorption and degradation. If the groundwater is shallow, the soil is permeable and rainfall exceeds the water-holding capacity of the soil; the travel time of the pesticide to reach the water table may be from a few days to a week.

Hydrogeologic conditions may dictate both the direction and rate of chemical movement. The presence of impermeable lenses in the soil profile may limit the vertical movement of pesticides but could contribute to the lateral flow of groundwater and the eventual discharge of groundwaters and pesticides into surface waters. The presence of karsts and fractured geologic materials generally allow for rapid transport of water and chemicals to the groundwater.

Climatic and weather conditions other than rainfall also affect the fate of pesticides. Higher temperatures tend to accelerate degradation. High winds and high evaporation rates may accelerate volatilization and other processes that contribute to gaseous losses of pesticides.

The slope will influence runoff and erosion rates. Increasing slope may increase runoff rate, soil detachment, and transport and increase effective depth for chemical extraction.

Soil crusting and compaction decrease infiltration rates and reduces time to runoff, resulting in increasing the initial concentration of soluble pesticides in runoff.

5.3   GROUNDWATER IMPACTS

5.3.1   Monitoring Studies

Numerous state, local, and multistate investigations have been carried out. Parsons and Witt[11] summarized data on the occurrence of pesticides in groundwater in 35 states. A more comprehensive database on pesticides in groundwater is the Pesticides in Groundwater Database (PGDB) compiled by the U. S. Environmental Protection Agency (EPA), which contains data from 45 states and 68,824 wells from 1971 to 1991.[12] The only study that has measured pesticides in groundwater in all 50 states is the EPA National Pesticide Survey (NPS).[13] Other multistate studies include the Mid Continent Pesticide Study (MCPS)[14] by the U.S. Geological Survey (USGS), Cooperative Private Well Testing Program[15] (PGWDB), National Alachlor Well Water Survey,[16] Metolachlor Monitoring Study,[17] and the USGS National Water Quality Assessment Program (NWQAP).[18]

Statewide monitoring surveys that have been conducted include Kansas, Iowa, Ohio, New York, Wisconsin, Massachusetts, Minnesota, Nebraska, Illinois, Louisiana, Indiana, Oregon, Arizona, and Connecticut.[19] All statewide and multistate surveys sampled existing community or domestic wells. The most extensive monitoring of groundwater has been carried out in California, Florida, New York, most of the states in New England, the central Atlantic Coastal Plain, and the central and northern midcontinent. The types of pesticides analyzed have been largely determined by the extent of use or concern at the time of sampling. Most site-specific studies that involve the application of one or more pesticides under controlled conditions are usually analyzed only for the pesticides applied and perhaps some of their transformation products. The principal objective of most monitoring studies, on the other hand, is to determine which pesticides are present in groundwater in the areas of interest, thereby requiring a broad spectrum of pesticides to be analyzed. With the increase in the use of triazine and acetanilide herbicides over the past three decades, more recent studies have increased the attention devoted to them. Ongoing concern over pesticides whose use had been discontinued, but that still persist in groundwater where former use was heavy, is reflected in the considerable number of recent studies of the long-term subsurface fate of the fumigants DBCP and 1,2-dibromoethane (EDB).

The MCPS study conducted in 12 states involved preplanting sampling in 1991 and postplanting sampling in July and August in 1991 and 1992. In total, 55% of compounds and eight degradation products were analyzed in 1992. Sixty-two percent of the wells sampled had detectable amounts of parent compound pesticides or their breakdown products in 1992. In 1991, only 11 pesticides were analyzed and 27.8% of the wells had detectable amounts of pesticides. In 1991, none of the pesticide concentrations were above the maximum contaminant level (MCL), whereas in 1992, 0.1% of the samples had concentrations above the MCL. Atrazine dominated the MCPS herbicide detections with 43% of the samples having atrazine concentrations above the detection limit of 0.005 µg/L in 1992. Simazine and metolachlor were also detected in more than 10% of the samples in 1992 along with the alachlor transformation products ethanesulfonic acid and 2-6-diethylaniline. Atrazine detections were generally more frequent in areas with heavier atrazine use, except in much of Ohio and Indiana, where atrazine was detected infrequently.

In the NPS program, atrazine and cyanazine were the most frequently detected pesticides.[13] Atrazine was also detected in 11.7% of the samples of the National Alachlor Well Water Survey; alachlor was detected in only 0.78%.[16]

The USGS NAWQA study was derived from 2227 wells and springs in 20 major hydrologic basins across the U.S. from 1993 to 1995. In total, 55 pesticides were analyzed, but the major emphasis was on the herbicides atrazine, cyanazine, simazine, alachlor, metolachlor, prometon, and acetochlor. All of these herbicides except acetochlor were detected in shallow groundwater (groundwater recharged within the past 10 years) in a variety of agricultural and nonagricultural areas, as well as in several aquifers that are sources of drinking water supply.[18]

Acetochlor was detected at two of 953 sites in the NAWQA study and in shallow groundwater in a statewide USGS study in Iowa in 1995 and 1996. Because acetochlor was first registered for use in 1994, the results are in agreement with those from previous field studies in that some pesticides may be detected in the shallow groundwater within 1 year following their application. More than 98% of pesticide detections in the NAWQA study were at concentrations of less than 1.0 μg/L. Frequencies of detection at or above 0.01 μg/L in shallow groundwater beneath agricultural areas were significantly correlated at the 0.05 level with agricultural use for atrazine, cyanazine, alachlor, and metolachlor, but not simazine.

Barbash and Resik[19] found no significant correlation between total pesticide use per unit area and the overall pesticide detection frequencies in states with data from 100 or more wells in the PGWDB. Of the herbicide classes examined in the PGWDB, the numbers of triazines and acetamilides detected in individual states appear to show the closest relations with use. In contrast, less of a geographic correspondence between occurrence and use is apparent for the chlorophenoxy acid, urea, and miscellaneous herbicides. The most frequently detected herbicides were atrazine, cyanazine, simazine, propazine, metribuzen, alachlor, metolachlor, propachlor, trifluralin, dicamba, DCPA, and 2-4-D. The most frequently detected insecticides were aldicarb and its degradates and carbofuran, whereas the most widely detected fumigants were 1,2-dibromo-3-chloropropance (DBCP), 1,2-dibromoethane (EDB) and 1,2-dichloropropane. Because of the health risks associated with the presence of these three fumigants in groundwater, their agricultural use has been cancelled in the U.S.

In a number of state studies, direct relations between the frequency of pesticide detection and pesticide use have been reported. Kross et al.[20] reported lower frequencies of atrazine detection in wells located on Iowa farms where herbicides had not been applied during the recent growing season, compared with farms where they had been applied. LeMasters and Doyle[21] also reported a direct relationship between atrazine use and occurrence in groundwater beneath various areas on Wisconsin grade A dairy farms across the state. Koterba et al.,[22] in a study of the groundwater beneath the Delmarva Peninsula, found that the pesticides detected in wells located near areas planted in corn, soybeans, or small grains were (with one exception) compounds that were commonly applied to those crops in that region. The single exception was hexazinone, an herbicide used to control brush and weeds in noncrop areas.

Wade et al.[23] sampled 97 wells in the surficial aquifer in areas that were more vulnerable to contamination in North Carolina. Twenty-three pesticides or pesticide degradates were detected in 26 of the 97 wells. Nine of the pesticides or degradates were no longer registered for use; dibromochloropropane and methylene chloride had concentrations above the state groundwater quality standards. They also found that areas with a high soil leaching potential index based on the pesticide DRASTIC model were no more likely to have pesticides detected in groundwater than areas with low soil-leaching potential index value.

5.3.2 WATERSHED AND FIELD-SCALE STUDIES

Atrazine and some of the other triazine herbicides have also been detected frequently in groundwater in many plot and watershed studies. Hallberg[24] reported that in the Big Springs watershed, the flow-weighted mean atrazine concentrations for groundwater discharge increased steadily from 1981 to 1985. Maximum concentrations of atrazine in the groundwater from 1981 to 1985 ranged from 2.5 to 10.0 µg/L.

Atrazine has also been found in the groundwater in Delaware.[25] Atrazine was detected in the groundwater in the Appoquinimink watershed in New Castle County in 11 of 23 monitoring wells in a Matapeake silt loam soil at depths of 6–9 m. Concentrations ranged from 1 to 45 µg/L.

Hallberg[24] also found cyanazine and alachlor in the groundwater in the Big Springs watershed. Maximum concentrations from 1981 to 1985 ranged from 0.5 to 4.6 µg/L,[24] and alachlor concentrations as high as 16.6 µg/L were measured.

Pionke et al.[26] detected atrazine, simazine, and cyanazine in groundwater in an agricultural watershed in Pennsylvania; the soils on the watershed ranged from coarse to fine textured. Atrazine was detected in 14 of 20 wells ranging in concentration from 0.013 to 1.1 µg/L. Simazine was detected in 35% of the wells at concentrations ranging from .01 to 1.7 µg/L and cyanazine was detected only in one well (0.09 µg/L).

Brinsfield et al.[27] studied pesticide leaching on no-till and conventional tillage watersheds on a silt loam Coastal Plain soil in Maryland. Over a 3-year period, atrazine was detected in the groundwater more frequently than simazine, cyanazine, or metolachlor. Pesticides were detected more frequently in the groundwater on the no-till watershed than on the conventional tillage watershed.

Dillaha et al.[28] found atrazine had the highest mean concentration of 20 pesticides detected in the groundwater on an agriculture watershed with a Rumford loamy sand soil in Virginia. The average concentration of 129 samples was 0.46 µg/L with concentrations ranging from 0 to 25.6 µg/L.

Isensee et al.[29] found atrazine in nearly all of their monitoring wells for a 3-year period in both conventional tillage and no-till plots. The wells were from 1.5 to 3.0 m deep. Atrazine concentrations ranged from 0.005 to 2.0 µg/L. Alachlor was detected in fewer than 5% of the wells.

In 1990, the Management Systems Evaluation Areas (MSEA) Program was initiated in eight states in the Midwest by USDA[30] to study the impact of prevailing and modified farming systems on groundwater and surface water quality. Many reports have been published on the results. In the Walnut Creek watershed in Iowa, annual atrazine losses in tile drainage water ranged from 0.02 to 2.16 g/ha in a corn and soybean rotation during the 4-year study.[31] Fewer than 3% of the groundwater samples contained atrazine concentration exceeding 3 µg/L. Metribrizan, which was applied to soybeans, was also found in groundwater, but only half as frequently as atrazine.

A number of researchers have found pesticides can move rapidly to the groundwater by macropore flow. Steenhuis et al.[32] found atrazine in the groundwater 1 month after it was applied in conservation tillage but did not detect any atrazine in the groundwater in conventional tillage until late fall. They concluded atrazine moved to the groundwater under conservation tillage by macropores that were connected to the surface, but under conventional tillage most of the atrazine was adsorbed in the root zone.

Ritter et al.[33] studied the movement of alachlor, atrazine, simazine, cyanazine, and metolachlor on an Evesboro loamy sand soil that had a water table near the surface. Over a period of 9 years in four different experiments, they found these pesticides may move to shallow groundwater by macropore flow if more than 30 mm of rainfall occurs shortly after they are applied. They found no large difference in pesticide transport between conventional tillage and no-tillage.

Gish et al.[34] found that average field-scale solute phase atrazine concentrations at 1 m resulting from 48 mm of rainfall 12 h after application on a loam soil were 243 µg/L for no-tillage and 59 µg/L for conventional tillage. Cyanazine concentrations were 184 µg/L for no-tillage and 69 µg/L on conventional tillage. They concluded these high concentrations were a result of preferential flow.

5.3.3 MANAGEMENT EFFECTS

Management practices such as tillage and method of application can influence the amount of pesticide leached to groundwater. The attempts by researchers to discern the influence of tillage practices on pesticide movement to groundwater are beset by a number of complicating factors. First, the effects of tillage on infiltration capacity are seasonal. Conventional tillage leads to transient increases in soil permeability relative to an untilled soil. Over the course of an entire growing season, however, long-term infiltration rates tend to be higher under reduced tillage than under conventional tillage.[35] Second, both the placement of pesticides during application and the magnitude of individual recharge events may influence the effect of tillage on pesticide transport.

The results of the effect of tillage practices on pesticide concentrations in the subsurface have not always been consistent among different investigations. In general, reduced tillage gives rise to pesticide distributions in the subsurface that are markedly different from those observed under conventional tillage. Although pesticide concentrations are typically higher in surficial soils under conventional tillage than under reduced tillage, the reverse is often observed at greater depths in the soil.

In addition, pesticides are usually detected more frequently and at higher concentrations in groundwater beneath no-till and reduced-tillage areas than beneath conventionally tilled fields.[36,37,38,39] There have been a number of cases where pesticide concentrations in the groundwater have displayed inconsistent relations with tillage practices. In some cases, pesticide concentrations have been higher or lower in the groundwater than those in reduced tillage, depending on the compound or the year examined.[36,40,41] Different trends observed in different years for the same compound may arise from variations in several key parameters related to tillage and recharge from year to year.

The available data on comparing the fluxes of pesticides leached to groundwater through conventional tillage and no-tillage are more consistent than those on pesticide concentrations. The majority of the research suggests that, all factors being equal, reduced tillage increases the mass loading of pesticides to groundwater compared with conventional tillage. Kanwar et al.[42] observed the amount of the applied herbicides alachlor, atrazine, cyanazine, and metribuzen entering tile drainage water from a fine loam soil in Iowa were generally higher for ridge tillage and no-tillage regimes than when the soil was worked with a moldboard plow or chisel plow. Hall and co-workers[36] reported increased fluxes of several pesticides through a silty clay loam in soil in Pennsylvania under no-tillage compared with conventional tillage. The proportions of applied herbicides recovered in pan lysimeters were three to eight times higher beneath conventional tillage areas for atrazine, simazine, cyanazine, and metolachlor.[36] The differences were even more pronounced for dicamba.[43] Difference in tillage practices may have much less impact on pesticide transport through low-permeability soils compared with more permeable soils. Logan et al.[49] observed no discernible difference between the losses of the herbicides atrazine, alachlor, metolachlor, and metribuzen in tile drainage from conventional tillage and no-tillage plots on a poorly drained silty clay soil in Ohio.

A number of studies have examined the effects of pesticide application strategies on pesticide residue levels and leaching to groundwater. It has been demonstrated that the incorporation of pesticides into controlled-release formulations diminishes the rate at which the active ingredient enters the soil solution. Hickman et al.[45] found a starch-encapsulated controlled-release atrazine formulation reduced atrazine concentrations in tile drainage significantly compared with commercial formulations of atrazine in a silt loam soil. Williams et al.[46] also found starch encapsulation of atrazine reduced leaching of atrazine through a calcareous soil. Encapsulation has also been shown to reduce the impact of preferential flow on alachlor.[47] Although a number of studies indicate that different formulations influence the rate at which active pesticide ingredients are released to soil and groundwater, not enough data are available to predict the results of different formulations of different compounds.[18]

Limiting the area of land surface to which pesticides are applied appears to reduce pesticide concentrations and depth of migration in the subsurface. Baker et al.[48] found herbicide concentrations were lower in tile drainage following banding compared with broadcast application for atrazine, alachlor, metolachlor, and cyanazine for five different tillage systems. Clay et al.[49] concluded that banding of pesticides along ridge tops compared with the troughs in a ridge-tillage system will reduce the transport of applied chemicals to the subsurface. In a ridge-tillage system with a sandy soil in Minnesota, they found alachlor concentrations were highest at the soil surface and decreased with depth under ridge application, whereas under trough application the opposite pattern was observed.

5.4 SURFACE WATER IMPACTS

5.4.1 MONITORING STUDIES

Since the 1950s, the most common pesticides monitored in U.S. surface waters have been the organochlorine insecticides, organophosphorus insecticides, triazine herbicides, acetanilide herbicides, and phenoxy acid herbicides. The use of organochlorine insecticides began in the 1940s and continued until the 1970s until most were banned or their use severely restricted. The organophosphate insecticides came into wide use in the late 1960s and 1970s and the total used in agriculture has remained relatively stable over the last two decades but declined from the 1970s.

In a comprehensive review of pesticides in surface water, Larson et al.[50] targeted 98 pesticides and 20 pesticide transformation products. Of these 118 compounds, 76 have been detected in one or more surface water bodies in at least one study. In terms of pesticide classes, 31 of 52 targeted insecticides, 28 of 41 herbicides, 2 of 5 fungicides, and 15 of 20 pesticide transformation products were detected in surface waters.

From 1957 to 1968, the Federal Water Quality Administration collected samples from about 100 rivers in the U.S. for analysis for pesticides and other organic compounds.[51,52] This was the first comprehensive multistate monitoring program. All rivers were sampled in September each year except in 1968 when samples were collected in June. Dieldrin, DDT, and heptachlor were the most frequently detected pesticides; dieldrin was detected in 47% of the samples with a maximum concentration of 0.1 µg/L.

The USGS and EPA examined pesticides in water and bed sediments of rivers throughout the U.S. from 1975 to 1980.[53] They examined 21 pesticides and transformation products at more than 150 sites. They observed pesticides in less than 10% of the samples but the detection limits were high. Most of the detections were for organophosphorus insecticides.

Starting in 1975 and continuing through the 1980s, Ciba-Geigy Corporation monitored atrazine concentrations at a number of sites throughout the Mississippi River basin.[54] Atrazine was detected frequently at nearly all the sites sampled, with a detection frequency of 60 to 100% of samples, depending upon the site. Annual mean atrazine concentrations were less than the EPA drinking water standards of 3 µg/L at 94% of the sites over the entire sampling period.

In 1989 and 1990, the USGS sampled 147 sites throughout the Midwest in spring (preplanting), summer (postplanting), and fall (postharvest, lower river discharge).[55] Samples were analyzed for 11 triazine and acetanilide herbicides and 2 atrazine transformation products. Herbicides were detected at 98 to 100% of the sites in the postplanting samples. Atrazine, alachlor, and metolachlor were the most frequently detected herbicides in both years, with detection at 81 to 100% of the sites in the post-planting samples. Concentrations in most postplanting samples ranged from 1 to 10 μg/L for atrazine, alachlor, metolachlor, and cyanazine. Maximum concentrations in the 1989 postplanting samples were 108 μg/L for atrazine, 40 to 60 μg/L for alachlor, metolachlor, and cyanazine; and 1 to 8 μg/L for simazine, propazine, and metribuzen. Concentrations were much lower in the preplanting and postharvest samples.

In 1992, the USGS conducted a survey of 76 reservoirs in the midwestern U.S.[56] The reservoirs were sampled in late April to mid-May, late June to early July, late August to early September, and late October to early November for 11 triazine and acetanilide herbicides and 3 selected transformation products; at least 1 of the 14 herbicides and transformation products were detected in 82–92% of the 76 sampled reservoirs during the four sampling periods. Atrazine was detected in 92% of the samples. Herbicides were detected most frequently in reservoirs where herbicide use was the highest.

In 1991 and 1992, the USGS sampled three sites on the mainstem of the Mississippi River and sites on the major tributaries (Platte, Missouri, Minnesota, Illinois, Ohio, and White Rivers) one to three times per week for 18 months.[55] The samples were analyzed for 27 high-use pesticides (15 herbicides and 12 insecticides). The triazine and acetanilide herbicides were observed most frequently, but the organophosphates and other compounds were rarely observed.

Water samples from 58 streams and rivers across the U.S. were analyzed for pesticides as part of the NWQA Program of the USGS.[57] The sampling sites represented 37 diverse agricultural basins, 11 urban basins, and 10 basins with mixed land use. Forty-six pesticides and pesticide degradation products were analyzed in approximately 2200 samples collected from 1992 to 1995. The targeted compounds account for approximately 70% of national agricultural pesticide use. All the targeted compounds were detected in one or more samples. The herbicides atrazine, metolachlor, prometon, and simazine were detected most frequently. Among the insecticides, carbaryl, chlorpyrifos, and diazinon were detected most frequently. Atrazine concentrations exceeded the EPA drinking water standard of 3 μg/L at 16 sites, and alachlor concentrations exceeded the EPA drinking water standard of 2 μg/L at 10 sites. Relatively high concentrations of atrazine, alachlor, metolachlor, and cyanazine occurred as seasonal pulses in corn-growing areas.

From the data reviewed, there is a clear relationship between agricultural use of the triazines and acetanilide herbicides and their occurrence in surface waters. The concentrations of these compounds in rivers are seasonal, with a sharp increase in concentrations shortly after application followed by a relatively rapid decline in concentration. These seasonal peaks in concentrations are influenced strongly by the timing of rainfall relative to application. The Lake Erie tributaries study, which is the longest and most complete continuous record of triazine and acetanilide concentrations, shows this variability from 1983 to 1991.[58] Much lower concentrations of alachlor, atrazine, and metolachlor were observed in the drought year of 1988.

The most widely used phenoxy herbicide, 2-4-D, was a relatively common contaminant in surface waters in the 1970s and 1980s.[50] Recent monitoring data are sparse. Most observed concentrations were below 1 µg/L. Little information has been published about monitoring for MPCA, the other phenoxy compound with significant agricultural use.[50]

5.4.2 WATERSHED AND FIELD-SCALE STUDIES

There have been numerous studies since the 1970s on measuring pesticide losses from field plots or watersheds. In some cases, losses were measured under natural rainfall conditions and in other studies, rainfall simulators were used. Hall et al.[59] studied the runoff losses of atrazine applied at seven different rates. Losses in runoff water ranged from 1.7 to 3.6% of the amount applied for the different application rates. No correlation was seen between application rate and percentage lost in runoff water. Losses in runoff suspended sediment ranged from 0.03 to 0.28% of the amount applied, with higher percentage lost at the higher application rates; the first runoff occurred 23 days after application. Ritter et al.[60] found up to 15% of the applied atrazine and 2.5% of the applied propachlor were lost in runoff water and sediment in a runoff event 7 or 8 days after application in Iowa from a small surface-contoured watershed.

Wu et al.[61] measured atrazine and alachlor from eight watersheds ranging in size from 16 to 253 ha in the Rhode River watershed in Maryland. Atrazine loadings represented from 0.05 to 2% of the amount applied. Alachlor loadings were less than 0.1% of the amount applied. Forney et al.[62] measured losses of atrazine, melotachlor, cyanazine, alachlor, metribuzin, nicosulfuron, tribenuron methyl, and thifensulfuron methyl from 1994 to 1996 from four different farming systems on small watersheds ranging in size from 2.1 to 9.0 ha in the Chesapeake Bay watershed. Atrazine losses were higher than any of the herbicides. On one of the watersheds, atrazine losses ranged from 1.25 to 15.43% of the amount applied for continuous no-till corn. Alachlor losses were less than 1.0% of the amount applied each year, and the highest amount of nicosulfuron lost was 7.3%. For all herbicides, the average annual runoff losses ranged from 0.82 to 5.08%. If significant runoff occurred shortly after the herbicides were applied, larger amounts of herbicides were lost.

In the Midwest and other areas, subsurface drainage is a common agricultural water management practice. During parts of the year, tile drainage flow may be a large percentage of stream flow in some streams. Pesticides discharged in subsurface drainage can influence surface water quality. There have been numerous studies to evaluate pesticide concentrations in tile drains. Masse et al.[63] found tile effluent represented a small fraction of atrazine and metolachlor applied for no-tillage and conventional tillage treatments in eastern Ontario on a loam soil. Atrazine losses ranged from 0.05 to 0.15% in no-tillage and 0.02 to 0.12% for conventional tillage; metolachlor losses were 0.02% or less for both tillage systems. Bengston et al.,[64] on a clay loam soil, found 97% of the atrazine lost was in surface drainage and 3% in subsurface drainage in Louisiana. In total, 1.4% of the atrazine applied was lost in surface runoff and subsurface drainage. For metolachlor, surface runoff contributed 89% and subsurface discharge contributed 11% of the total losses. Total metolachlor losses were 1.2% of the amount applied. When losses from the subsurface drainage plots were compared with plots with only surface drainage, subsurface drainage reduced atrazine losses by 55% and metolachlor losses by 51%. Based upon numerous studies, it appears subsurface drainage losses of pesticides to surface waters will be much smaller than surface runoff losses. In fact, subsurface may reduce pesticide losses to surface waters by reducing the amount of surface runoff.

5.4.3 MANAGEMENT EFFECTS

The amount of pesticide in the active zone at the soil surface at the time of runoff is probably the most important variable affecting amounts and concentrations in runoff. The effects of erosion control practices on pesticide runoff depends upon the adsorption characteristics of the pesticide and the degree of fine-sediment transport reduction. As sediment yield is reduced, pesticides adsorbed in runoff are reduced, but not necessarily in proportion because erosion control practices tend to reduce transport of coarse particles more than fine particles.[65] Smith et al.[66] compared pesticide runoff from terraced watersheds to runoff from watersheds with no planned conservation practices. Paraquat, which was strongly found to sediment, was reduced in proportion to sediment reduction. Terraces did not reduce runoff volumes and therefore losses of atrazine, diphenamid, cyanazine, propazine, and 2-4-D were not affected because they were transported primarily in the aqueous phase. Ritter et al.[60] showed that conservation practices that reduce runoff volumes also reduce losses of propachlor and atrazine. Baker and Johnson[67] and Baker et al.[65] related runoff and soil loss to crop residues in some tillage practices. Crop residues reduced runoff volumes in some soils, but not the losses of alachlor and cyanazine because concentrations tended to increase with increasing crop residue.

Over the years there has been considerable interest in pesticide transport and conservation tillage systems, and whether pesticide losses in runoff may be enhanced or reduced. Triazines and other soluble herbicides are easily removed from crop surfaces by rainfall and runoff,[65] and this washoff may be a source of enhanced concentrations in runoff as observed by Baker and Johnson[67] and Baker et al.[68] However, Baker et al.[69] reported that runoff concentrations were not affected by herbicide placement above or below the crop residue but were negatively correlated with time to runoff. Baker and Laflen[70] earlier reported that wheel tracks reduced time to runoff, increased initial herbicide concentrations in runoff and total runoff volumes, and, therefore, total herbicide losses.

Watanabe et al.[71] studied the effect of tillage practice and method of chemical application on atrazine and alachlor losses through runoff and erosion on four sites in Kansas and Nebraska. The five treatments evaluated were no-tillage and pre-emergent, disk and pre-emergent, plow and pre-emergent, disk and preplant incorporated, and plow and preplant incorporated. In total, 63.5 and 127 mm of rainfall were applied 24–36 hours after chemical application. The no-tillage, pre-emergent treatments had the highest losses of atrazine and alachlor, and the plow and the preplant incorporated treatments had the lowest losses. In the no-tillage treatments, 94% of the atrazine and 97% of the alachlor losses occurred in the runoff.

Baker[72] discussed three reasons that less strongly absorbed pesticide losses may be greater from conservation tillage systems than from moldboard plow tillage systems. One reason is that, on an individual storm basis, fields that have been recently tilled often have less runoff from the first storm after tillage, and pesticides soil-applied in the spring are usually applied at the time of or shortly after tillage is done. The second reason is that mechanical soil incorporation of pesticides has been shown to significantly reduce pesticide runoff losses by reducing the amount of pesticide in the surface-mixing zone. The degree of incorporation is normally directly related to the severity of tillage and inversely related to the crop residue remaining after tillage. In no-tillage systems, incorporation is not possible. The third reason is that surface crop residue will intercept sprayed pesticides such that a 30% crop residue condition would result in about 30% of a broadcast-sprayed pesticide found on crop residues after application. Washoff studies have shown that herbicides commonly used for corn can be easily washed off corn residue with up to 50% of the intercepted herbicide washed off with the first 10 mm of rain occurring shortly after application.[73]

As mentioned previously, pesticide application methods can have an effect on the amount of pesticide lost in runoff. In some cases, one of the reasons for higher pesticide losses in no-tillage is the lack of incorporation.[72] Pesticide formulation also can affect edge-of-field losses. Wettable powder formulations applied to the soil surface are among the most runoff-acceptable pesticides, and soil emulsifiable concentrates are among the least susceptible.[74] Wauchope,[75] in an extensive review of pesticide losses from cropland, estimated that seasonal losses of 2–5% for wettable powders could be expected. Because the bulk of a pesticide may be lost in the first storm, he defined "catastrophic" events as those in which runoff losses exceed 2% of the application. He also concluded that the first critical event must occur within 2 weeks of application with at least 10 mm of rainfall, 50% of which becomes runoff. Kenimer et al.[76] found that a microencapsulated formulation of alachlor and a controlled-release formulation of terbufos yielded higher surface losses than did the emulsifiable concentrate or granular formulations. They attributed greater losses of the microencapsulated and controlled-release formulations to transport of discrete particles of pesticide with eroded sediment.

Vegetative filter strips or riparian forest buffer systems to remove pesticides have received increased emphasis in recent years. Lowrance et al.[77] studied the effects of a riparian forest buffer system on the transport of atrazine and alachlor in the Coastal Plain of Georgia. Over a 3-year period, atrazine concentrations were reduced by a magnitude and alachlor concentrations by a factor of six. The riparian buffer system consisted of a bermuda grass and bahia grass strip (8 m wide) adjacent to the field, a pine forest strip (40–55 m wide), and then a hardwood forest (10 m wide) with a stream channel. The load reductions for the system relative to what was leaving the field was 97% for atrazine and 91% for alachlor.

Mikelson and Baker[78] conducted a rainfall simulation on the reduction of atrazine as it passed through a vegetative filter strip consisting of 59% smooth brome, 35% bluegrass, and 6% tall fescue. Cropping to filter strip areas of 5:1 and 10:1, no-tillage, and conventional tillage were evaluated. The 5:1 ratio plots were able to reduce the atrazine losses to a greater degree than the 10:1 plots. There was no significant difference between reductions of atrazine with the no-tillage runoff versus the conventional tillage runoff.

From a review of a number of studies, Baker et al.[79] concluded that buffer strips can be effective in reducing pesticide transport in runoff from treated fields, particularly if covered with close-grown vegetation. These buffers can take the form of grassed waterways, contour buffer strips, vegetative barriers, and tile inlet buffers within fields, or as field-borders, filter strips, set-backs, and riparian forest buffers at the field edge or offsite. The two major factors determining the effectiveness of buffers are the field runoff source area to buffer strip area and the pesticide adsorption potential for soil and sediment. For weakly to moderately adsorbed pesticides, the major carrier is runoff, and infiltration of runoff into the buffer strip is a major removal mechanism. As the field area to strip area increases, the effectiveness of the buffer strips in retaining pesticides decreases.

5.5   SUMMARY

Atrazine and alachlor are the two most widely used pesticides. Pesticide properties, soil properties, and site conditions influence the fate and transport of pesticides. Chemical characteristics that influence transport include strength (cationic, anionic, basic, or acidic), water solubility, vapor pressure, hydrophobic/hydrophilic character, partition coefficient, and chemical, photochemical, and biological activity. Soil properties influencing the fate and transport of pesticides include soil organic matter, hydraulic conductivity, infiltration capacity, pH, and soil structure. The most important site conditions include depth to groundwater, slope, hydrogeologic conditions, soil compaction, and climatic conditions.

Numerous state, local and multistate studies of pesticides in groundwater have been carried out. The most recent studies have been devoted mostly to the triazine and acetanilide herbicides. Atrazine has been the most widely detected herbicide in groundwater. A number of studies have indicated pesticides may be rapidly leached to shallow groundwater by preferential flow if significant rainfall occurs after the pesticides are applied. Management practices such as tillage and method of application influence the amount of pesticide leaches to groundwater. The effects of tillage on pesticide concentrations in groundwater have not always been consistent. Reduced tillage gives rise to pesticide distributions in the subsurface that are markedly different from those observed under conventional tillage. Reduced tillage in most studies increases the mass loading of pesticides to groundwater.

There is a clear relationship between agricultural use of the triazine and acetanilide herbicides and their occurrence in surface waters in the U.S. The concentrations in streams and rivers are seasonal, with a sharp increase in concentrations shortly after application. Pesticides in tile drainage appear to contribute small amounts of pesticides to surface waters compared with direct surface runoff. The amount of pesticide in the active zone at the soil surface at the time of runoff is the most important variable affecting pesticide amounts and concentrations in runoff. Pesticide concentrations and the amounts removed in runoff may be greater in conservation tillage than conventional tillage. One of the reasons is washoff of the pesticides from the residue by rainfall. This may be especially true for less strongly adsorbed pesticides. Mechanical incorporation of pesticides in conventional tillage also reduces the amount of pesticide in the surface-mixing zone. Vegetative filter strips have been shown to be effective in removing pesticides in surface runoff. The major factors in determining the effectiveness of buffers are the ratio of field runoff area to buffer area and the pesticide adsorption potential for soil and sediment.

REFERENCES

1. Steinheimer, J. R., Ross, L. J., and Spittler, J. D., Agrochemical movement: perspective and scale-of-study overview, in *Agrochemical Fate and Movement, Perspective and Scale of Study,* Steinheimer, J. R., Ross, L. J., and Spittler, J. D., Eds., ACS Symp. Series 751, Chem. Soc., Washington, DC, 2000, Chap. 1.
2. National Agricultural Statistics Service, 1998 agricultural chemical use estimates for field crops, USDA, NASS, ERS, Washington, D.C., 1999.
3. Himel, C. M., Loats, H., and Barley, G. W. Pesticide source to the soil and principles of spray physics, in *Pesticides in the Soil Environment: Processes, Impacts, and Modeling,* Cheng, H. H., Ed., Soil Sci. Soc. Am., Madison WI, 1990, Chap. 2.
4. Taylor, A. W. and Spencer, W. F., Volatilization and vapor transport processes, in *Pesticides in the Soil Environment: Processes, Impacts and Modeling,* Cheng, H. H., Ed., Soil Sci. Soc. Am., Madison, WI, 1990, Chap. 7.
5. Green, R. E. and Karickhoff, S. W., Sorption estimates for modeling, in *Pesticides in the Soil Environment: Processes, Impacts and Modeling,* Cheng, H. H., Ed., Soil Sci. Soc. Am., Madison, WI, 1990, Chap. 4.
6. Koskinen, W. C. and Harper, S. S., The retention process: mechanisms, in *Pesticides in the Soil Environment: Processes, Impacts and Modeling,* Cheng, H. H., Ed., Soil Sci. Soc. Am., Madison, WI, 1990, Chap. 3.
7. Karickhoff, S. W., Semi-empirical estimation of sorption of hydrophobic pollutants on natural sediments and soils, *Chemosphere,* 10, 833, 1981.
8. Rao, P. S. C. and Hornsby, A. G., Behavior of pesticides in soils and waters, Soil Sci. Fact Sheet SL 40, University of Florida, Gainesville, FL, 1989.
9. Bollag, J. M. and Liu, S. Y., Biological transformation processes of pesticides, in *Pesticides in the Soil Environment: Processes, Impacts and Modeling,* Cheng, H. H., Ed., Soil Sci. Soc. Am., Madison, WI, 1990, Chap. 6.
10. U.S. Environmental Protection Agency, Pesticides in groundwater, background document, EPA, Washington, DC, 1986.
11. Parsons, D. W. and Witt, J. M., Pesticides in groundwater of the United States of America: A report of the 1988 survey of lead state agencies, Report EM 8401, Oregon State University Extension Service, Corvallis, OR , 1989.
12. U.S. Environmental Protection Agency, Pesticides in groundwater database 1. A compilation of monitoring studies, 1971–1991, Report EPA734-12-92-001, EPA, Washington, DC, 1992.
13. U.S. Environmental Protection Agency, Another look: National survey of pesticides in drinking water wells, phase II report, Report No. EPA 579/09-91-021, EPA, Office of Pesticides and Toxic Substances, Washington, DC, 1992.
14. Koplin, D. W., Goolsby, D. A., Aga, D. S., Iverson, J. L., and Thurman, E. M., Pesticides in near surface aquifers: results of mid-continent United States groundwater reconnaissance: 1991–1992, in *Selected Papers in Agricultural Cultural Chemicals in Water Resources of the Midcontinental United States,* Goolsby, D. A., Boyer, L. L., and Mallard, G. E., Eds., USGS Open File Report 93, 0418, USGS, Denver, CO, 1993, 64.

15. Richards, R. P., Baker, D. B., Creamer, N. L., Kramer, J. W., Ewing, D. E., Merryfield, B. J., and Wallrabenstein, L. K., Well water quality, well vulnerability, and agricultural contamination in the midwestern United States, *J. Environ. Qual.,* 26, 935, 1992.

16. Holden, L. R., Graham, J. A., Whitmore, R. W., Alexander, W. J., Pratt, R. W., Liddle, S. K., and Piper, L. L., Results of the national alachlor well survey, *Environ. Sci. Tech.,* 26, 935, 1992.

17. Roux, P. H., Balu, K., and Bennett, R., A large-scale retrospective groundwater monitoring study for metolachlor, *Groundwater Monitoring Rev.,* 11(3), 104, 1991.

18. Barbash, J. E., Thelin, G. P., Kolpin, D. W., and Gilliom, R. J., Distribution of major herbicides in groundwater of the United States, Water Resources Investigation Report 98-4245, USGS, Denver, CO, 1998.

19. Barbash, J. E. and Resek, E. A., *Pesticides in Groundwater: Distribution, Trends and Governing Factors,* Ann Arbor Press, Inc., Chelsea, MI, 1996, Chap. 2.

20. Kross, B. C., Hollberg, G. R., Bruner, D. R., Libra, R. D., Rex, K. D., Weik, L. M. B., Vermace, M. E., Burmeister, L. F., Hall, N. H., Cherryholmes, K. L., Johnson, J. K., Selim, M. J., Nations, B. K., Seigly, L. S., Quade, D. J., Dudler, A. G., Sesker, K. D., Culp, M. A., Lynch, C. F., Nicholson, H. F., and Hughes, J. P., The Iowa state-wide rural well water survey water quality data: initial analysis, Tech. Information Series 19, Iowa Department of Natural Resources, Des Moines, IA, 1990.

21. LeMasters, G. and Doyle, D. J., Grade A dairy farm well water quality survey, Wisconsin Dept. of Agriculture and Wisconsin Agricultural Statistics Service, 1989.

22. Koterba, M. J., Banks, W. S. L., and Shedlock, R. J., Pesticides in shallow groundwater in the Delmarva Peninsula, *J. Environ. Qual.,* 22, 500, 1993.

23. Wade, H. F., York, A. C., Morey, A. E., Padmore, J. M., and Rudo, K.M., The impact of pesticide use on groundwater in North Carolina, *J. Environ. Qual.,* 27, 1018, 1998.

24. Hallberg, G., Agricultural chemicals and groundwater in Iowa: status report, 1985, Circular CE-2158q, Cooperative Extension Service, Iowa State Univ., Ames, IA, 1985.

25. Ritter, W. F., Chirnside, A. E. N., and Lake, R., Best management practices impacts on water quality in Appoquinimink Watershed, Paper No. 88-2034, ASAE, St. Joseph, MI, 1988.

26. Pionke, H. B., Glotfeltz, D. E., Lucas, A. D., and Urban, J. B., Pesticide contamination of groundwaters in the Mahantango Creek watershed, *J. Environ. Qual.,* 17, 76, 1988.

27. Brinsfield, R. B, Staver, K. W., and Magette, W. L., Impact of tillage practices on pesticide leaching in Coastal Plains Soils, Paper No. 87-2631, ASAE, St. Joseph, MI, 1987.

28. Dillaha, T. A., Mostaghimi, S., Reneau, R. R, McClellan, P. V., and Shanholtz, V. O., Subsurface transport of agricultural chemicals in the Nomine Creek Watershed, Paper No. 87-2629, ASAE, St. Joseph, MI, 1987.

29. Isensee, A. R., Nash, R. G., and Helling, C. S., Effect of conventional and no-tillage on pesticide leaching to shallow groundwater, *J. Environ. Qual.,* 19, 434, 1990.

30. Hatfield, J. L., Bucks, D. A., and Horton, M. L., The Midwest water quality initiative: research experiences at multiple sites, in *Agrochemicals Fate and Movement Prospective and Scale of Study,* Steinheimer, T. R., Ross, L. J., and Spittler, T. D., Eds., Am. Chem. Soc., Washington, DC, Symp. Series 751, 2000, Chap. 16.

31. Moorman, T. B., Jaynes, D. B., Cambardella, C. A., Hatfield, J. L., Pfeiffer, R. L., and Morrow, A. J., Water quality in Walnut Creek watershed: herbicides in soils, subsurface drainage, and groundwater, *J. Environ. Qual.,* 28, 35, 1999.

32. Steenhuis, T. S., Stanbitz, W., Andreini, M. S., Surface, J., Richard, T., Paulsen, R, Pickering, N. B., Hagerman, R., and Geohring, L. D., Preferential movement of pesticides and traces in agricultural soils, *J. Irrig. Drain. Eng.,* 1/6, 50, 1990.

33. Ritter, W. F., Chirnside, A. E. M., and Scarborough, R. W., Movement and degradation of triazines, alachlor and metolachlor in sandy soils, *J. Environ. Sci. Health,* A31, 2699, 1996.

34. Gish, T. J., Helling, C. S., and Mojasevic, D. L., Preferential movement of atrazine and cyanazine under field conditions, *Trans. ASAE,* 34, 1699, 1991.

35. Baker, J. L., Hydrologic effects of conservation tillage and their importance relative to water quality, in *Effects of Conservation Tillage in Groundwater Quality: Nitrates and Pesticides,* Logan, T. J., Davidson, T. J., Baker, J. L., and Overcash, M. R., Eds., Lewis Publishers, Chelsea, MI, 1987, Chap. 6.

36. Hall, J. K., Murry, M. R., and Hartwig, N. L., Herbicide leaching and distribution in tilled and untilled soil, *J. Environ. Qual.,* 18, 439, 1989.

37. Ritter, W. F., Scarborough, R. W., and Chirnside, A. E. M., Contamination of groundwater by triazines, metolachlor and alachlor, *J. Contam. Hydrol.,* 15, 73, 1994.

38. Ritter, W. F., Chirnside, A. E. M., and Scarborough, R. W., Leaching of dicamba in a Coastal Plain soil. *J. Environ. Sci. Health,* A31, 505, 1996.

39. Gish, T. J., Isensee, A. R., Nash, R. G., and Helling, C. S., Impact of pesticides on shallow groundwater quality, *Trans. ASAE,* 34, 1745, 1991.

40. Kanwar, R. S. and Baker, J. L., Tillage and chemical management effects on groundwater quality, in Agricultural Research to Protect Water Quality, *Proc. of Conf. Minneapolis, MN, Soil and Water Conservation Society,* Ankeny, IA, Vol. 1, 1994, 455.

41. Shirmohammadi, A., Magette, W. L., Brinsfield, R. B., and Staver, K., Ground water loadings of pesticides in the Atlantic Coastal Plain, *Ground Water Monitoring Rev.,* 9(4), 141, 1989.

42. Kanwar, R. S., Stoltenberg, D. E., Pfeiffer, R., Karlen, D., Colvin, T. S., and Simpkins, W. W., Transport of nitrate and pesticides to shallow groundwater systems as affected by tillage and crop rotation practices, in *Research to Protect Water Quality, Proc. of the Conf., Minneapolis, MN, Soil and Water Conservation Society,* Ankeny, IA, Vol. 1, 1994, 270.

43. Hall, J. K. and Mumma, R. O., Dicamba mobility in conventionally tilled and no-tilled soil, *Soil Tillage Res.,* 30, 3, 1994.

44. Logan, T. J., Eckert, D. J., and Beak, D. G., Tillage, crop and climatic effects on runoff and tile drainage losses of nitrates and four herbicides, *Soil Till. Res.,* 30, 75, 1994.

45. Hickman, M. V., Schreiber, M. M., and Vail, G. D., Role of controlled release herbicides formulations in reducing groundwater contaminations, in *Clean Water—Clean Environment—21st Century, Vol. 1: Pesticides,* ASAE, St. Joseph, MI, 1995, 89.

46. Williams, C. F., Nelson, S. D., and Gish, T. J., Release and mobility of starch-encapsulated atrazine in calcareous soils, in *Clean Water—Clean Environment—21st Century, Vol. 1: Pesticides,* ASAE, St. Joseph, MI, 1995, 173.

47. Gish, T. J., Shirmohammadi, A., and Wienhold, B. J., Field-scale mobility and persistence of commercial and starch-encapsulated atrazine and alachlor, *J. Environ Qual.,* 23, 355, 1994.

48. Baker, J. L., Colvin, T. S., Erbach, D. C., Kamwar, R. S., and Lawlor, P. A., Herbicide banding to reduce inputs and environmental losses, in *Clean Water—Clean Environment—21st Century, Vol. 1: Pesticides,*ASAE, St. Joseph, MI, 1995, 13.

49. Clay, S. A., Clay, S.E., Koskinen, W. C., and Malzer, G. L., Agrichemical placement impacts on alachlor and nitrate movement through soil in a ridge tillage system, *J. Environ. Sci. Health,* B27, 125, 1992.

50. Larson, S. J., Capel, P. D., and Majewski, M. S., *Pesticides in Surface Waters,* Ann Arbor Press, Inc., Chelsea, MI, 1997, Chap. 3.

51. Weaver, B. C. E., Gunnersen, C. G., Breidenbach, A. W., and Lichtenberg, J.J., Chlorinated hydrocarbon pesticides in major U.S. river basins, *Publ. Health Rev.,* 80, 481, 1965.

52. Lichtenberg, J. J., Eichelberger, J. W., Dressman, R. C., and Longbottom, J. E., Pesticides in the surface waters of the United States—a 5-year summary, 1964–68, *Pest. Monit. J.,* 4(2), 71, 1970.

53. Gilliam, R. J., Alexander, R. B., and Smith, R. A., Pesticides in the nation's rivers, 1975–1980, and implications for future monitoring, Water Supply Paper 2271, USGS, Denver, CO, 1985.

54. Ciba-Geigy, A review of historical surface water monitoring for atrazine in the Mississippi, Missouri, and Ohio Rivers, Tech. Report 6-92, Ciba-Geigy Corporation, Agricultural Division, Environ. and Public Affairs Dept., 1992.

55. Goolsby, D. A. and Battaglin, W. A., Occurrence, distribution, and transport of agricultural chemicals in surface waters of the midwestern United States, in *Selected Papers in Agricultural Chemicals in Water Resources of the Midcontinental United States,* Goolsby, D. A., Boyer, L. L., and Mallard, G. E., Eds., Open File Report 93-418, USGS, Denver, CO, 1993, 1.

56. Goolsby, D. A., Battaglin, W. A., Fallon, W. A., Aga, D. S., Kaplin, D. W., and Thurman, E. M., Persistence of herbicides in selected reservoirs of the mid-continent United States: some preliminary results, in *Selected Papers in Agricultural Chemicals in Water Resources of the Midcontinental United States,* Goolsby, D. A., Boyer, L. L. and Mallard, G. E., Eds., Open File Report 93-418, USGS, Denver, CO, 1993, 51.

57. Larson, S. J., Gilliom, R. J., and Capel, P. D., Pesticides in streams of the United States—initial results from the national water-quality assessment program, Water Resources Investigation Report 98-422, USGS, Sacramento, CA, 1999.

58. Richards, R. P. and Baker, D. B., Pesticide concentration patterns in agricultural drainage networks in the Lake Erie Basin, *Environ. Toxicol. Chem.,* 12, 13, 1993.

59. Hall, J. K., Paulus, M., and Higgins, E. R., Losses of atrazine in runoff and soil sediment, *J. Environ. Qual.,* 1, 172, 1972.

60. Ritter, W. F., Johnson, H. P., Lovely, W. G., and Molnau, M., Atrazine, propachlor, and diazinon residues on small agricultural watersheds: runoff losses, persistence and movement, *Environ. Sci. Technol.,* 8, 38, 1974.

61. Wu, T. L., Correll, D. L., and Remenapp, H. E. H., Herbicide runoff from experimental watersheds, *J. Environ. Qual.,* 12, 330, 1983.

62. Forney, D. R., Strahan, J., Rankin, C., Steffin, D., Peter, C. J., Spittler, T. D., and Baker, J. L., Monitoring pesticide runoff and leaching from four farming systems in field scale Coastal Plain watersheds in Maryland, in *Agrichemical Fate and Movement, Perspective and Scale of Study,* Steinheimer, T. R., Ross, L. J., and Spittler, T. D., Eds., ASC Symp. Series 751, Am. Chem. Soc., Washington DC, 2000, Chap. 2.

63. Masse, L., Patni, N. K., Jui, P. Y., and Clegg, B. S., Tile effluent quality and chemical losses under conventional and no tillage—part 2: atrazine and metolachlor, *Trans. ASAE,* 39, 1673, 1996.

64. Bengston, R. L., Southwick, L. M., Willis, G. H., and Carter, C. E., The influence of subsurface drainage practices on herbicide losses, *Trans. ASAE,* 32, 415, 1990.

65. Leonard, R. A., Movement of pesticides into surface waters, in *Pesticides in the Environment: Processes, Impacts and Modeling,* Cheng, H. H., Ed., Soil Sci., Soc. Am., Madison, WI, 1990, Chap. 9.

66. Smith, C. N., Leonard, R. A., Langsdale, G.W ., and Bailey, G. W., Transport of agricultural chemicals from small upland Piedmont watersheds, EPA, 600/3-78-056, EPA, Washington, DC, 1978.

67. Baker, J. L. and Johnson, H. P., The effect of tillage systems on pesticides in runoff from small watersheds, *Trans. ASAE,* 22, 554, 1979.
68. Baker, J. L., Laflen, J. M., and Johnson, H. P., Effect of tillage systems on runoff losses of pesticides, a rainfall simulation study, *Trans. ASAE,* 21, 886, 1978.
69. Baker, J. L., Laflen, J. M., and Hartwig, R. O., Effects of corn residues and herbicide placement on herbicide runoff losses, *Trans. ASAE,* 25, 340, 1982.
70. Baker, J. L. and Laflen, J. M., Runoff losses of surface applied herbicides as affected by wheel tracks and incorporation, *J. Environ. Qual.,* 8, 602, 1979.
71. Watanbe, H., Steichen, J., Barnes, P., Watermeier, N. L., Jasa, P. J., Shelton, D. P., and Dickey, E. C., Water quality aspects of tillage, soil type and slope,—Part II: atrazine and alachlor losses, ASAE Paper No. 92-0010, ASAE, St. Joseph, MI, 1992.
72. Baker, J. L., Effects of tillage and crop residues on field losses of soil—applied pesticides, in *Fate of Pesticides and Chemicals in the Environment,* Schnoor, J.L., Ed., John Wiley and Sons, Inc., New York, NY, 1992, Chap. 11.
73. Baker, J. L. and Shiers, L. E., Effects of herbicide formulation and application method on washoff from corn residue, *Trans. ASAE,* 32, 830, 1989.
74. Wauchope, R. D. and Leonard, R. A., Maximum pesticide concentrations in agricultural runoff. A semi-empirical prediction formula, *J. Environ. Qual.,* 7, 459, 1978.
75. Wauchope, R. D., The pesticide content of surface water drainage from agricultural fields: a review, *J. Environ. Qual.,* 7, 459, 1978.
76. Kenimer, A. L., Mitchell, J. K., Felsot, A. S., and Hissehi, M.C., Pesticide formulation and application technique effects on surface pesticide losses, *Trans. ASAE,* 40, 1617, 1997.
77. Lowrance, R., Vellidis, G., Wauchope, R. D., Gay, P., and Bosch, D.D., Herbicide transport in a managed riparian forest buffer system, *Trans. ASAE,* 40, 1047, 1997.
78. Mickelson, S. K. and Baker, J. L., Buffer strips for controlling herbicide runoff losses, ASAE Paper No. 93-2084, ASAE, St. Joseph, MI, 1993.
79. Baker, J. L., Mickelson, S. K., Arora, K., and Missa, A. K., The potential of vegetated filter strips to reduce pesticide transport, in *Agrochemical Fate and Movement, Perspective and Scale of Study,* Steinheimer, J. R., Ross, L .J., and Spittle, T. D., Eds., ACS Symp. Series 751, Am. Chem. Soc., Washington, DC, 2000, Chap. 18.

6 Nonpoint Source Pollution and Livestock Manure Management

W. F. Ritter

CONTENTS

1-56670-222-4/01/$0.00+$.50
© 2001 by CRC Press LLC

6.1   INTRODUCTION

Man has used animals for food and as a source of labor throughout history. In the 1960s and 1970s, there were major changes in livestock and poultry production. As the consumer demand for meat and animal products increased, so also did mechanization of production. There was a major trend toward the production of confinement livestock and poultry. Poultry broilers and layers led the way with housing systems with increasingly large numbers of animals. Large beef cattle feedlots became common in the 1960s. With the introduction of confinement facilities and the increase in livestock and poultry in individual enterprises, the quantity of manure to be disposed of became a problem. During the late 1960s and 1970s, livestock waste management evolved as a field of engineering to protect the environment and make livestock production systems more cost effective. Overcash et al.[1] summarized the state-of-the-art of livestock waste management up until 1980.

Over the years, the number of farms has decreased, but they have become larger. Production efficiency has also increased, as indicated by the dairy industry. In 1950, New York state had 60,000 farms with 1.36 million dairy cows with an average annual milk production of 2405 kg/cow. In 1994 there were 10,700 dairy farms in New York with 718,000 cows and an average annual milk production of 7218 kg/cow.[2] The hog industry is also changing dramatically. In the last 15 years, the number of hog farms in the U.S. has plunged from nearly 600,000 to 157,000. Fewer than 8% of the farms in the U.S. now have hogs. Meanwhile, the total U.S. hog inventory has declined only 4.3%. Livestock and poultry production occurs in every state; however, the livestock and poultry industries are concentrated in various regions because of favorable climate, feed availability, proximity to market, labor availability, etc. Iowa and North Carolina are the two largest hog producing states with 12.2 and 9.3 million head, respectively. California and Wisconsin are the leading dairy states, and Texas and Kansas have the largest concentration of cattle feedlots. Arkansas and Georgia are the two leading broiler production states, and Ohio and Indiana are the leading egg production states.

Livestock production became regulated at the federal level with the passage of the amendments to the Federal Water Pollution Control Act (PL-92-500) in 1972. Concentrated animal feeding operations above a certain size were treated as a point source under the National Pollutant Discharge Elimination System (NPDES) and required a permit. Effluent guidelines require no discharge of runoff, manure, or process-generated wastewater from rainfall less than a 25-year frequency, 24-hour duration storm event. The Coastal Zone Management Act (CZMA) of 1972 was reauthorized and amended by the Coastal Zone Act Reauthorization Amendments (CZARA) in 1990.[3] Section 6217 of the CZARA is to address nonpoint source pollution of coastal waters, portions of 24 states are subject to CZARA. Nonpoint source pollution control related to the livestock industry that is covered by the Act includes large- and small-animal confinement facilities, plant nutrients, and pasture and range.[4] All states affected by the Act must develop management plans for controlling nonpoint source pollution. Although federal guidelines may control pollution from animal agriculture, in some states, federal regulations are superseded by state regulations that are more stringent. Just recently, EPA and USDA finalized a national strategy for confined animal feeding operations (CAFOs).[5] The goal of the policy is to minimize water quality impacts from large animal agriculture operations.

6.2   MANURE CHARACTERISTICS

Both ASAE[6] and the Natural Resources Conservation Service (NRCS)[7] have published standard values for physical and chemical properties of manure for livestock and poultry. Physical properties of manure that are important in planning and designing manure management systems are weight, volume, total solids, and moisture content. The most important chemical properties are nitrogen (N), phosphorus (P), and potassium (K). These parameters are used in planning manure land application plans. Some of the physical and chemical properties of manure for beef, dairy, swine, and poultry are presented in Tables 6.1 and 6.2.[6,7] ASAE data was last revised in 1988 to reflect the latest research data. In most cases, average values of dry manure and nutrients were revised upward, and standard deviations were calculated to reflect the degree of variability. The NRCS characteristics are based upon the ration, feed digestibility, and 5% feed waste.[8] If the waste feed is more than 5%, NRCS manure characteristic values should be increased.

Values in Tables 6.1 and 6.2 are as excreted, which are the most reliable data. Manure properties resulting from other situations, such as flushed manure, feedlot manure, and poultry litter are the result of certain "foreign" materials being added or some manure components being lost from the excreted manure. Characteristics of stored or treated manure are strongly affected by actions such as sedimentation, flotation, and biological degradation. When possible, on-site manure sampling and testing should be done to plan manure management systems.

Manure can be handled as a solid, semisolid, slurry, or liquid.[7] In general, manure of less than 4–5% solids can be handled as a liquid, manure of 5–10% solids can be handled as a slurry, and manure of 10–15% solids can be handled as a semisolid. Above 20% solids, most manures can be handled as a solid.

6.3   WATER QUALITY IMPACTS

6.3.1   SOURCES

Livestock production can affect both groundwater and surface water. Surface waters can be impacted by runoff from feedlots and barnyards, from manure land application

TABLE 6.1
Fresh Manure Production and Characteristics per 1000 kg Live Animal Mass per Day[6]

Parameter	Units[a]	Dairy 640 kg[b]	Beef 360 kg	Swine 61 kg	Layer 1.8 kg	Broiler 0.9 kg	Turkey 6.8 kg
					Typical Live Animal Masses		
Total manure[c]	kg						
mean[d]		86	58	84	64	85	47
std. deviation		17	17	24	19	13	13
Urine	kg						
mean		26	18	39	***[7]	**	**
std. deviation		4.3	4.2	4.8	**	**	**
Density	kg						
mean		990	1000	990	970	1000	1000
std. deviation		63	75	24	39	**	**
Total solids	kg						
mean		12	8.5	11	16	22	12
std. deviation		2.7	2.6	6.3	4.3	1.4	3.4
Volatile solids	kg						
mean		10	7.2	8.5	12	17	9.1
std. deviation		0.79	0.57	0.66	0.84	1.2	1.3
BOD	kg						
mean		1.6	1.6	3.1	3.3	**	2.1
std. deviation		0.48	0.75	0.72	0.91	**	0.46
COD	kg						
mean		11	7.8	8.4	11	16	9.3
std. deviation		2.4	2.7	3.7	2.7	1.8	1.2
pH							
mean		7.0	7.0	7.5	6.9	**	**
std. deviation		0.45	0.34	0.57	0.56	**	**
Total Kjeldahl N	kg						
mean		0.45	0.34	0.52	0.84	1.1	0.62
std. deviation		0.096	0.073	0.21	0.22	0.24	0.13
Ammonia N	kg						
mean		0.079	0.086	0.29	0.21	**	0.080
std. deviation		0.083	0.052	0.10	0.18	**	0.018

Total P	kg	mean[d]	0.094	0.092	0.18	0.30	0.30	0.23
		std. deviation	0.024	0.027	0.10	0.081	0.053	0.093
Ortho phosphorus	kg	mean	0.061	0.030	0.12	0.092	**	**
		std. deviation	0.0058	**	**	0.016	**	**
Potassium	kg	mean	0.29	0.21	0.29	0.30	0.40	0.24
		std. deviation	0.094	0.061	0.16	0.072	0.064	0.080
Calcium	kg	mean	0.16	0.14	0.33	1.3	0.41	0.63
		std. deviation	0.059	0.11	0.18	0.57	**	0.34
Magnesium	kg	mean	0.071	0.049	0.070	0.14	0.15	0.073
		std. deviation	0.016	0.015	0.035	0.042	**	0.0071
Sulfur	kg	mean	0.051	0.045	0.076	0.14	0.085	**
		std. deviation	0.010	0.0052	0.040	0.066	**	**
Sodium	kg	mean	0.052	0.030	0.067	0.10	0.15	0.066
		std. deviation	0.026	0.023	0.052	0.051	**	0.012
Chloride	kg	mean	0.13	**	0.26	0.56	**	**
		std. deviation	0.039	**	0.052	0.44	**	**
Iron	kg	mean	12	7.8	16	60	**	75
		std. deviation	6.6	5.9	9.7	49	**	28
Manganese	kg	mean	1.9	1.2	1.9	6.1	**	2.4
		std. deviation	0.75	0.51	0.74	2.2	**	0.33
Boron	kg	mean[d]	0.71	0.88	3.1	1.8	**	**
		std. deviation	0.35	0.064	0.95	1.7	**	**

(continued)

TABLE 6.1 (*continued*)

Parameter	Units[a]		Dairy 640 kg[b]	Beef 360 kg	Swine 61 kg	Layer 1.8 kg	Broiler 0.9 kg	Turkey 6.8 kg
Molybdenum	kg	mean	0.074	0.042	0.028	0.30	**	**
		std. deviation	0.012	**	0.030	0.057	**	**
Zinc	kg	mean	1.8	1.1	5.0	19	3.6	15
		std. deviation	0.65	0.43	2.5	33	**	12
Copper	kg	mean	0.14	0.31	1.2	0.83	0.98	0.71
		std. deviation	0.14	0.12	0.84	0.84	**	0.10
Cadmium	kg	mean	0.0030	**	0.027	0.038	**	**
		std. deviation	**	**	0.028	0.032	**	**
Nickel	kg	mean	0.28	**	**	0.25	**	**
		std. deviation	**	**	**	**	**	**
Lead	kg	mean	**	**	0.084	0.74	**	**
		std. deviation	**	**	0.012	**	**	**

[a] All values wet basis.

[b] Typical live animal masses for which manure values represent. Differences within species according to exist, but sufficient fresh manure data to list these differences were not found.

[c] Feces and urine as voided.

[d] Parameter means within each animal species are composed of varying populations of data. Maximum numbers of data points for each species are dairy, 85; beef, 50; veal, 5; swine, 58; 39; 3; horse, 31; layer, 74; broiler, 14; turkey, 18.

[e] All nutrients and metals values are given in elemental form.

[f] Data not found.

TABLE 6.2
Fresh Manure Production and Characteristics per 1000 kg Live Weight[7]

Parameter	Unit	Dairy		Beef Feeder[a]	Swine Grower[b]	Layer	Broiler
		Lactating	Dry				
Total manure	kg	80	82	59	63	61	46
Density	kg/m^3	977	1001	987	1006	1032	99
Total solids	kg	10.0	9.5	6.8	6.3	15.1	11.4
Volatile solids	kg	8.5	8.1	6.0	5.4	10.8	9.7
BOD	kg	1.6	1.2	1.4	2.1	3.7	3.3
COD	kg	8.9	8.5	6.1	6.1	13.7	12.2
Total N	kg	0.45	0.36	0.31	0.42	0.83	0.62
Total P	kg	0.07	0.05	0.11	0.16	0.31	0.24
Potassium	kg	0.26	0.23	0.24	0.22	0.34	0.26

[a]Beef feeder on high forage diet of 340–500 kg.

[b]Grower pig, 18–100 kg.

sites, and from pastures where livestock are grazing. Overflows from manure storage and treatment systems can also contaminate surface waters. Where animals have direct access to streams, animal urine and feces may be directly discharged to streams. Organic matter, nutrients, microorganisms, and salts are the major pollutants found in manure that may contaminate surface waters.

The major concern with groundwater contamination is NO_3 leaching. Potential sources of groundwater contamination from manure include seepage from manure storage basins and lagoons and leaching of nutrients from land application sites.

6.3.2 ORGANIC MATTER

Whenever organic matter enters a stream, lake or pond, it is degraded by aquatic microorganisms by the following generalized reaction:

Organic matter + microorganisms + $O_2 \rightarrow CO_2 + H_2O$ + more microorganisms.

The organic matter is used as an energy source for synthesis of new cell material, and the microorganisms use the oxygen in the water to break down the organic matter. As a result, the dissolved oxygen is decreased in the water. Dissolved oxygen is critical to the survival of fish and other desirable aquatic organisms. Organic matter also contains organic N which is converted to NH_3 during the degradation process. Fish are sensitive to NH_3; nonionic NH_3 concentrations as low as 0.2 mg N/L may prove toxic to fish.

The biodegradable organic matter concentration can be measured by the biochemical oxygen demand test (BOD). The BOD is determined by measuring the quantity of dissolved oxygen utilized by microorganisms under aerobic conditions in stabilizing the carbonaceous organic matter during a specified period of time and at a constant temperature, usually 5 days and 20°C. The carbonaceous or first-stage reaction is assumed to follow first-order kinetic and can be represented by the following equation:

$$\frac{dy}{dt} = K(L - y) \tag{6.1}$$

where y is the BOD concentration up to time t, mg/L, L is the total first stage or carbonaceous BOD, mg/L, t is time in days, and K is the rate constant in days^{-1}.

Another measure of organic matter is the chemical oxygen demand test (COD). Instead of microorganisms, the COD test uses a strong chemical oxidizing agent, usually potassium dichromate in an acid solution. The COD test is run more quickly than the BOD test with a digestion time of from 1 to 2 hours.

6.3.3 NUTRIENTS

Nitrogen and P can cause eutrophication in lakes and estuaries. Eutrophication can be defined as an increase in the nutrient status of natural waters that causes growth of algae or other vegetation, depletion of dissolved oxygen, increased turbidity, and a degradation of water quality. A body of water may be N- or P-limited. If the N:P ratio is >15:1, the water body is P-limited; if the ratio is <10:1 it is N-limited. The eutrophication threshold for most P-limited systems is from 10 to 100 μP/L. For N-limited systems, the threshold is 0.5 to 1.0 mg N/L.[9]

Nitrate contamination of groundwater is a global concern. Strebel et al.[10] stated that the major causes of NO_3 contamination of groundwater in Europe were (1) intensified plant production and increased use of N fertilizers, (2) intensified livestock production with high livestock densities that cause enormous production of manure on an inadequate land base, and (3) conversion of large areas of permanent grassland to usable land. Livestock production is concentrated in certain areas of the U.S., which can result in a surplus of manure that can cause groundwater contamination. Ninety percent of the 6.2 billion broilers produced in 1995 were grown in 15 states and 55 percent of the eggs were produced in eight states.[11] Two areas of concentrated poultry production with documented environmental problems are the Delmarva Peninsula and northwestern Arkansas. Ritter and Chirnside[12] sampled more than 200 wells in southern Delaware. More than 34% of the wells tested in Sussex County had NO_3 concentrations above 10 mg N/L. They cited intensive agricultural activity, particularly land application of poultry manure, as the cause. Scott et al.[13] reported that application of poultry litter on pasture in northwestern Arkansas adversely impacted groundwater and springs.

When manure is used as a fertilizer, application rates are based mostly upon the N requirements of the plants. The efficiency of applied N in terms of the amount applied and what is taken up by the crop is always less than one because of: (1) N uptake in the nonharvested parts of the plant, (2) denitrification in the soil, (3) NH_3 volatilization, and (4) leaching into deeper soil horizons. It is more difficult to predict the amount of manure to apply to meet the crop N requirements than with commercial fertilizer. Most of the N in manure is in the organic and NH_3 forms. If the manure

TABLE 6.3
Percent of Nitrogen Losses During Land Application[14]

Application Method	Type of Waste	Nitrogen Lost, %
Broadcast	Solid	15–30
	Liquid	10–25
Broadcast with immediate cultivation	Solid	1–5
	Liquid	1–5
Knifing	Liquid	0–2
Sprinkler irrigation	Liquid	15–40

is not incorporated shortly after it is applied, most of the NH_3 may be lost by volatilization. Total N losses from broadcast manure may be as high as 30% (Table 6.3).[14] Nitrogen losses also occur during treatment or storage. Seventy to eighty percent of the N from fresh excreted manure may be lost if lagoons are used, while an anaerobic pit may lose only 15 to 30% of the N (Table 6.4).[14]

Organic N is mineralized to NH_3 and NO_3 when manure is applied to soil. Factors such as how the manure has been treated or stored, soil temperature, and soil moisture can affect the mineralization rate. Deciding on what mineralization rate to use is important in determining manure application rates for N. Mineralization rates may vary from 25 to 60% the first year depending upon the type of manure (Table 6.5).[14] Organic N released during the second, third, and fourth cropping years after initial application is usually 50, 25, and 12.5%, respectively, of that mineralized during the first cropping year.[14]

When N is used to determine manure application rates, for most manure types P is generally applied at rates beyond crop removal in the harvested biomass except in

TABLE 6.4
Nitrogen Losses from Storage and Treatment[14]

System	Nitrogen lost, %
Solid	
Daily scrape and haul	20–35
Manure pack	20–40
Open lot	40–55
Deep pit (poultry)	25–50
Litter	25–50
Liquid	
Anaerobic pit	15–30
Above-ground storage	10–30
Earth storage	20–40
Lagoon	70–85

TABLE 6.5
Organic Nitrogen Mineralization Rates the First Year
After Application[14]

Manure Type	Manure Handling	Mineralization Factor
Swine	Fresh	0.50
	Anaerobic liquid	0.35
	Aerobic liquid	0.30
Beef	Solid without bedding	0.35
	Solid with bedding	0.25
	Anaerobic liquid	0.30
	Aerobic liquid	0.25
Dairy	Solid without bedding	0.35
	Solid with bedding	0.25
	Anaerobic liquid	0.30
	Aerobic liquid	0.25
Sheep	Solid	0.25
Poultry	Deep pit	0.60
	Solid with litter	0.60
	Solid without litter	0.60
Horses	Solid with bedding	0.20

extremely P-deficient soils. If manure is applied year after year with N-based manure management, soil P levels will continue to increase. Soil test results from 1991 to 1992 for Sussex County, Delaware, showed that 77% of the samples from agricultural fields had high or excessive levels of soil test P.[15] Sussex County has the most concentrated broiler production in the U.S. Soils with high P levels that are susceptible to erosion will cause high levels of eutrophication. Inorganic phosphates are mainly Fe and Al phosphates in acid soils and Ca phosphates in alkaline soils. Any P added as fertilizer or released in decomposition of organic matter rapidly is converted to one of these compounds. All forms of inorganic P in soils are extremely insoluble. Because of the high adsorptive capacity of P by clays, the Fe and Al oxides leaching of P to groundwater is rare.[16] The situation where P leaching may occur is in well-drained, deep, sandy soils.[17]

6.3.4 MICROORGANISMS

Livestock manure contains large quantities of microorganisms from the intestine of the animal. Manures are a potential source of approximately 150 diseases. Illnesses that may be transmitted by bacterial diseases include typhoid fever, gastro-intestinal disorders, cholera, tuberculosis, anthrax, and mastitis. Transmittable viral diseases are hog cholera, foot and mouth disease, polio, respiratory diseases, and eye infections. Although the potential for disease transmission from livestock manures is present, the incidence of human disease attributable to manure contact has been infrequent.

Manure applied to land or lagoon and storage basin overflows pose public health hazards. Numerous factors such as climate, soil types, infiltration rates, topography, animal species, animal health, and presence of carrier organisms influence the nature and amount of disease-producing organisms that will reach a stream. When manure is applied to land on hot, sunny days, harmful bacteria die rapidly. Rain falling on freshly applied manure or manure applied to frozen ground increases the potential for harmful organisms to reach watercourses.

Fecal coliform are used as an indicator organism to test for organic pollution. They are nonpathogenic and reside in the intestine of warm-blooded animals, including humans. The fecal coliform to fecal streptococcus ratio can be used to differentiate waste origin or source in fresh water.

In recent years cryptosporidium, which is a protozoan found in surface waters, has become a concern. It can cause cryptosporidosis, a severe diarrhea, in humans and animals. Runoff from fields receiving livestock manure have been blamed for contributing to outbreaks in recent years. In 1993, 400,000 people were infected in Milwaukee. In Ontario, Fleming and McLellan[18] measured cryptosporidium in 20 surface water sites, of which 10 received livestock manure and 10 were nonlivestock areas. Of 60 samples collected in total, only 9 tested positive for cryptosporidium and only at relatively low levels.

6.3.5 SALTS

Animal manures contain salts that can be harmful to soils and crops if the manure is applied at too high an application rate. Sodium chloride (NaCl) is supplemented in swine diets at the rate of 0.025 to 0.5% to prevent deficiency symptoms, 0.25–0.30% are most common.[19] In anaerobic swine manure storage pits, Na ranges from 5000 to 9000 mg/L on a dry-weight basis for dietary NaCl additions of 0.2 to 0.5%.[20]

Feedlot runoff held in evaporation ponds may have extremely high salt concentrations with electrical conductivity of over 20 mmhos/cm.[21] Dilution of feedlot runoff may be needed when used for irrigation with dilution ratios of 3:1–10:1 depending upon soil texture and characteristics of the effluent and irrigation water.[22] Salt tolerance has been established for most crops.[23] High salt-tolerant crops include sorghum, barley, wheat, rye, and bermuda grass. Corn is less salt tolerant but is a high user of N and a good crop to use on manure or feedlot runoff application sites. Research in Kansas showed that about 250 mm of undiluted feedlot runoff applied per year produced peak yields of corn silage, but beyond that level it began to reduce yields. Liebhardt[24] found grain corn yields were reduced if broiler litter was applied at an application rate of greater than 22.4 mg/ha.

Sweeten et al.[25] found that application of 100 to 235 mm/yr of undiluted feedlot runoff in level border irrigation maintained a good stand of wheat over a 4-year period in Texas. Final soil electrical conductivity levels were 1.4, 1.8, and 1.3 mmhos/cm for 100, 170, and 235 mm of application of feedlot runoff, respectively, compared with control treatments of 0.4 mmhos/cm.

6.4 BARNYARD AND FEEDLOT RUNOFF

Runoff from feedlots contains high concentrations of nutrients, salts, pathogens, and oxygen-demanding organic matter. Some typical cattle feedlot runoff characteristics are presented in Table 6.6.[26] Feedlots in the Great Plains and southwestern U.S. begun in the late 1960s and 1970s were required to control discharges. Texas and several other cattle-feeding states instituted individual permit programs by the early 1970s that are still in effect. In 1974, the EPA adopted feedlot effluent guidelines requiring no-discharge and a federal permit system for feedlots of more than 1000 head that discharge less than a 25-year, 24-hour duration storm event.[27]

In 1987, the Texas Natural Resources Conservation Service developed a set of regulations that stated there shall be no discharge from livestock feeding facilities, but the animal waste material must be collected and used or disposed of on agriculture land. Beef feedlots with more than 1000 head on feed need a permit, but with less than 1000 beef cattle on feed, they do not need a permit but still must meet the no-discharge policy. In 1993, EPA adopted a general permit for Concentrated Animal Feeding Operations (CAFOs) in Texas, Louisiana, Oklahoma, and New Mexico.[28] The general permit requires CAFOs with more than 1000 animal units to come under the general permit. Also, operations with 300 or more animal units come under the general permit if they discharge wastewater through a manmade conveyance structure. The general permit requires the following: (1) design, implementation, and maintenance of best management practices (BMPs) for control of rainfall runoff manure and processing wastewater including overflow cattle drinking water, (2) prevention of hydrologic connection to surface waters, (3) and application of manure and wastewater onto land at agronomic nutrient loading rates.

In recent years EPA has been working on an animal feeding operation (AFO) strategy that was finalized in 1998. The objectives of the strategy are to expand compliance and enforcement efforts, improve Clean Water Act (CWA) permits, focus on priority watersheds, review existing regulations, and increase EPA/USDA coordination. The vast majority of 450,000 animal feeding operations in the U.S. will not be the focus of compliance and enforcement by EPA. The focus for compliance and enforcement activities will be on the larger operations that meet the regulatory definition of CAFOs and other facilities designated as CAFOs because of their impact on the environment. It is the goal of the strategy to issue CWA permits to all CAFOs by 2005 consistent with any new regulations EPA will have promulgated.

Early research in cattle feedlot runoff was directed to characterizing the runoff for pollutants and to develop runoff versus rainfall relationships for designing runoff holding ponds. Gilbertson et al.[29] found it takes about 13 mm of rainfall to induce runoff from a cattle feedlot. Rainfall versus runoff relationships predict less runoff per unit of rainfall in dry climates than in wetter climates.[26] It is recommended holding ponds be designed using a NRCS runoff curve number of 90, which would provide a conservative estimate of runoff in the Great Plains.[27] In the Great Plains cattle feeding regions, the annual amount of runoff expected is about 20–33% of rainfall. With a NRCS runoff curve number of 90, a 40-ha feedlot in a 450-mm rainfall area will produce an average of 42,000 m^3 of runoff per year.

Groundwater quality may be impacted by seepage from runoff holding ponds or by the feedlot itself. Standards for seepage control for runoff holding ponds generally require them to be built in (or lined with) at least 30 cm compacted thickness of soil material with 30% or more passing a No. 200 mesh sieve, a liquid limit of 30% or

TABLE 6.6
Average Chemical Characteristics of Runoff from Beef Cattle Feedyards in the Great Plains[26]

Location	Total Solids	Chemical Oxygen Demand	Total Nitrogen	Total Phosphorus	Potassium	Sodium	Calcium	Magnesium	Chloride	Electrical Conductivity
					ppm					mmhos cm^{-1}
Bellville, TX	9,000	4,000	85	85	340	230	—	—	410	—
Bushland, TX	15,000	15,700	1,080	205	1,320	588	449	199	1,729	8.4
Ft. Collins, CO	17,500	17,800	—	93	—	—	—	—	—	8.6
McKinney, TX	11,430	7,210	—	69	761	408	698	69	450	6.7
Mead, NE	15,200	3,100	—	300	1,864	478	181	146	700	3.2
Pratt, KS	7,500	5,000	—	50	815	511	166	110	—	5.4
Sioux Falls, SD	2,990	2,160	—	47	—	—	—	—	—	—

more, and a plastic index of 15 or more.[27] These three criteria require a sandy clay loam, clay loam, or clay soil and should attain a hydraulic conductivity of 1×10^{-7} cm/sec, which is required in most permits. A clay liner 45 cm thick with materials having a hydraulic conductivity of 1×10^{-7} cm/sec is specified as one method for establishing "no hydrologic connection" to waters of the U.S.

Norstadt and Duke[30] measured soil NO_3 levels that decreased from 80 mg N/kg at the top of the feedlot soil profiles to less than 10 mg N/kg at 1.0 to 1.5 m depth. The same results were obtained from a clay loam soil and a layered soil that consisted of 0.75 m of sand over 0.75 m of clay loam.

In some feedlot soil profiles, denitrification may take place. Schuman and McCalla[31] measured NO_3 concentrations of 7.5 mg N/kg in the top 100 mm of a Nebraska feedlot. Below 200 mm, NO_3 concentrations were below 1.0 mg N/kg because of denitrification. Elliott et al.[32] collected soil water samples at 0.45, 0.70, and 1.1 m beneath a level cattle feedlot on a silt loam/sand soil profile. Nitrate concentrations were generally less than 1.0 mg N/L compared with 0.3 to 101 mg N/L in the top 75 mm.

The feedlot profile usually contains a compacted interfacial layer of manure and soil that provides a biological seal that reduces water infiltration rates to less than 0.05 mm/hr and reduces leaching of salts, NH_3, and NO_3.[34,31]

6.5   MANURE STORAGE AND TREATMENT

Manure may be stored in earthen, concrete, steel, or fiberglass structures or treated by physical, chemical, or biological methods. Biological treatment of manure is the most commonly used method. Anaerobic lagoons have found widespread application in the treatment of animal wastes because of their low initial cost, ease of operation, and convenience of loading by gravity flow from the livestock buildings.[34] Aerobic and aerated lagoons are not widely used. Feedlot runoff is collected mostly in holding ponds. Manure may be stored as a solid, semi-solid, or liquid. The greatest potential for water pollution from manure storage and treatment systems is by seepage from anaerobic lagoons, earthen manure storage basins, or feedlot runoff holding ponds. There is also the potential for lagoons and manure storage basins to overflow or the berm of the lagoon or storage basin to break. Leachate may also occur from solid-manure storage systems.

Some studies have shown that lagoons can cause groundwater contamination, and other studies indicate biological sealing takes place. In a study of unlined lagoons in the Coastal Plain soils in Virginia, Ciravolo et al.[35] found that two anaerobic swine lagoons caused measurable (but minimum) groundwater contamination. A third lagoon hold contaminated groundwater with Cl and NO_3 in excess of drinking water standards. Sewell[36] found that NO_3 and Cl concentrations in groundwater taken from wells 15 m from an unlined anaerobic dairy lagoon increased rapidly during the first six months of lagoon operation, and later decreased to levels similar to those before the lagoon was loaded. Median NO_3 concentrations of all the test wells were below 10 mg N/L. The lagoon was located in an area with silt loam and sandy loam soils to a depth of 1 m and a quartz sand horizon at 1-4 m. Nordstedt et al[37] found that NO_3 concentrations were above background levels in the groundwater in wells at a depth of 3.0 m and a distance of 15 m from a dairy lagoon in a clay soil that had been in operation for 8 months. At a distance of 15 m, the average NO_3 concentration in the wells was 14.3 mg N/L.

Ritter et al.[38] found that an unlined anaerobic lagoon for swine wastes had some impact on groundwater quality. During the first year of operation, NO_3, NH_3, and organic N concentrations increased in some of the monitoring wells but decreased to lower levels after the first year. None of the monitoring wells had NO_3 concentrations above 10 mg N/L. In a second study, Ritter and Chirnside[39] monitored groundwater quality for three years at two sites around clay-lined anaerobic lagoons. A swine waste lagoon located in an Evesboro loamy sand soil (excessively well drained) was having a severe impact on groundwater quality. Ammonium N concentrations above 1000 mg N/L were measured in shallow monitoring wells around the lagoon. Chloride and total dissolved solids (TDS) concentrations were also high. At the second site, which has three lagoons and a settling pond in poorly drained soils, some seepage was occurring. Ammonium N, NO_3, Cl, and TDS were above background concentrations in some of the monitoring wells. There was a strong correlation between NO_3 and Cl concentrations in the monitoring wells. The results indicated that clay-lined animal waste lagoons located in sandy loam or loamy sand soils with high water tables may lead to degradation of groundwater quality.

Westerman et al.[40] found that seepage losses from older unlined lagoons in North Carolina were much higher than previously believed. Two swine lagoons that had received swine waste from 3.5 to 5 years had high NH_3 and NO_3 concentrations in the shallow groundwater. The variation with time, with spatial location, and with depth in the groundwater were substantial. They concluded that the variations made it very difficult to develop groundwater transport models to accurately predict transport and transformations of NH_3 and NO_3 resulting from seepage from anaerobic lagoons. In a follow-up study, Huffman[41] evaluated 34 swine lagoons for impacts to shallow groundwater from lagoon seepage. About two-thirds of the sites showed seepage contamination exceeding drinking water standards at 38 m down gradient.

Numerous studies have shown holding ponds, manure storage basins, and treatment lagoons have a tendency to be partially self-sealing. Research in Canada showed that clogging of soil pores by bacterial cells and organic matter is the mechanism responsible for partial self-sealing.[42] The initial freshwater infiltration rate in 4.5-m deep holding ponds was 10^{-2}, 10^{-3}, and 10^{-4} cm/sec for sand, clay, and loam, respectively. After only 2 weeks of storage, the infiltration rates of dairy lagoon effluent were reduced to only 10^{-6} cm/sec in loam and sandy soils compared with $0-1.8 \times 10^{-6}$ cm/sec after a year for all three soils. Miller et al.[43] also found an unlined earthen storage basin in a sandy soil became effectively sealed to infiltration within 12 weeks after the addition of beef cattle manure.

Clay liners help reduce the movement of chemicals below manure storage ponds. Phillips and Culley[43] found NO_3 concentrations at 1.5 to 4.5 m below a dairy manure storage pond were 0.4 mg N/L for a clay soil, 1.2 mg N/L for a loam soil, and 17 mg N/L for a sandy soil. Gangbazo et al.[44] concluded that all manure storage basins with a hydraulic conductivity of less than 10^{-5} cm/sec had no contamination from NH_3 or NO_3.

6.6   LAND APPLICATION OF MANURES

An efficient manure management and application system meets, but does not exceed, the needs of the crop and thereby minimizes pollution. Any farm enterprise that applies manure to land should have such a system.

Certain farming practices will help prevent the loss of nutrients from manure and manured fields, thus reducing fertilizer expenses and water pollution. The key to conserving manure P and K is to reduce erosion and runoff from fields. Conserving manure N also requires erosion and runoff control, proper handling, storage, treatment, and timing of manure applications and incorporation into the soil; and other practices that reduce leaching.

6.6.1   APPLICATION METHODS

The goal of any manure application system is to apply manure to land and minimize environmental change, community relations problems, damage to the land, cost, and frustration, and to maximize the use of nutrients in the manure.[45]

Manure may be applied to the surface, incorporated, or injected. If manure is simply applied to the surface of the soil, much of the unstable, rapidly mineralized organic N from the urine will be lost through the volatilization of NH_3 gas. Volatilization increases with time, temperature, wind, and low humidity. Loss from runoff, and the resulting water pollution, are particularly great when manure is spread on frozen or snow-covered ground or on fields that are flooded. Incorporating manure into the soil, either by tillage or subsurface injection, increases the amount of manure N available for use by crops and can reduce water pollution. A soaking rain of 1.5 cm with no runoff has the same effect as incorporating manure. When tillage tools such as moldboard plows, chisel plows, and heavy discs are used to incorporate the manure, care must be taken to incorporate the manure completely before it dries, usually within two days or less.

Injection is probably the best method for incorporating manure in reduced-till or no-till cropping systems because crop residues are left on the surface to act as a mulch, and exposed soil surface is minimal. Injection requires a liquid manure spreader and equipment to deposit manure below the soil surface. To be effective, the openings made by the injectors must be closed over the manure following application. It may be possible to inject manure into a growing row crop to supply nutrients closer to the time when the crop needs them.

Manure can be handled as a solid, semisolid, or liquid. Solid manure generally has from 15 to 23% solids content, depending upon the livestock type, and can be handled with a fork or front end loader with tines. It is applied to land with a box-type spreader. Other types of equipment used for applying solid manure include flail-type spreaders, dump trucks, earth movers, or wagons.

Semisolid manure (from 4 to 15% solids) can be pumped and handled with liquid manure handling equipment. It can also be handled with a front-end loader and a box-type or flail-type spreader. Piston, helical rotor, submerged centrifugal, and positive displacement gear type pumps can handle heavy semisolids against high pressures. Submerged centrifugal, piston, or auger pumps are used to pump heavy semisolids against low pressures.

If the manure contains fibrous material, such as bedding, hair, or feed, a chopper pump to cut the fibrous material should be used. Piston pumps readily handle manure with bedding.

A liquid tanker spreader is the best choice for handling semisolid manure up to 10% solids. Big gun sprinklers are required to handle semisolid manure by irrigation.

Manure with less than 4% solids is classified as liquid manure. If large quantities of liquid manure are handled, a pipeline and irrigation system is preferred to a tank wagon for transporting and applying the manure.

6.6.2 SURFACE WATER QUALITY

The main factors influencing the impact of land application of manure on surface water quality are the fate of N and P in surface soil and manure management. Phosphorus is adsorbed by soil particles, so loss of P in surface runoff is of greater concern than leaching. It may be lost in both the particulate and dissolved forms. Because P is adsorbed by the soil fraction most susceptible to erosion (clays, oxides of Fe and Al), it is important to reduce soil erosion to control particulate P losses. Phosphorus often accumulates in the upper few centimeters of the soil, particularly under minimum tillage conditions where manures and fertilizers are not incorporated. Hence, dissolved phosphate levels can be quite high in the upper few centimeters of soil that are most interactive with surface runoff.

When animal manures are applied at rates based on crop N requirements, P levels can build up rapidly in the soil. Sharpley et al.[46] indicated in a P balance and efficiency of plant and animal uptake of P the surplus for the U.S. was 26 kg/ha and for the Netherlands was 88 kg/ha. Poultry manure is higher in P than other manures. Broiler manure has an approximate N:P ratio of 40:16.9 with a plant-available N value of 50%, the ratio becomes 20:16.9. As a result of this ratio, in areas with intensive poultry production such as the Delmarva Peninsula and Arkansas, many soils have high levels of soil test P.

Poultry litter is a common source of nutrients for forage crops in poultry growing areas. Research has shown that it increases yields for forage crops such as fescue, orchard grass, and bermuda grass.[47] One of the concerns with applying poultry litter to forages is the impact on surface water quality. A number of researchers have found that runoff concentrations of various litter constituents are higher from litter-treated areas than from untreated areas for simulated rainfall events occurring soon (1–3 days) after application.[48,49] In addition to N and P concerns with poultry manure, the growth hormones testosterone (0.8 to 2.9 ng/L) and estrogen (1.2 to 4.1 ng/L) have been found in several streams of the Conestoga River Valley of the Chesapeake Bay watershed,[50] surface runoff from manured fields contained 215 ng/L testosterone and 19 ng/L estrogen.

The rate, method, and timing of manure application will influence the amount of N and P lost in surface runoff. Edwards and Daniel[49,51] found that concentrations of total N, NH_3, dissolved P, and total P increased linearly with increased poultry litter and swine manure when applied to fescue in northeast Arkansas.

Incorporating manure into the soil profile, either by tillage or subsurface injection, reduces the potential for N and P losses in runoff. Mueller et al.[52] showed incorporation of dairy manure by chisel plowing reduced total P loss in runoff from corn 20 times compared with no-till areas receiving surface applications. Some of the decrease was caused by the reduced volume of runoff with chisel plowing compared with no-till. Infiltration rates increased with the incorporated manure. They also found there was no significant relationship between soil test P and the mass of dissolved P lost in runoff.

Timing of manure application relative to rainfall also affects N and P losses. Westerman and Overcash[53] found concentrations of total N and P in runoff were reduced approximately 90% when simulated runoff was delayed from 1 hr to 3 days after poultry manure or swine manure was applied to fescue in North Carolina. Edwards and Daniel[51] found little effect of time on N and P loss in runoff with longer periods between swine manure application to fescue and rainfall runoff initiation in Arkansas. These two studies suggest intervals of more than 3 days between manure application and runoff will not greatly affect N and P loss in runoff. The type of manure does not appear to affect the amount of N and P lost in surface runoff. A number of studies are summarized in Table 6.7. Nitrogen and P losses are highly variable.

Crane et al.[63] concluded that land application of wastes can significantly increase bacterial concentrations in runoff if safety precautions and wise management are not taken. Robbins et al.,[64] studying various livestock operations in North Carolina, determined 2–23% of the fecal coliform deposited on fields by manure application were lost in runoff on an annual basis. McCaskey et al[65] found bacteria losses were highest for solid-spread dairy manure and lowest for liquid-spread manure when they compared liquid, semisolid, and solid dairy manure application with a minimally sloped sandy loam soil with bermuda grass cover. For solid manure application, the maximum annual removal of applied total coliforms, fecal coliforms, and fecal streptococci was 0.06, 0.007, and 0.008%, respectively. These rates were much lower than those cited by Robbins et al.[64]

6.6.3   SUBSURFACE DRAINAGE WATER QUALITY

Subsurface drainage waters may be impacted by liquid manure application. Dean and Foran[66] reported numerous incidents of bacterial contamination from tile drains in Ontario, Canada. Of 12 monitored liquid manure spreading sites under a variety of field conditions and soil types, 8 resulted in water quality degradation within 20 minutes to 6 hours following application. One site resulted in a 725,000 times increase in bacteria levels within 2 hours, and two other sites showed increases in tile flow in response to the application. In southwestern Ontario, Fleming and Bradshaw[67] also

TABLE 6.7
Proportion of N and P Added in Manure Transported in Surface Runoff

	Amount Added		Study Period	Percent Loss		Reference and Location
	N	P		N	P	
	kg ha^{-1} yr^{-1}			%		
Dairy manure						
Corn	451	108	3 months	11.1	8.1	Klausner et al.,[54] NY
C. bermuda grass	807	175	4 years	1.6	—	Long,[55] AL
Fescue	133	142	4 events	2.1	1.3	McLeod and Hegg,[56] SC
Corn	—	100	2 events	—	6.2	Mueller et al.,[52] WI
Fescue - dry[a]	415	104	8 events	2.8	7.9	Reese et al.,[57] AL
Fescue - slurry[a]	403	112	8 events	4.1	12.1	Reese et al.,[57] AL
Alfalfa - spring[b]	205	21	1 year	10.7	12.1	Young and Mutchler,[58] MN
Alfalfa - fall[b]	285	55	1 year	13.2	13.3	Young and Mutchler,[58] MN
Corn - spring[b]	205	21	1 year	1.0	2.4	Young and Mutchler,[58] MN
Corn - fall[b]	285	55	1 year	0.8	4.7	Young and Mutchler,[58] MN
Poultry litter						
C. bermuda grass	1177	—	2 years	4.3	—	Dudinsky et al.,[59] GA
	699	—	5 years	4.6	—	Dudinsky et al.,[59] GA
	1397	—	5 years	10.7	—	Dudinsky et al.,[59] GA
Fescue	218	54	1 event	4.0	2.2	Edwards and Daniel,[49] AR
	435	108	1 event	4.2	2.3	Edwards and Daniel,[49] AR
Fescue	450	150	1 year	0.3	1.9	Heathman et al.,[60] OK
Fallow	287	165	1 event	20.0	19.0	Westerman et al.,[48] NC
Poultry manure						
Fescue	220	76	1 event	3.1	2.6	Edwards and Daniel,[61] AR
	879	304	1 event	3.3	3.2	Edwards and Daniel,[61] AR
Fescue	149	85	4 events	4.2	2.4	McLeod and Hegg,[56] SC
Fallow	428	95	1 event	5.0	12.6	Westerman et al.,[48] NC
Swine manure	217	19	1 event	2.6	7.4	Edwards and Daniel,[62] AR
Fescue	435	38	1 event	2.9	8.4	Edwards and Daniel,[62] AR

[a]Applied as dry manure or as a slurry.

[b]Manure applied in the spring and fall.

observed tile water contaminated as a result of applying liquid manure. They used NH$_3$ loadings as an indicator of manure entry into tile drains and found that injection of liquid manure contributed to tile water degradation at least as much or even more than simply broadcasting the liquid manure onto the soil surface. Bacteria contamination of the tile water also occurred.

In a long-term study in Ontario, Patni[68] found that high manure application rates (500 kg N/ha/yr) lead to high NO$_3$ concentrations in tile effluent that tend to persist for a few years after applications are reduced or stopped. The yearly and cumulative loss of N in the tile effluent was insignificant compared with the applied manure N.

Geohring[69] discussed control methods to reduce the environmental impacts of the drainage effluent from manure spreading. He discussed controlled drainage, time and rate of manure application, and tillage as viable control methods. When tiles are flowing, liquid manure application should be avoided or low applications of 0.3 to 0.8 cm should be applied. Tillage before the application of liquid manures will reduce and delay the opportunity for preferential flow, minimizing the incidence of high concentrations of bacteria and NH_3 entering the drains.

Kanwar et al.[70] studied the effects of liquid swine manure application on corn and soybean production and shallow groundwater quality. The experiment was on a Kenyon silt-clay loam soil with 3–4% organic matter in northeastern Iowa. The manure was applied to 0.4-ha plots that were tile-drained. Nitrogen applications for the swine manure for the continuous corn and corn-soybean rotation plots varied from 82 kg/ha in 1993 to 486 kg/ha in 1995. The swine manure applications were compared with other N management practices that included strip-cropping, late spring N test, and a single N fertilizer application. No N was applied to soybeans. In 1994 the NO_3 concentrations were below 10 mg N/L for all N management practices except for manure-applied plots. In 1995, much higher NO_3 concentrations were observed from continuous corn manured plots than in 1993 and 1994 because of the much higher manure application rates in 1995. The authors had difficulty in applying the intended N application rate with swine manure, which had an impact on groundwater quality. The strip cropping (corn-soybean-oats-hay) and the forage crop (alfalfa) had the lowest groundwater NO_3 concentrations.

6.6.4 GROUNDWATER QUALITY

Over-application of manure will cause NO_3 leaching into the groundwater. Ritter and Chirnside[71] found that 32% of 200 wells sampled in Sussex County, Delaware, had NO_3 concentrations above 10 mg N/L. The major cause of NO_3 contamination was poultry manure. Adams et al.[72] evaluated NO_3 leaching in soils fertilized with both poultry litter and hen manure at 0, 10, and 20 Mg/ha. They found that the amount of NO_3 leaching into the groundwater was a function of litter application rate.

Westerman et al.[73] applied swine lagoon effluent at rates of 380–440 kg N/ha of estimated available N to coastal bermuda grass to two fields for 3 years in North Carolina. One field had intensive grazing of beef cattle and the other was harvested for hay. The soil was a Cainhoy sand. In the third year of the study, elevated NH_3, NO_3, and Cl levels were found in the shallow groundwater beneath each field. The hay plot in year two also had potentially dangerous NO_3 levels in the hay (1% N). The results imply lower effluent application rates are needed to prevent NO_3 leaching because of the rapid leaching in the sandy soils.

A number of studies have shown excessive applications of liquid dairy manure can cause NO_3 leaching. Hubbard et al.[74] found NO_3 concentrations exceeded drinking water standards on a Georgia Coastal Plain plinthic soil when dairy manure was applied to coastal bermudagrass at rates of 44 and 91 kg N/ha per month. Davis et al.[75] found 600 kg N/ha/yr of liquid dairy lagoon effluent applied to a year-round forage production system resulted in maximum yields but increased soil and water NO_3

concentrations to a depth of 1.5 m on a Coastal Plain soil. The system consisted of rye planted in the fall in bermudagrass sod and cut twice in winter and early spring, followed by corn planted in the grass sod in March and harvested for silage in July, before three bermuda grass cuttings in the summer and fall.

Doliparthy et al.[76] found that liquid dairy manure applied to alfalfa for three years in Massachusetts significantly increased NO_3, concentrations in the soil water when applied at a rate of 336 kg N/ha/yr to a sandy loam soil. When applied at a rate of 112 kg N/ha/yr NO_3, concentrations in the soil water were no higher than in unmanured alfalfa.

6.7 PRACTICES TO REDUCE NONPOINT SOURCE POLLUTION

6.7.1 BARNYARD AND FEEDLOT RUNOFF

Runoff from cattle feedlots, other unroofed animal enclosures, and manure storage areas requires collection and diversion to storage or treatment areas. To minimize the quantity of water that comes in contact with manure, all relatively clean water from roof drainage and rainfall on driveways and adjacent cropland or pasture should be diverted away from the feedlot.

Components of a runoff control system include a clean water diversion system, runoff collection system, solids retention facility, runoff retention basin, and runoff application area. Common components of a diversion facility include roof gutters, downspouts, concrete gutters, earthen channels, and culverts. Curbs and terraces may also be used to divert the clean water.

The runoff collection system generally consists of a series of canals, ditches, and flow ways designed to collect runoff from the individual pens in an orderly fashion. When designing collection facilities, consideration should be given to keeping animals dry and protecting traffic ways for ease of servicing.

A solids retention facility is used to entrap the solids and prevent rapid filling of the runoff retention basin with solids that feedlot runoff commonly carries. The principle of a solids retention basin is to reduce the velocity sufficiently for the solids to settle, removing the liquid without disturbing the settled solids, allowing the solids to dry as much as possible, and provide a means to remove the solids. Settling tanks, basins, or channels are used for settling, with the latter two options being the most common. A 10-yr, 1-hr storm is usually used for designing settling facilities.[14]

A runoff retention basin provides storage for feedlot runoff from the time it leaves the lot until it is applied to land. Typically, runoff retention basins are designed to hold a 25-yr, 24-hr storm.[14] In some cases, storage basins may be designed to hold up to 180 days of runoff depending on local regulations and conditions, or an infiltration area (or vegetative filter) may be used as an alternative to holding ponds for runoff control.

The most common management method for feedlot runoff is application to cropland. Nutrients in the runoff are utilized by the crop. Application rates are generally determined by the N content. Detailed design information for all components of a runoff control system can be found in a number of references.[6,14]

6.7.2 MANURE STORAGE AND TREATMENT SYSTEMS

Manure storage basins and lagoons may overflow, or seepage can occur from them. Site selection is important in preventing seepage.[77] Areas with very permeable soils, high water tables, or underlying rock fissues should be avoided. The bottom of earthen manure storage basins should be at least 1.0 m above bedrock and 0.6 m above the water table.[14] Sites should be avoided where the bottom of a lagoon is less than 6.0 m above limestone. Lagoons and earthen storage basins require sealing on highly permeable soils. Sealing may be accomplished with clay, soil cement, or a membrane liner. Liners are the most expensive and difficult to install. Before constructing a lagoon or earthen manure storage basin, regulations should be checked as to the location of the facility relative to wells.

To keep lagoons from overflowing, they must be managed properly and constructed with sufficient freeboard. Surface water should be diverted away from the lagoon. Lagoons should be pumped on a regular basis down to the minimum design operating level.

6.7.3 LAND APPLICATION

Erosion and runoff may occur from land application sites that contain N, P, organics, and bacteria. Nitrogen may also be leached to groundwater. The main approach to addressing pollution today is to implement best management practices (BMPs) on land application sites. All BMPs can be classified as managerial or structural. Many BMPs are discussed in Chapter 10. *The National Handbook of Conservation Practices of the Natural Resources Conservation Service*[78] provides detailed descriptions of many BMPs. Only some of the BMPs associated with nutrient management are discussed in this section.

6.7.3.1 Application Timing

The longer manure is in the soil before crops take up its nutrients, the more those nutrients, especially N, can be lost through volatilization, denitrification, leaching, and erosion. Therefore, application timing and site selection are important considerations.

Spring application is best for conserving nutrients. Spring is the time nearest to nutrient utilization that manure application is practical.

Summer application of manure is suitable for small-grain stubble, noncrop fields, or little-used pastures. Manure should not be spread on young stands of legume forage because legumes fix atmospheric N, and additional fertilizer N will stimulate competitive grasses and broadleaf weeds. It can be applied effectively to pure grass stands or to old legume-grass mixtures with low legume percentages (less than 25%).

Fall application of manure generally results in greater nutrient loss than does spring application, regardless of the application method, but especially if the manure is not incorporated into the soil. If manure is incorporated immediately, the soil will immobilize some of the nutrients, especially at soil temperatures below 50°F. In fall, manure is best applied at low rates to fields that are to be planted in winter grains or cover crops. If winter crops are not to be planted, manure should be applied to the fields containing the most vegetation or crop residues. Sod fields to be plowed the next spring are also acceptable, but fields where corn silage was removed and a cover crop is not to be planted are undesirable sites.

Winter application of manure is the least desirable, from both a nutrient utilization and a pollution point of view, because the frozen soil surface prevents rain and melting snow from carrying nutrients into the soil. The result is nutrient loss and pollution through leaching and runoff. If daily winter spreading is necessary, manure should be applied to the fields with the least runoff potential, and it should be applied to distant or limited-access fields in early winter, then to nearer fields later in the season when mud and snow make spreading more difficult.

6.7.3.2   Application Rate

Manure should be applied to fields at the rate that supplies only the amount of nutrients that the crop will use. Supplying an excess of nutrients is essentially a waste of valuable resources, may even depress yields, and may result in ground- and surface-water pollution. Determining the rate at which nutrients, and thus manure, would be applied requires careful calculation of crop need and the amount of residual nutrients already present in the soil.

Manure nutrients, especially N, are used more efficiently by corn and cereal grains than by legumes. In general, if manure is applied to meet the N needs of a grain crop, P and K eventually build up to excessive levels in the soil. Planting forage crops in rotation with grain crops will help remove the excess P and K and keep the three nutrients in balance.

6.7.3.3   Realistic Crop Yield Goals

The nutrient needs of a crop are determined by the expected yield. An important factor in setting realistic yield expectations is the yield potential of the soil, which is a function of soil depth and drainage independent of manure or fertilizer application. Realistic yield goals are best calculated as the average yield (using proven yield estimates) for the past five to seven growing seasons. In this way, yield goals would be adjusted to account for many variables such as weather, management, and economics.

6.7.3.4   Soil Testing for Residual Nutrients

The rate at which manure should be applied depends in part on the amount of nutrients already present in the soil and available to the crop. Soil tests are essential for indicating the levels of available P and K in the soil. Soil tests show where P and K are present in excess and where applying manure containing these two nutrients will have a profitable effect on yields.

Once N enters soils, its availability cannot be measured, so residual N in a field must be calculated on the basis on the N supplied. All sources must be considered, such as manure applied over the past several years, N supplied by previous legume crops, and any fertilizer applications.

6.7.3.5 Manure Testing

There are many variables in animal production systems that can affect manure quality at the time of application. Management factors can cause a wide range in nutrient content applied to land.[62] It is not only important to test the soil, but also, the manure should be analyzed for N and P before it is applied to land. Manure should be analyzed as close as possible to the application site and the analysis should be used only as a guideline in determining application rates. The N meter can provide a rapid on-farm approximation of available N in the manure and compares favorably with laboratory analysis. The N meter has been tested by a number of researchers to estimate the plant-available N content of liquid slurry manure.[79,80]

6.7.3.6 Calibrating Manure Spreading Equipment

It is important to calibrate applicator equipment for liquid and solid manure. The task is simple and easy. Nutrients in manure can be utilized more efficiently when a farmer knows how much manure the spreader is applying per unit area. Details on calibrating manure spreaders can be found in a number of publications.[81]

6.7.3.7 Early-Season Soil and Plant Nitrate Tests

Early-season soil and plant NO_3 tests have been developed for estimating available N contributions from soil organic matter, previous legumes, manure under the soil, and climatic conditions that prevail at specific production locations.[82,83] These tests are performed 4 to 6 weeks after the corn is planted. Early-season soil NO_3 tests involve taking soil samples in the top 30 cm of the soil profile from 4 to 6 weeks after the corn is planted. Early-season plant NO_3 testing involves determining the NO_3 concentration in the basal stem of young corn plants approximately 30 days after emergence. One disadvantage of the early-season soil and plant NO_3 testing is that there must be a rapid turnaround between sample submitted and fertilizer recommendations from the soil testing laboratory. If side-dress N fertilizer is being used in conjunction with manure, the early-season NO_3 test should help reduce the potential for over-fertilization.

6.7.3.8 Nitrification Inhibitors

Nitrification inhibitors are available to stabilize N in the NH_4 form. Stabilizing the N in manure by inhibiting nitrification should increase its availability for crop uptake later in the season, reduce its mobility in soil, and reduce its pollution potential under both conventional and conservation tillage.[84] Sutton et al.[80] found that stabilized swine manure had an efficiency for crop production similar to that of anhydrous NH_3.

6.7.3.9 Winter Cover Crops

Small-grain cover crops can be used to remove residual N from the soil profile following a grain crop such as corn. The cover crop not only reduces NO_3 leaching but also can increase evapotranspiration. Winter cover crops that can be used are wheat, barley, rye, and oats. Brinsfield and Staver[85] have found that rye offered the most potential for rapid N uptake as a winter cover crop. Nitrate leachate concentrations were consistently lower when a rye cover crop was present than in previous years when no cover crops on two Coastal Plain watersheds were present.

6.7.3.10 Alfalfa as a Nutrient Scavenging Crop

Legumes will fix N from the atmosphere but will take up residual inorganic N from the soil in preference to fixing N. Alfalfa often utilizes N below the rooting depth of other crops. Mather et al.[86] found that significant removal of NO_3 from the soil profile occurred to a depth of 1.8 m during the first year of an alfalfa stand. Vocasek and Zupancic[87] found alfalfa reduced initial 3.5 m profile NO_3 accumulations by 88–92%, reaching background levels during the first 48 to 60 months after seeding when it was used at two land application sites.

6.7.3.11 Alteration of Feed

Increasing dietary P levels may decrease the P levels in manure and increase the N/P ratio of the manure. Sutton et al.[88] has found that by adding the enzyme phytase to a low P diet for swine increased P digestibility in pigs from 4 to 21% units and reduced the P content of the manure by 18–36% compared with pigs fed a low phosphorus diet without phytase. Cantor et al.[89] supplemented broiler diets with different phytase products that increased available P in the diet from 0.10 to 0.12%. There have been other studies since the late 1960s showing P supplement levels can be reduced in both poultry and swine diets by adding the enzyme phytase.[90]

Another method that has been used to lower the amount of mineral phosphate supplements needed in poultry diets is the use of grains in which a greater proportion of the P exists as available P. A low-phytate corn variety has been developed by USDA-ARS and licensed by Pioneer Seed. This corn has only about 10% of the P tied up as phytate, compared with 65% for normal corn.

Moore et al.[90] evaluated the effect of low-phytase corn and on adding the enzyme phytase to the diet on soluble and total P in the litter. They also conducted a runoff study using a rainfall simulator to measure P in the runoff for the various treatments. There were no significant differences in soluble P concentrations in the runoff among litter types. The low-phytase corn and low-phytase corn plus phytase treatments lowered P runoff by 2 and 26%, respectively.

6.7.3.12 Alum Addition

Aluminum sulfate ($Al_2(SO_4)_3 \cdot 14H_2O$), commonly called alum, is an acid when it dissolves in water. If alum is added to litter it should reduce the NH_3 volatilization and reduce the amount of soluble P. Alum will react with P in the following manner:

$$Al_2(SO_4)_3 \cdot 14H_2O + 2H_3PO_4 \rightarrow 2AlPO_4 + 6H^+ + 3SO_4^{2-} + 14H_2O$$

to form insoluble aluminum phosphate. Moore and Miller[91] conducted a laboratory study where 100 different treatments with various Al, Ca, and Fe compounds were added to broiler litter. Many of the compounds reduced soluble P from 2000 mg/kg to 1 mg/kg. In a small-plot study with a rainfall simulator, it was shown alum could reduce P concentrations in runoff water by 87%.[92]

In a 3-year paired watershed study, alum was added to poultry litter at a rate of 0.09 kg/bird. Litter applications rates were 5.6, 6.7, and 9.0 mg/ha for the 3 years, respectively. Alum applications reduced soluble P concentrations in runoff water by 75% over a 3-year period.[90] Long-term studies of alum-treated litter on tall fescue plots were initiated in 1995. Treatments included an unfertilized control, four rates of normal broiler litter, four rates of alum-treated litter, and four rates of ammonium nitrate. Litter application rates were 2.2, 4.5, 6.7, and 9.0 mg/ha. After three years, large differences in soil test P were observed. Normal litter-fertilized plots had increased levels of soluble P, but the alum-treated litter plots had soluble P concentrations similar to the unfertilized plots. Alum-treated litter shows great promise for reducing P concentrations in runoff. Aluminum phosphates are more stable than Fe or Ca phosphates under a wide range of soil conditions.[90]

6.8   LIVESTOCK GRAZING IMPACTS

Water quality impacts from pastured livestock areas and rangelands depends in part on the stocking density, length of grazing period, average manure loading rate, manure spreading rate, manure spreading uniformity by grazing animals, and disappearance of manure with time.[93] Normally, pasture areas have not presented appreciable water quality problems except under special circumstances.[64] Smeins[94] studied the effect of various rangeland livestock-grazing management programs on the quantity and quality of surface runoff. The highest total N concentration from a heavily and continuously grazed pasture was 0.94 mg/L, whereas a pasture with a defined rotation grazing scheme had a total N concentration of 0.64 mg/L on the same date. Nutrient losses appeared to be more related to sediment loss than to animal waste loadings. Olness et al.[95] found that rangelands where animals were continuously grazed contributed at least four times more N and P in runoff compared with rotationally grazed rangelands.

Sewell and Alphin[96] studied problem areas associated with unconfined animal production systems. Average NO_3 concentrations in runoff from two sites on a heavily grazed dairy pasture system exceeded those from all other sites, including those from an aerobic lagoon and drainage from cultivated lands. Mean ortho P concentrations in runoff from the dairy pasture were exceeded only by those of aerobic lagoon waters.

Correll et al.[97] compared discharge loads for organic carbon, total N, and total P for a completely forested watershed, a cropland/riparian forest watershed, and a pasture-dominated watershed for a 4-year period in the Rhodes River of the Chesapeake Bay basin. On average, less total organic carbon and total N and P were discharged from the pasture than from either the forest or cropland-dominated watershed. They also measured baseflow water quality in 47 other sub-watersheds of the Chesapeake Bay watershed in the Piedmont and Appalachian physiographic provinces. Nitrate concentrations in the pasture-dominated watershed were 40 times higher than in the Rhodes River pasture watershed, but dissolved NH_3 concentrations were somewhat lower than in the Rhodes River pasture watershed. They generalized that the high NO_3 could have been as a result of fertilization of the pasture watersheds, or the livestock had access to the stream channels.

Reese et al[57] found that total coliform and fecal coliform levels on an unfertilized pasture were higher than the permissible drinking water supply standards in South Carolina. Crane et al.[89] concluded from a review of the literature that there is little difference in the bacterial concentrations in runoff between areas used as pastures and controlled areas where manure had not been applied. This suggests that the low manure loading associated with low-density pasture systems presents a minimal contribution of microorganisms to surface runoff from these areas.

Nitrogen and P loads and microbial contamination from pastures is not related to the number of animals involved but is related to the hydrologic and management practices. If livestock is not allowed access to streams, microbial loads will be much lower. If runoff and erosion and sediment transport are controlled, nutrient loads will be lower.

6.9   SUMMARY

Livestock production can affect both groundwater and surface water. The major pollutants that may contaminate surface water are organics, N, P, microorganisms, and salts. Nitrates are a major concern in groundwater contamination. Potential sources of nonpoint source pollution are runoff from feedlots and barnyards, manure land application sites, livestock grazing, and manure storage and treatment units. Runoff from cattle feedlots, other unroofed animal enclosures, and manure storage areas should be collected and diverted to storage or treatment areas. All clean water should be diverted away from the feedlot.

To prevent seepage from manure storage basins and lagoons, site selection is important. Research has shown unlined manure storage basins and lagoons can contaminate groundwater. Areas with high water tables, very permeable soils, or underlying rock fissures should be avoided.

Nutrient management practices should be used on manure application sites along with runoff and erosion control practices. Some of the nutrient management practices include calibration of application equipment, timing of application, applying only enough manure to meet the crop nutrient requirements, soil testing, and manure testing. Phosphorus is becoming more of a concern in manure application. Increasing the dietary P levels by adding the enzyme phytase to feed and the development of a low-phytate corn show great promise in reducing manure P levels. Alum added to broiler litter reduces the amount of soluble P in surface runoff.

REFERENCES

1. Overcash, M. J., Humenik, F. J., and Miner, J. R., Livestock Waste Management, Vol. I & II, CRC Press Inc., Boca Raton, FL, 1983.
2. Golton, D. M., and Knoblauch, W. A., Why do farms expand, in *Proceedings from the Animal Agriculture and the Environment Conference,* Northeast Agricultural Engineering Service, Cornell University, Ithaca, NY, NRAES 96, 1, 1996.
3. U.S. Congress, Coastal zone act reauthorization amendments of 1990, Public Law, 101–508, Washington, DC, 1990.
4. U.S. Environmental Protection Agency, Guidance specifying management measures for sources of nonpoint pollution in Coastal Waters. Report No. 840-B-92-002, EPA, Office of Water, Washington, DC, 1993.
5. U.S. Environmental Protection Agency, USDA and EPA unified national strategy for animal feeding operations, http://www.epa.gov/own/finalfost.htm, 1999.
6. American Society of Agricultural Engineers, Manure production and characteristics, in *ASAE Standards 1993,* 40th ed., ASAE, St. Joseph, MI, 530, 1993.
7. Soil Conservation Service, Agricultural Waste Management Field Handbook, SCS, USDA, Washington, DC, Chap. 4, 1992.
8. Barth, C. L., Livestock waste characterization—a new approach, in *Proc. Fifth Int. Symp. on Agric. Wastes,* ASAE, St. Joseph, MI, 286, 1985.
9. Mason, C. F., *Biology of Freshwater Pollution,* 2nd ed., John Wiley & Sons, New York, NY, Chap. 2, 1991.
10. Strebel, O., Duynisveld, W. H. M., and Bottcher, J., Nitrate pollution of groundwater in Europe, *Agric. Ecosyst. Environ.* 26, 189, 1989.
11. U.S. Department of Agriculture, Agricultural statistics, 1998, U.S. Gov. Printing Office, Washington, DC, 1998.
12. Ritter, W. F., and Chirnside, A. E. M., Impact of land use on groundwater quality in southern Delaware, *Ground Water,* 22, 38, 1984.
13. Scott, H. D., Smith, P. A., Mauromoustakos, A., and Limp, W. F., Geographical and statistical relationships between landscape parameters and water quality indices in an Arkansas watershed. *Bull. Ark. Agric. Exp. Stn.,* 933, 1992.
14. Midwest Plan Service, Livestock waste facilities handbook, 3rd ed., Iowa State University, Ames, IA, MWPS-18, 1993.
15. Sims, J. T., and Wolf, D. C., Poultry waste management: agricultural and environmental issues, *Adv. Agron.,* 52, 1, 1994.
16. Logan, T. J., Phosphorus losses by runoff and leaching, in *Future Directions for Agricultural Phosphorus Research,* National Fertilizer and Environmental Research Center, Tennessee Valley Authority, Muscle Shoals, AL, Bul Y2343, 16, 1991.
17. Gerritse, R. G., Simulation of phosphate leaching in acid sandy soils, *Aust. J. Soil Res.* 27, 1989.
18. Fleming, R. J., McLellan, J., Alves, D., and Pintar, K., Cryptosporidium in livestock manure and surface water, Tech. Bul., University of Guelph, Ridgetown, Ont., Canada, 1997.

19. Reese, D. E., Miller, P. S., Lewis, A. J., Brumm, M. C., and Ahlschwede, W. T., University of Nebraska swine diet suggestions, Nebraska Coop. Extension Pub. 92-210A, U. of Nebraska, Lincoln, NB, 1992.

20. Sutton, A. L., Mayrose, V. B., Nye, J. C., and Nelson, D. W., Effect of dietary salt level and liquid handling systems on swine waste composition, *J. Anim. Sci.,* 43, 1129, 1976.

21. Sweeten, J. M., Feedlot runoff characteristics for land application, in *Agricultural and Food Processing Wastes, Proceedings of the 6th International Symposium on Agricultural and Food Processing Wastes,* ASAE, St. Joseph, MI, 168, 1990.

22. Sweeten, J. M., Dilution of feedlot runoff, MP-1297, Texas Agricultural Extension Service, Texas A&M University, College Station, TX, 1976.

23. Wallingford, G. W., Murphy, L. S., Powers, W. L., and Manges, H. L., Effect of beef feedlot lagoon water on soil chemical properties—growth and composition of corn forage. *J. Environ. Qual.,* 3, 74, 1974.

24. Liebhardt, W. C., Soil characteristics and corn yield as affected by previous applications of poultry manure, *J. Environ. Qual.,* 5, 459, 1976.

25. Sweeten, J. M., Wolfe, M. L., Chasteen, E. S., Sanderson, M., Auvermann, B. A., and Alston, G. D., Dairy lagoon effluent irrigation: effects on runoff quality, soil chemistry and forage yield, in *Animal Waste and the Land-Water Interface,* Steel, K., Ed., CRC Press, Boca Raton, FL, 99, 1995.

26. Clark, R. N., Gilbertson, C. B., and Duke, H. R., Quantity and quality of beef feedyard runoff in the Great Plains, in *Managing Livestock Wastes, Proc. of the 3rd Int. Symp. on Livestock Wastes,* ASAE, St. Joseph, MI, 289, 1975.

27. Sweeten, J. M., Cattle feedlot manure and wastewater management practices, in *Animal Waste Utilization: Effective Use of Manure as a Soil Resource,* Hatfield, J. L. and Stewart, B. A., Eds., Ann Arbor Press, Chelsea, MI, 125, 1998.

28. U.S. Environmental Protection Agency, National pollutant discharge elimination system general permit and reporting requirements for discharges from concentrated animal feeding operations, *Federal Register,* February, 7610, 1993.

29. Gilbertson, C. B., Clark, R. N., Nye, J. C., and Swanson, N. P., Runoff control for livestock feedlots: state of the art, *Trans. ASAE,* 23, 1207, 1980.

30. Norstadt, F. A., and Duke, H. R., Stratified profiles: characteristics of simulated soils in a beef cattle feedlot, *Soil Sci. Soc. Am. J.,* 46, 827, 1982.

31. Schuman, G. E., and McCalla, T. M., Chemical characteristics of a feedlot soil profile, *Soil Sci.,* 119, 113, 1975.

32. Elliott, L. F., McCalla, T. M., Mielke, L. N., and Travis, T. A., Ammonium, nitrate and total nitrogen in the soil water of feedlot and field soil profiles, *Appl. Microbiol.,* 23, 810, 1972.

33. Mielke, L. N., Swanson, N. P., and McCalla, T. M., Soil profile conditions of cattle feedlots, *J. Environ. Qual.,* 13, 14, 1974.

34. Day, D. L. and Funk, T. L., Processing manure: physical, chemical and biological treatment, in *Animal Waste Utilization: Effective Use of Manure as a Soil Resource,* Hatfield, J. L., and Stewart, B. A., Eds., Ann Arbor Press, Chelsea, MI, 243, 1998.

35. Ciravolo, T. G., Martens, D. C., Hallock, D. L., Collins, E. R., Jr., Kornegay, E. J., and Thomas, H. R., Pollutant movement to shallow groundwater tables from swine waste lagoons, *J. Environ. Qual.,* 8, 126, 1979.

36. Sewell, J. R., Dairy lagoon effects on groundwater quality, *Trans. ASAE* 21, 948, 1978.

37. Nordstedt, R. A., Baldwin, L. B., and Hortenstine, C. C., Multistage lagoon systems for treatment of dairy farm waste, in *Livestock Waste Management and Pollution Abatement, Proc. Int. Symp. Livestock Wastes,* ASAE Publication PROC-271, ASAE, St. Joseph, MI, 77, 1971.

38. Ritter, W. F., Walpole, E. W., and Eastburn, R. P., Effect of an anaerobic swine lagoon on ground-water quality in Sussex County, Delaware, *Agric. Wastes,* 10, 267, 1984.
39. Ritter, W. F., and Chirnside, A. E. M., Impact of animal waste lagoons on ground-water quality, *Biol. Wastes,* 34, 39, 1990.
40. Westerman, P. W., Huffman, R. L., and Feng, J. S., Swine lagoon seepage in sandy soil, *Trans. ASAE,* 38, 1749, 1995.
41. Huffman, R. L., Evaluating the impacts of older swine lagoons on shallow groundwater, in *1999 Animal Waste Management Symposium,* Havenstein, G.B., ed., North Carolina State University, Raleigh, NC, 92, 1999.
42. Barrington, S. F., and Jutras, P. J., Soil sealing by manure in various soil types, Paper No. 83-4571, ASAE, St. Joseph, MI, 1983.
43. Phillips, P. A., and Culley, J. L. B., Groundwater nutrient concentrations below small-scale earthen manure storages, *Trans. ASAE,* 13, 672, 1985.
44. Gangbazo, G.D., Cluis, D., and Vallieres, M., Nitrogen seepage from earthen built manure storage tanks, Paper No. 89-2002, ASAE, St. Joseph, MI, 1989.
45. Hilborn, D., Choosing a liquid manure application system, in *Proc. from the Animal Agriculture and the Environment Conference,* Northeast Agricultural Engineering Service, Cornell University, Ithaca, NY, 205, 1996.
46. Sharpley, A., Meisinger, J. J., Breeuwsma, A., Sims, J. T., Daniel, T. C., and Schepers, J. S., Impacts of animal manure management on ground and surface water quality, in *Animal Waste Utilization: Effective Use of Manure as a Soil Resource,* Hatfield, J. L., and Stewart, B. A., Eds., Ann Arbor Press, Chelsea, MI, 173, 1998.
47. Edwards, D. R. and Daniel, T. C., Quality of runoff from fescuegrass plots treated with poultry litter and inorganic fertilizer, *J. Environ. Qual.,* 23, 379, 1994.
48. Westerman, P. W., Donnelly, T. L., and Overcash, M. R., Erosion of soil and poultry manure—a laboratory study, *Trans. ASAE,* 26, 1070, 1983.
49. Edwards, D. R. and Daniel, T. C., Effects of poultry litter application rate and rainfall intensity on quality of runoff from fescuegrass plots, *J. Environ. Qual.,* 22, 361, 1993.
50. Shore, L. S., Correll, D., and Chakraborty, P.K., Sources and distribution of testosterone and estrogen in the Chesapeake Bay Watershed, in *Impact of Animal Manure and the Land Water Interface,* Steel, K., ed., Lewis Pub., CRC Press, Boca Raton, FL., 155, 1995.
51. Edwards, D. R. and Daniel, T. C., Drying interval effects on runoff from fescue plots receiving swine manure, *Trans. ASAE,* 36, 1673, 1993.
52. Mueller, D. H., Wendt, R. C., and Daniel, T. C., Phosphorus losses as affected by tillage and manure application, *Soil Sci. Soc. Am. J.,* 48, 901, 1984.
53. Westerman, P. W., and Overcash, M. R., Short-term attenuation of runoff pollution potential for land-applied swine and poultry manure, in *Livestock Waste—A Renewable Resource, Proc. 4th Int. Symp. on Livestock Wastes,* ASAE, St. Joseph, MI, 289, 1980.
54. Klausner, S. D., Zwerman, P. J., and Ellis, D. F., Nitrogen and phosphorus losses from winter disposal of dairy manure, *J. Environ. Qual.,* 5, 47, 1976.
55. Long, F. L., Runoff water quality as affected by surface-applied dairy cattle manure, *J. Environ. Qual.,* 8, 215, 1979.
56. McLeod, R. V. and Hegg, R. O., Pasture runoff water quality from application of inorganic and organic nitrogen sources, *J. Environ. Qual.,* 13, 122, 1984.
57. Reese, L. E., Hegg, R. O., and Gantt, R. E., Runoff water quality from dairy pastures in the Piedmont region, *Trans. ASAE,* 25, 697, 1982.

58. Young, R. A. and Mutchler, C. K., Pollution potential of manure spread on frozen ground, *J. Environ. Qual.,* 5, 174, 1976.

59. Dudinsky, M. L., Wilkinson, S. R., Dawson, R. N., and Barnett, A. P., Fate of nitrogen from NH$_4$ NO$_3$ and broiler litter applied to coastal bermudagrass, in *Nutrient Cycling in Agricultural Ecosystems,* Lowrance, R., Todd, R., Asmussen, L., and R. Leonard, Eds., Georgia Agric. Exp. Sta. Spec. Bull. 23, Athens, GA, 373, 1983.

60. Heathman, G. C., Sharpley, A. N., Smith, S. J., and Robinson, J. S., Poultry litter application and water quality in Oklahoma, *Fert. Res.,* 37, 165, 1995.

61. Edwards, D. R. and Daniel, T. C., Potential runoff quality effects of poultry manure slurry applied to fescue plots, *Trans. ASAE,* 35, 1827, 1992.

62. Edwards, D. R. and Daniel, T. C., Runoff quality impacts of swine manure applied to fescue plots, *Trans. ASAE,* 36, 81, 1993.

63. Crane, S. R., Moore, J. A., Grismer, M. E., and Miner, J. R., Bacterial pollution from agriculture sources: a review, *Trans. ASAE* 26, 858, 1983.

64. Robbins, J. W., Kriz, G. J., and Howells, D. H., Quality of effluent from farm animal production sites, in *Proc. 2nd Int. Symp. on Livestock Wastes,* Pub. 271, ASAE, St. Joseph, MI, 166, 1971.

65. McCaskey, T. A., Rollins, G. H., and Little, J. A., Water quality of runoff from grassland applied liquid, semi-liquid and solid dairy manure, in *Proc. 2nd Int. Symp. on Livestock Wastes,* Pub. 271, ASAE, St. Joseph, MI, 239, 1971.

66. Dean, D. M. and Foran, M. E., The effect of farm liquid waste application on tile drainage, *J. Soil Water Conserv.,* 47, 368, 1992.

67. Fleming, R. J. and Bradshaw, S. H., Contamination of subsurface drainage systems during manure spreading, Paper No. 92-2618, ASAE, St. Joseph, MI, 1992.

68. Patni, N., Nitrate in agricultural drainage water from a long-term mixed farm operation—a case study, in *Proc. 7th Int. Symp. on Agricultural and Food Processing Wastes,* Ross, C.C., Ed., ASAE, St. Joseph, MI, 564, 1995.

69. Geohring, L. D., Controlling environmental impact in tile-drained fields, in *Proc. of Liquid Manure Application Systems Conference,* Northeast Agricultural Engineering Service, Cornell University, Ithaca, NY, 175, 1994.

70. Kanwar, R. S., Melvin, S. W., Karlen, D. L., Cambardella, C. A., Steenheimer, T. R., Moorman, T. B., and McFadden, V., Impact of liquid swine manure application on agricultural productivity sustainability and water quality, in *1996 Research Investment Report,* National Pork Producers Council, Des Moines, IA, 325, 1996.

71. Ritter, W. F. and Chirnside, A. E. M., Influence of agricultural practices on nitrates in the water table aquifer, *Biol. Wastes* 19, 168, 1987.

72. Adams, P. L., Daniel, T. C., Edwards, D. R., Nichols, D. J., Pote, D. H., and Scott, H. D., Poultry litter and manure contributions to nitrate leaching through the vadose zone, *Soil Sci. Soc. Am. J.,* 58, 1206, 1994.

73. Westerman, P. W., Huffman, R. L., and Barker, J. C., Environmental and agronomic evaluation of applying swine lagoon effluent to coastal bermudagrass for intensive grazing, in *Proc. 7th Int. Symp. on Agricultural and Food Processing Wastes,* Ross, C.C., Ed., ASAE, St. Joseph, MI, 162, 1995.

74. Hubbard, R. K., Thomas, D. L., Leonard, R. A., and Butler, J. L., Surface runoff and shallow ground water quality as affected by center pivot applied dairy cattle wastes, *Trans. ASAE,* 30, 430, 1987.

75. Davis, J. G., Vellidis, G., Hubbard, R. K., Johnson, J. C., Newton, G. L., and Lowrance, R. R., Nitrogen uptake and leaching in a no-till forage rotation irrigated with liquid dairy

manure, in *Animal Waste and the Land-Water Interface,* Steele, K., Ed., Lewis Publishers, CRC Press, Boca Raton, FL, 405, 1995.

76. Daliparthy, J., Herbert, S. J., and Veneman, P. L. M., Dairy manure application to alfalfa: crop response, soil nitrate and nitrates in soil water, *Agron. J.,* 86, 927, 1994.

77. Doughtery, M., Geohring, L.D., and Wright, P., Liquid Manure Application Systems Design Manual, NRAES-89, Northeast Regional Agricultural Engineering Service, Cornell University, Ithaca, NY, 1998.

78. Natural Resources Conservation Service, National handbook of conservation practices, NRCS-USDA, Web Site, http://www.ncg.nrcs.usda.gov/nhcp2.html., 1998.

79. Westerman, R. W., Safely, L. M., Barker, J. C., and Chescher, G. M., III, Available nutrients in livestock wastes, in *Proc. 5th Int. Symp. in Agricultural Wastes,* ASAE, St. Joseph, MI, 295, 1985.

80. Sutton, A. L., Huber, D. M., Jones, B. C., and Jones, D. D., Management of nitrogen in swine manure to enhance crop production and minimize pollution, in *Proc. 7th Int. Symp. on Agricultural and Food Processing Wastes,* Ross, C.C., Ed., ASAE, St. Joseph, MI, 532, 1995.

81. Pennsylvania Department of Environmental Resources, Manure management for environmental protection and supplements, Graves, R. E., Ed., Commonwealth of Pennsylvania, Harrisburg, PA, 1986.

82. Magdoff, F. R., Ross, D., and Amadon, J., A soil test for nitrogen availability to corn, *Soil Sci. Soc. Am. J.,* 48, 1301, 1984.

83. Iversen, K. V., Fox, R. H., and Piekielek, W. P., The relationships of nitrate concentrations in young corn stalks to nitrogen availability, *Agron. J.,* 77, 927, 1985.

84. Sutton, A. L., Huber, D. M., Jones, D. D., and Kelly, D. T., Use of nitrification inhibitors with summer application of swine manure, *J. Appl. Eng. Agric.,* 6, 296, 1990.

85. Brinsfield, R.B. and Staver, K.W., Role of cover crops in reduction of cropland nonpoint source pollution, Technical Report, University of Maryland, Maryland Agric. Exp. Sta., College Park, MD, 1991.

86. Mathers, A. C., Stewart, B. A., and Blair, B., Nitrate-nitrogen from soil profiles by alfalfa, *J. Environ. Qual.,* 4, 403, 1975.

87. Vocasek, F. F., and Zupancic, J. W., Field evaluation of soil profile nitrate removal by alfalfa, Paper No. 95-2618, ASAE, St. Joseph, MI, 1995.

88. Sutton, A. L., Adeola, O., Schinckel, A. P., Kelly, D. T., and Meyerholtz, K. A., Improving dietary phosphorus utilization swine to reduce phosphorus levels in manure, in *Proc. 7th Int. Symp. on Agricultural and Food Processing Wastes,* Ross, C.C., Ed., ASAE, St. Joseph, MI, 107, 1995.

89. Cantor, A., Moore, E., Pescatore, A., Straw, M., and Ford, M., Evaluation of phytase products on phosphorus utilization in broilers, *Poultry Sci.,* 73, 78, 1994.

90. Moore, P. A., Miller, D. M., Shreve, B. R., and Daniel, T. C., Use of high available phosphorus corn and phytase enzyme additions to broiler diets to lower phosphorus levels in poultry litter, in *Proc. of 1998 National Poultry Waste Management Sympsoium,* Blake, J. P. and Patterson, P. H., eds., 346, 1998.

91. Moore, P. A. and Miller, D. M., Decreasing phosphorus solubility in poultry litter with aluminum, calcium, and iron amendments, *J. Environ. Qual.,* 23, 293, 1994.

92. Shreve, B. R., Moore, P. A., Daniel, T. C., and Edwards, D. R., Reduction of phosphorus in runoff from field-applied poultry litter using chemical amendments, *J. Environ. Qual.,* 24, 106, 1995.

93. Sweeten, J. M., and Reddel, D. L., Nonpoint sources: state-of-the-art overview, *Trans. ASAE,* 21, 474, 1978.

94. Smeins, T. E., Influence of vegetation management on yield and quality surface runoff, Annual Report No. C-6310, Texas Water Resources Institute, Texas A&M University, College Station, TX, 1976.

95. Olness, A., Smith, S. J., Rhoades, E.D., and Menzel, R.G., Nutrient and discharge from agricultural watersheds in Oklahoma, *J. Environ. Qual.,* 4, 331, 1975.

96. Sewell, J. J. and Alphin, J. M., Effects of agricultural land uses on runoff quality, in *Animal Waste Management Facilities and Systems,* Bull. 548, Univ. of Tennessee, Ag. Exp. Station, Knoxville, TN, 1975.

97. Correll, D. L., Jordon, T. E., and Weller, D. E., Livestock and pasture land effects on the water quality of Chesapeake Bay watershed streams, in *Animal Wastes and the Land-Water Interface,* Steele, K., Ed., Lewis Pub., CRC Press, Boca Raton, FL, 107, 1995.

7 Irrigated Agriculture and Water Quality Impacts

Blaine R. Hanson and Thomas J. Trout

CONTENTS

7.1   INTRODUCTION

Nonpoint source pollution of groundwater and surface water from irrigated agriculture is a major concern in many areas of the western United States and elsewhere. Pesticides cause water quality impairment in rivers and streams in California, and nitrate causes groundwater pollution.[1] Nitrate and pesticide contamination of groundwater are serious threats in New Mexico.[2] Nebraska reports that pollutants such as pesticides, ammonia, nutrients, siltation, organic enrichment, and total dissolved solids are found in many surface waters, and that, in addition to nitrate residues, 15 pesticides occur in groundwater, the most common being atrazine.[3] Nitrate pollution of groundwater is a concern in Texas,[4] and agricultural activities are the leading cause of impairment of rivers, lakes, and streams in Colorado, with total dissolved solids being a particularly serious problem for the Colorado River.[5] Sediment pollution is a serious concern on the Snake River in Idaho.[6]

7.2   WHY IRRIGATION CAUSES NONPOINT SOURCE POLLUTION

In arid areas, irrigation is necessary for crop production because little or no rainfall occurs during the growing season. Types of irrigation methods commonly used are surface irrigation (furrow, border, basin), sprinkler irrigation (periodic-move, solid-set, continuous-move), and microirrigation (microsprinklers, drip emitters, and drip tape).

Water applied by irrigation infiltrates the soil and sometimes runs off the field. The infiltrating water replenishes the soil moisture depleted by crop water use or evapotranspiration. Infiltrated amounts exceeding soil moisture depletions drain below the root zone. Sources contributing to this drainage include nonuniform application of irrigation water and excessive irrigation times (the time that irrigation water is applied to a field). Nonuniform water applications, which occur in all irrigation methods, mean some parts of the field receive more water than others. Drainage can occur in those parts receiving more water, even for a properly designed and managed irrigation system. Excessive irrigation times result in too much water applied throughout the field.

Irrigation water infiltrating the soil dissolves chemicals in the soil. These chemicals include naturally occurring salts and trace elements, fertilizers, and pesticides. The infiltrating water carries these chemicals downward in the soil profile, and, if drainage below the root zone occurs, to the groundwater.

Surface runoff occurs when the application rate of the applied water exceeds the infiltration rate. Runoff usually occurs under surface irrigation but can occur under sprinkler irrigation. Runoff picks up sediments as it flows across the soil. Nutrients such as phosphorus and pesticides may be adsorbed to these sediments. These suspended materials can cause sedimentation and turbidity problems and detrimental concentrations of nutrients and pesticides in receiving waters.

Nonpoint source pollution from irrigation generally does not cause the elevated and localized concentration of pollutants frequently found from industrial activities. Pollution concentrations from irrigation are generally lower, but much larger volumes of water are affected compared with industrial pollution because of the large land areas used for agricultural production.

7.3 TYPES OF NONPOINT SOURCE POLLUTION CAUSED BY IRRIGATION

7.3.1 NITRATE

About 20–70% of applied nitrogen is used by crops.[7] The remaining nitrogen can be denitrified (a soilbased process that transforms nitrate into gases that escape into the atmosphere), incorporated into soil organic matter, or leached in the nitrate form.

Nitrate readily moves with water in soil because of anion repulsion. Anion repulsion occurs because most soil particles are negatively charged, as are nitrate ions.[8] This repulsion forces nitrate ions away from the soil particles where water velocity in the soil pore is the slowest and out into the pores where the water velocity is the fastest. Thus, nitrate ions move readily with water and are easily leached below the root zone during irrigation.

Potential nitrate leaching from irrigation is greatest in sandy soils and least in clay soils. Schmidt and Sherman[9] indicated that many areas with high nitrate concentrations in the groundwater correlate with surface sandy soils. Research has shown nitrate concentration in the root zone to decrease with increased clay content.[8] Letey et al.[10] found similar behavior at a site containing sandy soil with clay lenses. Lund and Wachtell[11] concluded that the denitrification was greater in finer-textured soils than in sandy soils because of greater soil moisture and organic carbon percentages in fine-textured soils. In general, McNeal and Pratt[8] feel little denitrification occurs below 2 m where submerged tile drains exist. Pratt[12] listed the criteria shown in Table 7.1 for assessing areas sensitive to quality degradation of receiving waters from nitrate leaching from irrigation. In general, excessive nitrate leaching can occur under the following conditions:

1. Crop conditions that create high potential for nitrate leaching.
   a. Nitrogen (N) removed in the harvestable portion of the crop is a small portion of the total N. About 25–35% or less is removed by fruit crops, about 35–45% or less is removed by vegetable crops, and about 45–60% is removed by grain crops.
   b. Quality or quantity of crop requires high N input and frequent irrigation to ensure rapid vegetative and fruiting growth.
   c. Crop gives a high dollar return per acre and N costs are small compared with total costs.
   d. Crop does not suffer reduced yield or reduced quality when more than adequate amounts of N are applied.
2. Soils with a high potential for nitrate leaching.
   a. High infiltration rates.
   b. Low denitrification potential—usually sandy soils.
   c. No layers restricting water movement.

Nitrate nonpoint source pollution normally occurs in groundwater. However, in some areas, nitrogen fertilizers are injected into irrigation water used for furrow irrigation. Surface runoff from these fields can have elevated levels of nitrate and ammonium. Discharging this surface runoff into off-farm receiving waters causes those waters to be polluted by the fertilizer.

7.3.1.1   Case Study: Nitrate Pollution of Groundwater in the Salinas Valley of California

The Salinas Valley is located along the central California coast. The valley, about 140 km long, runs northwest (starting at Monterey Bay) to southeast. Groundwater is the only source of water for agricultural and urban uses. The amount of annual rainfall varies from an annual average of about 254 mm along the upper part of the valley to about 406 mm along the lower part. Most of the rainfall occurs between November and April.

The west side of the lower part of the valley contains three major water-bearing strata separated by clay layers about 55–121 m deep. These strata, called the pressure zones, extend about 15 km up the valley. Recharge to these strata comes from adjacent unconfined aquifers, from adjacent hillsides, and from drainage below the root zone, stream flow, and rainfall percolation. The aquifer for the rest of the valley is considered to be unconfined, although varying degrees of semiconfinement may be caused by localized clay layers. Recharge of this aquifer is from the Salinas River, drainage from irrigated lands, percolation from precipitation, and runoff from the western slope of the Gabilan Mountains, which run along the east side of the valley. Major crops grown in the valley are lettuce, broccoli, cauliflower, celery, artichokes, and peppers.

TABLE 7.1

Guidelines or Criteria for Judging the Relative Sensitivity of an Area to Nitrate Leaching from Irrigated Lands

	Criteria or Guidelines		
	Low Sensitivity	Medium Sensitivity	High Sensitivity
Receiving water	Not a source requiring low NO_3 concentrations	Intermediate situations	Multiple uses, some requiring low NO_3 concentrations
	Already has such high NO_3 load that more will do no damage		Low dilution of drainage water
	High dilution of drainage waters		No alternate supplies Economic impact of NO_3 leaching is high
	Irrigated agriculture is an insignificant source of NO_3		Irrigated agriculture is significant source of NO_3
Soils	Clayey soils and soils having layers that restrict water flow limit drainage volume and promote denitrification	Loamy soils, intermediate in water flow characteristics	Sandy soils having no layers that restrict water flow
			Well aggregated soils that have high water-flow characteristics
Crops	Require low N inputs or have high N use efficiencies	Good mixture of crops requiring high N inputs with low efficiency of use with crops that are efficient and that require low N inputs	Vegetable and fruit crops of low N use efficiency requiring high N inputs
	Hay crops including legumes, grains, sugarbeets, grapes		No or low acreage of efficient crops in the area
Irrigation	Efficient systems and management that allow low drainage volumes.	Carefully managed surface irrigation systems where low drainage volume is expected.	Inefficient systems that promote large drainage volumes. Typically surface flow systems with long irrigation runs and large amounts of water used
	Typically well-managed sprinkler systems with controls on quantity of water used or drip systems	Mixture of efficient and inefficient systems	Heavy winter rains concentrated in a short period
	Low rainfall that creates no leaching hazard	Infrequent rains that occasionally promote leaching	Temperatures are sufficiently high for nitrification and winter crops are grown

In 1987, data from 300 wells were collected to determine the distribution of nitrate concentrations throughout the valley.[13] Twenty six percent of the wells exceeded the drinking water standard. A similar study, which found that 25% of the wells exceeded the standard, was conducted in 1993.[14] However, in some areas, nitrate concentrations increased following 1987, whereas in other areas, concentrations decreased. Sources contributing to the high levels of nitrate concentration include: (1) fertilizer applications on coarse-textured irrigated soils; (2) greenhouse, dairies, and cattle feedlots and chicken ranches; (3) leaking fertilizer tanks; (4) septic tanks, and (5) lack of backflow prevention devices on wells where fertilizer was injected into the irrigation water.

7.3.2 PESTICIDES

Mobility and persistence determine the pollution potential of a pesticide.[15] Mobility refers to the ease of movement in a soil, and persistence refers to the life of the chemical. Some factors affecting both mobility and persistence of pesticides include volatilization, transformations, adsorption, and solubility. Volatilization depends on the nature and concentration of pesticide, climatic conditions at the soil surface, depth of pesticide in the soil, pesticide adsorption (affected by soil water content, clay content, organic matter content, soil temperature), diffusion of pesticide from the soil, convection of pesticide by evaporating soil water, and pesticide movement caused by bulk flow of soil water to the surface.[16] Transformations involve the degradation of a pesticide by photodecomposition, chemical transformation, and microbiological transformations.[17] Adsorption depends on the nature and concentration of the chemical (surface charge of pesticide), pH of soil water, water solubility of pesticide, and soil characteristics such as type of clay, clay content, and organic matter content.[17] Factors affecting solubility include temperature, salinity (dissolved salts tend to decrease solubility), dissolved organic matter, and pH.[18] The higher the solubility, the higher the mobility, the single most important property influencing pesticide movement.[19]

Persistence is described by the half-life of a pesticide, or the time required for half of the amount of applied pesticide to be degraded and released as carbon dioxide.[19] A measure of the mobility of a pesticide is the partition coefficient. This coefficient is defined as the ratio of pesticide concentrations bound to soil particles to the pesticide concentrations in the soil water.[19] Pesticides with low partition coefficients are more likely to be leached than those with larger values.

Pesticides applied to the soil can be leached below the root zone and transported down to the groundwater. Pesticides also may be applied to the irrigation water as is done for rice production. Surface runoff from these fields can contain unacceptable levels of pesticide concentrations that contaminate the receiving waters used for disposal of surface runoff.

7.3.2.1 Case Study: Pesticides in Surface Runoff from Rice
Fields in the Sacramento Valley, California

About 90% (142,000 ha) of California's rice acreage is in the Sacramento Valley. Surface water is used for irrigation. High-quality irrigation water is distributed throughout the rice production area by a network of canals and ditches supplied by water, primarily from the Sacramento and Feather rivers.

A continuously ponded flow-through basin irrigation system historically has been used for rice irrigation in California. Rice fields are divided into a series of basins. The field is irrigated by supplying water to the uppermost basin. Outflow from this basin irrigates the next basin and so forth. Outflow from the bottom basin is discharged into drainage ditches and eventually to the river.

Herbicides applied to the rice fields for weed control have contaminated the return flows to the Sacramento River, creating a bitter taste in the municipal drinking water of the city of Sacramento.[20] Thus, starting in the early 1980s, measures to reduce herbicide discharges from rice fields have been implemented. These measures consist of the following:[21]

1. Holding the water in the field longer to allow dissipation of the pesticide. The longer the holding time, the more the dissipation. Holding times were increased from 4 to 14 days between 1983 and 1989 to achieve the water quality performance goals set by the state regulatory agency. Required holding time for all pesticides was 24 days in 1991 except for throbencarb, which required a 30-day holding time.
2. Ponding outflow from the last basin on fallow land. This requires the grower to dedicate land for ponding.
3. Improve irrigation water management. Measures used include the following:
   a. Better flow rate control of historical systems to reduce return flow of the last basin.
   b. Recirculation of outflow to the upper basins.
   c. Eliminate outflow by using level basins with no outflow, referred to as the static system.

A project demonstrating the effect of improved irrigation practices on pesticide discharges was initiated in 1991 at two locations.[20] The following rice irrigation approaches were used: (1) conventional irrigation—continual flow-through with surface runoff discharged into a regional system of surface drains, (2) recirculating system—water discharge from the last basin is recirculated to the first basin, (3) static—level basins are used with no water discharged from the basins.

Results of the demonstration projects show that considerable reductions in pesticide discharges can be achieved through better management of existing systems or through an improved irrigation system.[21] Pesticide discharges of static and recirculating systems averaged about 85% less than those of conventional flow-through systems. Overall, better irrigation practices have considerably reduced pesticide concentration in the surface water of the valley. For example, peak molinate concentration declined by about 96% between 1982 and 1991.

7.3.3 SALTS AND TRACE ELEMENTS

Soils in arid areas may contain substantial amounts of naturally occurring soluble salts and trace elements because rainfall in these areas has been insufficient to leach these materials throughout the ages. Irrigation of these soils leaches these materials from the root zone and carries them downward to the groundwater. Soluble salts consist mostly of calcium, magnesium, sodium, chloride, sulfate, and bicarbonate/carbonate. Concentrations of potassium and nitrate generally are very small compared with these other constituents. Trace elements of concern include arsenic, boron, cadmium, chromium, copper, molybdenum, nickel, selenium, and strontium.[22]

At the same time, drainage from irrigated land may create a shallow water table, resulting in subsurface drainage problems. Where shallow water tables exist, evaporation of the groundwater increases concentrations of salts and trace elements over time in the shallow groundwater. Subsurface drainage systems are normally used to reduce or prevent crop production problems caused by shallow groundwater. The drain water collected by these systems usually is discharged into a surface water system. If, however, large concentrations of salts and trace elements exist in the drainage water, these discharges may create downstream water quality problems.

7.3.3.1   Case Study: Subsurface Drainage Problem Along the West Side of the San Joaquin Valley

The San Joaquin Valley of California is a gently sloping alluvial plain about 400 km long and an average of 74 km wide. Its temperate climate, productive soils, and use of irrigation have made the valley one of the world's most important agricultural areas. The soils of the west side of the San Joaquin Valley were derived from marine sediments of the Coastal Range mountains, which are west of the valley. These soils contain the natural salts and trace elements found in the marine sediments. In contrast, the soil of the east side of the valley contain few soluble salts and trace elements, reflecting their origin from the granitic Sierra Nevada mountains, which lie east of the valley.

Irrigation along the west side of the valley was greatly accelerated in 1960 on completion of federal and state water projects that transported northern California water to the San Joaquin Valley. As a result, irrigation water applied to these soils has leached these naturally occurring salts and trace elements down to the groundwater and has also created a shallow water table throughout much of the lower-lying areas. Because of evapoconcentration of salts and trace elements in the shallow groundwater, elevated concentrations of salts and trace elements now exist. Many areas with shallow water tables have salinity levels exceeding 20 dS m^{-1} (electrical conductivity of the groundwater), selenium concentrations exceeding 200 ppb, boron concentrations exceeding 8 ppb, molybdenum concentrations exceeding 1000 ppb, and arsenic concentrations between 100 and 300 ppb.[23]

To deal with the subsurface drainage problem, a master drain (San Luis Drain) was to be built to collect drainage water from farm-installed drainage systems and discharge it into the San Francisco Bay. About 137 km of the drain were built by 1975. The drainage water was discharged into a regulating reservoir (Kesterson Reservoir) until completion of the master drain.

In 1983, deformities and deaths of aquatic birds in Kesterson Reservoir were attributed to the selenium in the drainage water. As a result, discharges to the reservoir were halted and the reservoir was closed. This in turn resulted in termination of discharges of farm drainage systems into the San Luis Drain. Currently, no drainage discharges into receiving waters are occurring from those areas served by the master drain. It is unlikely that the master drain will ever be completed.

Because of the lack of a discharge point for the drainage water, several in-valley approaches to drainage water disposal have been investigated. These include removing some of the trace elements through chemical and biological processes, deep-well injection, desalination, and farm and regional evaporation ponds. None of the approaches has proven to be technically, economically, and environmentally feasible at this time. Currently, using very salt-tolerant trees and shrubs is being investigated for drainage water disposal.

Improved irrigation practices have been implemented to reduce subsurface drainage, although no method for disposing of the remaining drainage water exists. Although drainage amount can be reduced by improved practices, the effect of these improvements on long-term salinity levels is uncertain.

7.3.4 Suspended Sediments in Surface Runoff

7.3.4.1 Effect of Surface Runoff on Water Quality

When irrigation water is applied to sloping land faster than it is infiltrated, a portion of the water runs off the field. In furrow irrigation, the water application rate must be sufficient to advance water across the field, and application time must be sufficient that a large portion of the field receives adequate infiltrated water. This usually results in water running off the tail end of the field. Twenty to fifty percent of the water applied to most furrow-irrigated fields with slopes greater than 0.5% runs off the tail end. Border irrigation on sloping fields may also produce runoff, but because irrigation times are usually short, runoff amounts are often small. When sprinkler application rate exceeds soil infiltration rate, water may run off, although sprinkler water seldom runs off the field in large quantities.

Runoff water is nearly always of lower quality than the irrigation water supply. Water running across the land surface can erode soil. The extent of irrigation-induced erosion is not well documented, although measurements in Idaho, Wyoming, Washington, and Utah show that it is a serious problem in some areas of the western U.S.[24] Runoff water carries part of the eroded sediment off the field. Annual sediment loads in runoff between 4 and 40 Mg ha^{-1} are commonly measured from furrow-irrigated fields with slopes greater than 1%.[24] Surface runoff or tailwater from irrigation is often used on other fields, and a portion of the sediment deposits in surface drains and channels, but the remainder eventually reaches rivers and lakes.[25,26]

Runoff water can also carry other constituents that can degrade downstream water quality. Nutrients, pesticides, and chemicals that are on the soil surface or attached to surface soil particles can leave the field with the sediment. Phosphorus, applied as an agricultural fertilizer, is strongly adsorbed to soil particles and is common in irrigation runoff that carries sediment.[27] Plant pathogens such as nematodes and fungal diseases may be transported with sediments. Sediments may also carry persistent agricultural chemicals that are adsorbed to surface soils. Runoff water from the west side of the San Joaquin Valley carries low concentrations of organochlorine (DDT family) pesticide residues.[28] Weed seeds and other organic matter float off the fields with the flow. Mobile chemicals such as nitrate, salts, and agricultural chemicals are leached below the soil surface by the infiltrating water and are usually not present in harmful quantities in surface runoff. Chemicals or nutrients that are applied in the irrigation water ("chemigation") will leave the field with runoff water, and can pollute the receiving waters.

The sediment and its adsorbed constituents negatively impact downstream water users. Sediment fills surface drains and downstream reservoirs and irrigation canals. Some irrigation companies spend a large portion of their annual maintenance budget mechanically removing sediment deposits from reservoirs, drains, and canals.[25] Runoff water often becomes the irrigation water supply for downstream farms. Sediment-laden irrigation water prevents farmers from adopting drip and even sprinkler irrigation and increases maintenance costs of ditches, pipelines, and ponds. Weed seeds and other soil-borne pests such as crop pathogens can be spread from farm to farm with runoff sediment.

Sediment from irrigated fields has degraded many western U.S. rivers, including the Yakima in Washington,[29] the Snake in Idaho,[30] and the San Joaquin in California.[31] Sediments in surface runoff are deposited in rivers and streams and cover fish-spawning beds and other natural habitats. Sediment accumulation in river beds is often severe because river flow rates (and thus carrying capacity) are usually low during the irrigation season in irrigated valleys, and traditional spring flushing flows are reduced by upstream irrigation storage facilities. Agricultural sediments usually carry sufficient phosphorus to promote plant growth in the river and lake deposits, further stabilizing them. Trace amounts of agricultural chemicals in sediments can accumulate in river and lake beds, vegetation that grows in the beds, and wildlife that eat the vegetation. Sediments that are transported through the rivers often accumulate in downstream reservoirs, reducing reservoir storage capacity, or at the river mouths, where they may interfere with shipping or recreation facilities.

7.3.4.2 Assessing the Potential for Erosion and Surface Runoff Quality Problems

Sediment discharge from irrigation is seldom a problem other than with furrow irrigation. However, irrigated agriculture can increase rainfall-induced erosion and runoff by permitting cultivation of areas that would otherwise have permanent cover, and by maintaining high soil-water contents in soils that would otherwise be dry. These indirect effects of irrigation on surface runoff quality are not discussed here.

Furrow erosion depends on the erosiveness of the flowing water and the erodibility of the soil.[32] Flow shear, or velocity, which determines the flow erosiveness, increases with flow rate and slope. Erosion is usually low where furrow slope is less than 0.5%, but erosion potential increases dramatically at slopes greater than 1%. Roughness created by residue on the soil surface decreases erosiveness. Thus, erosion is often low in close-growing crops or where reduced tillage or residue management is used.

In spite of extensive study, soil erodibility is still difficult to predict. Texture is important, with high-silt soils being most erodible, but variation is large for similarly textured soils. Soil erodibility also varies with time and tillage practices for a given soil. Freshly tilled soil is more erodible than soil with a stabilized, consolidated surface.

Although much is known about erosion and sedimentation processes, irrigation-induced erosion cannot yet be accurately modeled and predicted, primarily because of the inability to predict soil erodibility. The Universal Soil Loss equation, USLE (and its recent revision, RUSLE) does not apply to irrigation-induced erosion. The Water Erosion Prediction Project (WEPP) model has a furrow irrigation component not validated by field studies.

The furrow tail-end condition can strongly affect sediment discharge from the field. Where water tends to pond up and flow slowly at the field downstream end, much of the sediment in the water may be deposited before leaving the field. Where farmers cut a tailwater ditch across the lower end of the field to discharge runoff and prevent water ponding, serious erosion can occur in the tailwater ditch and at the end of the furrows, resulting in a downward sloping (convex) field end and greatly increased sediment discharge.

Sediment discharge is quantified by measuring the flow rate and sediment concentration. Flow rate is measured with flow measurement flumes or weirs and concentration is measured on volumetric grab samples.[33] Accurately assessing sediment discharge requires numerous measurements during an irrigation and measurements of several irrigation events. Both flow rate and sediment concentration from furrow-irrigated fields varies widely with time. Runoff rates from an irrigation are initially zero and increase with time after water reaches the end of the field. Sediment concentration is often highest initially, especially with freshly tilled soils, and decreases with time.[33] Sediment concentration in runoff is usually high following tillage and decreases after several irrigations because the soil surface stabilizes and consolidates.

All sediment discharging from farm fields does not usually reach downstream rivers or lakes. Some of the runoff water may be rediverted by downstream farmers, and some of the sediment may deposit in drains or sediment basins. Field measurements must be supplemented by return flow quality and quantity measurements to assess sediment and chemical inflows to water bodies. Assessing damages to rivers and lakes and their complex aquatic biological systems requires a thorough understanding of those systems and an accounting of the various sources of pollutants.

7.3.4.3 Case Study: The Rock Creek Rural Clean Water Project—Erosion and Sediment Control in Southern Idaho

Over 3 million acres are irrigated in the Snake River plain in southern Idaho and eastern Oregon. The combination of highly erodible silt loam soils, field slopes commonly varying from 0.5 to 2%, and furrow irrigation, have resulted in high irrigation-induced erosion. Sediment discharge measurements in the Middle Snake area near Twin Falls showed 10–100 mg ha^{-1} of sediment leaving row crop furrow-irrigated fields,[27] and an average of 0.5 mg ha^{-1} eventually reaching the Snake River.[26] Sediment in return flows has been identified as a major cause of serious sedimentation and water-quality problems in the Middle Snake River.

A major tributary to the Middle Snake River is Rock Creek, which had long been recognized as one of the most severely degraded streams in the state, with the primary problem being sediment from irrigation return flows. Rock Creek drains 32,000 ha in south central Idaho, 8,500 ha of which are irrigated, with 4,500 ha being critical for sediment production.

In 1980, Rock Creek was selected as one of 20 Rural Clean Water Program projects in the nation. The goals were to reduce sediment by 70% and phosphorus by 60% in subbasins where practices were applied, and to improve fish and wildlife habitat, aesthetics, and recreational uses of Rock Creek. Between 1981 and 1986, 182 contracts for a total of nearly $2,000,000 were written with farmers to install or adopt Best Management Practices on 3,400 designated critical hectares. Approved Best Management Practices included permanent vegetative cover; conservation tillage; sediment retention, erosion, and water control structures; irrigation system improvements; stream protection; and fertilizer and pesticide management.

Water quality monitoring showed that suspended sediment and phosphorous loading during the irrigation season decreased in most subbasin drains receiving treatment between 1982 and 1990.[34] Rock Creek contributions to the Snake River showed a 75% decrease in sediment loadings (from 20,000 to 5,000 Mg during the irrigation season) and a 68% decrease in total phosphorus loading (from 28 to 9 Mg).

Specific findings of the study included:[34]

1. Irrigation practices such as concrete ditch and gated pipe, although not the most cost-effective practices, are highly effective in obtaining farmer participation.
2. Sediment practices are effective in demonstrating to farmers the magnitude of the soil-erosion/water quality problem.
3. For long-term soil erosion and water quality benefits, emphasis should be placed on converting from surface irrigation to sprinklers.
4. Large sediment ponds are effective in reducing sediment and positively affecting fish habitat in Rock Creek.
5. Streambank erosion continues to be a major source of sediment impacting Rock Creek.
6. Instream beneficial uses, including salmonid spawning and primary contact recreation, remain impaired on lower Rock Creek, because of both sediment and phosphorous and nitrogen levels.

7.4 PERFORMANCE CHARACTERISTICS OF IRRIGATION SYSTEMS AFFECTING NONPOINT SOURCE POLLUTION

Nonpoint source pollution in arid and semiarid areas of the western United States and elswhere is the result of irrigating land in these areas. Thus, developing and implementing practical and effective measures to reduce these pollution problems through improved irrigation requires an understanding of the performance characteristics of irrigation systems. This is necessary to identify opportunities and limitations for reducing nonpoint source pollution through the improved practices.

The performance of an irrigation system is described by its uniformity and efficiency. Uniformity refers to the evenness of the infiltrated water throughout a field and depends on system design and maintenance. Efficiency refers to the amount of water needed for crop production compared with the amount applied to the field, and depends on system uniformity and management.

7.4.1 UNIFORMITY

A uniformity of 100% means the same amount of water infiltrates everywhere in a field. No irrigation system, however, can apply water at 100% uniformity. Regardless of the irrigation method, some parts of a field infiltrate more water than other areas. If an amount of water equal to that needed for crop production infiltrates the least-watered area of a field (referred to as a properly irrigated field), excess water will be applied to the remainder of the field. These excess amounts contribute to drainage below the root zone because more water infiltrated than was needed to replenish the soil moisture. The larger the nonuniformity, the larger the differences in infiltrated water throughout the field, and the more the drainage below the root zone.

Many indices have been used to describe uniformity.[35,36] The most common index is the distribution uniformity defined as:

$$DU = \frac{100\overline{X}_{LQ}}{\overline{X}} \qquad (7.1)$$

where $\overline{X}$ equals the average amount of infiltration and $\overline{X}_{LQ}$ equals the average of the lowest one-fourth of the measurements, commonly called the low quarter.

The emission uniformity sometimes is used for microirrigation, where $\overline{X}$ equals the average measured emitter discharge rate and $\overline{X}_{LQ}$ equals the average of the lowest one-fourth of the measured emitter discharge rates.

The field-wide uniformity should be determined when assessing the uniformity of an irrigation system. Frequently, however, system uniformity is assessed by measuring only one uniformity component such as emitter or sprinkler discharge rates along one lateral only instead of along three or four laterals spread throughout the field. Procedures for estimating the field-wide uniformity are in Burt et al.[37]

To illustrate the effect of nonuniform water applications on drainage, ratios of drainage to applied water, shown in Table 7.2, were developed using data from periodic-move sprinkler systems. These ratios, calculated for various sprinkler distribution uniformities, show that for a DU of 93%, about 10% of the applied water drains below the root zone, whereas for a DU of 74%, drainage is about 34% of the applied water.

Components contributing to nonuniform infiltration are discussed for each irrigation method.

7.4.1.1 Surface Irrigation

Surface irrigation uses the soil surface to flow water across the field. Thus, the uniformity of these systems is affected by soil characteristics such as infiltration rate and

TABLE 7.2
Ratio of Drainage (DP) to
Applied Water (AW) for
Various Distribution
Uniformities Developed
from Evaluations of Periodic-
Move Sprinkler Systems

DU (%)	Ratio (DP/AW)
93	0.10
83	0.23
74	0.34
63	0.50
48	0.68

surface roughness and field characteristics such as length, slope, and inflow rate. Some of these characteristics are easily measured, whereas others, such as the infiltration rate, are not. Thus, making reasonable estimates of the distribution uniformity may be difficult.

The main components contributing to field-wide nonuniformity of infiltration are varying infiltration opportunity times along the field length and variable infiltration rates. Varying opportunity times along the field length are caused by the time required for irrigation water to flow to the end of the field. Field-wide uniformity is also affected by different day and night irrigation times. Frequently, more water is applied at night because irrigation times tend to be longer at night than during the day to avoid changing irrigation sets in darkness. Other factors include varying inflow rates during the irrigation, water temperature differences between day and night irrigations, and infiltration differences caused by tillage and planting equipment. Detailed information is found in Hanson and Schwankl.[38]

Soil variability caused by soil texture differences can severely affect the uniformity of infiltration. Childs et al.[39] found most of the nonuniform infiltration to be caused by soil variability in a field with soil textures ranging from a clay loam to a sand. Infiltration variability caused by varying infiltration opportunity times along the field length was minor. Tarboton and Wallender[40] found that soil variability and varying infiltration opportunity times contributed about equally to nonuniform infiltration in a field with a relatively uniform soil texture.

Variable infiltration rates also can be caused by cultural practices. Infiltration rates in wheel furrows are usually less than those in nonwheel furrow. "Guess" furrows, which occur at the edge of the cultivation pattern, can have infiltration rates much greater than the other nonwheel furrows.

7.4.1.2 Sprinkler Irrigation

Sprinkler irrigation uniformity depends on the hydraulic characteristics of the system and the areal distribution of water applied between the sprinklers. Specific components include pressure changes throughout the field caused by friction losses and elevation changes, catch-can uniformity, and minor factors such as mixed nozzle sizes, worn nozzles, malfunctioning sprinkler heads, nonvertical sprinkler risers, and leaks. Catch-can uniformity describes the pattern of applied water between adjacent sprinklers. It depends mainly on sprinkler spacing, pressure, wind speed, and sprinkler head/nozzle type. Different day and night set times can also affect the field-wide uniformity.

7.4.1.3 Microirrigation

The uniformity of microirrigation systems also depends on the hydraulic characteristics of the system and on system maintenance. Nonuniformity in microirrigation systems is caused by field-wide variability in emitter discharge rates. The main components contributing to this variability are manufacturing variation in flow path dimensions, pressure changes caused by friction and elevation changes, and clogging of emitters or microsprinklers. Other components include mixing of emitter sizes and types, emitter wear and aging, leaks, pressure regulator variability, and different irrigation times throughout a field.

It is commonly assumed that the uniformity of microirrigation is much higher than that of other irrigation methods. However, an analysis of data on nearly 1000 irrigation system evaluations indicate otherwise.[41] This analysis showed the field-wide uniformity of microirrigation systems to be similar to that of other irrigation methods. The study also concluded that microirrigation has the potential for higher uniformities, but only if the systems are properly designed and maintained. However, little correlation between age of the system and field-wide uniformity was found, indicating that new systems were not designed to realize the potential of microirrigation.

7.4.2 IRRIGATION EFFICIENCY

Irrigation efficiency is defined as the ratio of the amount of water needed for crop production to the amount of water applied to the field. The amount of water needed for crop production is the beneficial use. Another term frequently used is the application efficiency, defined as the ratio of the amount of irrigation water stored in the root zone to the amount of applied water.

Crop evapotranspiration is the largest beneficial use of irrigation water. This is water that evaporates from the plant leaves and from the soil surface. More than 95% of the water uptaken by the plant is used as evapotranspiration. Other beneficial uses include leaching for salinity control, frost protection, and climatic cooling.

Major losses affecting irrigation efficiency are drainage and surface runoff. However, drainage needed for leaching to control salts in the root zone is beneficial use and is not considered a loss, although it may contribute to nonpoint source pollution. Surface runoff is also beneficially used if it is recirculated back onto the field being irrigated or used to irrigate other fields.

A relationship exists between distribution uniformity and irrigation efficiency. If the amount infiltrated in the low quarter equals the beneficial use, the distribution uniformity is an estimate of the potential maximum irrigation efficiency, assuming no

TABLE 7.3
Practical Maximum Potential Irrigation Efficiencies

Irrigation Method	Irrigation Efficiency (%)
Sprinkler	
Continuous-move	80–90
Periodic-move	70—80
Portable Solid-set	70–80
Microirrigation	80–90
Furrow	70–90
Border	70–85

surface runoff losses. An actual irrigation efficiency less than the distribution uniformity indicates overirrigation occurs throughout the entire field. An irrigation efficiency greater than the distribution uniformity indicates deficit irrigation in parts of the field.

Table 7.3 lists potential maximum practical irrigation efficiencies developed from data analyzed by Hanson.[41] A practical irrigation efficiency is one that is technically and economically feasible. These values assume that the least watered part of the field receives an amount equal to the beneficial use, and surface runoff is beneficially used. Because microirrigation has the potential for higher distribution uniformities, its potential irrigation efficiency is also higher.

Some have reported a potential irrigation efficiency of 95% for drip.[42] Such values are not realistic for an economical system, but they usually are based on special circumstances and may not reflect the field-wide uniformity.

7.5   REDUCING DRAINAGE FROM IRRIGATED LAND: A CONCEPTUAL APPROACH

Reducing nonpoint source pollution from irrigated land involves reducing the amount of irrigation water that drains below the root zone or runs off the field. Drainage can be substantially reduced by simply decreasing applied water, which, however, may severely reduce crop yield. Thus, an integrated approach must be used in developing and implementing measures for reducing pollution that considers the effectiveness of measures, their cost, and their effect on both crop yield and farm-level economics. This, in turn, requires understanding how crop yield and drainage below the root zone are affected by uniformity and amount of irrigation water.

Crop yield is directly related to crop evapotranspiration. Maximum yield occurs when the evapotranspiration is maximum, whereas reduced evapotranspiration caused by deficit irrigation decreases crop yield. Many crops including alfalfa, processing tomato, grape, almond, sugar beet, wheat, and corn exhibit a linear relationship between yield and evapotranspiration, but other crops may show a curvilinear relationship.

Figure 7.1 shows alfalfa yield versus evapotranspiration and alfalfa yield versus applied water for several distribution uniformities. The alfalfa yield/evapotranspiration relationship was obtained from Grimes et al.[43] A linear relationship (solid line) exists between yield and evapotranspiration with a maximum evapotranspiration of 1001 mm and a maximum yield of 26.3 mg ha^{-1}.

For irrigation water applied at a uniformity of 100%, the yield/applied water relationship is the same as the yield/ET line (solid line) until an amount of applied water equal to maximum evapotranspiration is reached. For amounts greater than maximum evapotranspiration, the yield/applied water relationship is a straight horizontal line (dotted line in Figure 7.1). The difference between the amount of applied water and the maximum evapotranspiration is drainage below the root zone.

A different yield/applied water relationship occurs for smaller uniformities. Yield/applied water is the same as yield/evapotranspiration until a threshold value is reached. The yield-applied water curve then deviates from the yield/evapotranspiration relationship for amounts of applied water exceeding the threshold value. This deviation means that, for a given yield, more water must be applied than that needed at 100% uniformity. The lower the uniformity, the more the deviation from the yield/ET line, and the more applied water needed to obtain a given yield.

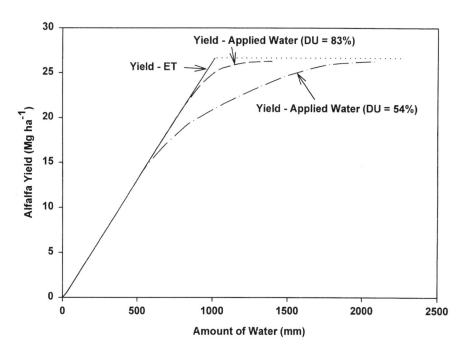

FIGURE 7.1   Relationships between alfalfa yield and evapotranspiration and alfalfa and applied water.

This deviation is caused by nonuniform water application. Once the threshold value is exceeded, some parts of the field receive more water than needed to replenish the soil moisture depletion, resulting in drainage below the root zone. Drainage is small when the applied water slightly exceeds the threshold value. As the amount of applied water increases, more and more drainage occurs.

The effect of both uniformity and applied water on drainage is shown in Figure 7.2. No drainage occurs until applied water exceeds 559 mm for DU = 54% and 800 mm for DU = 83%. As applied water continues to increase, drainage amounts also increase. Thus, for a given amount of applied water, more drainage occurs as the uniformity of the applied water decreases. From a nonpoint source pollution viewpoint, the more the drainage, the greater the leaching of chemicals from the root zone

Figure 7.3 shows the effect of this drainage on the irrigation efficiency. For amounts of applied water less than the threshold value, irrigation efficiency equals 100%. Once the threshold is exceeded, irrigation efficiency decreases with applied water. The smaller the distribution uniformity, the smaller the irrigation efficiency for a given amount of applied water, which reflects nonuniform water application.

The behaviors described in Figures 7.2 and 7.3 indicate that several factors can affect drainage below the root zone. First, even though the uniformity is 100%, overirrigation can cause nonpoint source pollution from drainage. Second, the

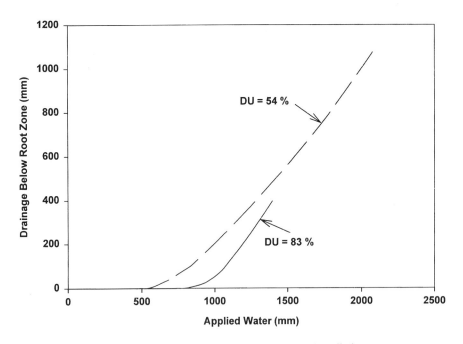

FIGURE 7.2   Relationships between drainage and applied water.

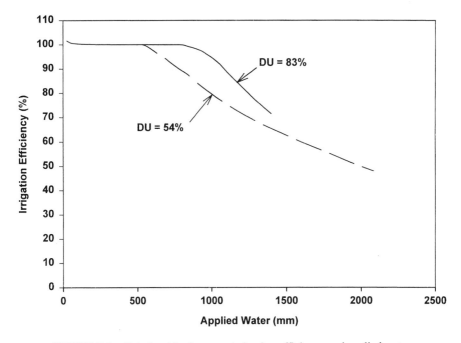

FIGURE 7.3 Relationships between irrigation efficiency and applied water.

smaller the uniformity of infiltrated water, the greater the potential for nonpoint source pollution because of increased drainage.

As the drainage increases, more and more leaching of chemicals such as nitrate and pesticides occurs from the root zone. This leaching deprives the crop of the positive benefits of the chemical and transports the material to the groundwater. This leaching may reduce crop yield unless additional fertilizer is applied, which in turn may contribute even more to nonpoint source pollution.

The interaction between leaching of nitrate, uniformity, and amounts of applied water is illustrated by Pang et al.[44,45] Using a computer growth model verified with field data, they modeled the effect of uniformity and amounts of applied water and applied nitrogen on corn growth and nitrate leaching. Their results showed the following:

1. The lower the uniformity of the applied water, the smaller the yield for a given amount of applied water.
2. Yields increased with applied water to some maximum value and then decreased. The water application at which the decrease starts to occur was larger for the larger nitrogen applications.

3. Maximum yield was never reached for the lowest uniformity, probably because of the nitrogen leaching in the parts of the field receiving the most irrigation.
4. The lower the uniformity of irrigation, the larger the nitrogen leaching, with more nitrogen leaching occurring for the larger nitrogen application.

Tanji et al.[46] conducted a similar study using lettuce grown in the Salinas Valley of California. They found that seasonal irrigation amounts larger than about 300 mm had little effect on crop yield but that nitrate leaching was greatly increased by the larger water applications. Maximum yield and profit occurred for 300 mm of irrigation. They concluded that decreasing the irrigation amounts was more effective in reducing nitrate leaching than reducing the applied nitrogen fertilizer.

7.6 MEASURES FOR REDUCING DRAINAGE

This conceptual approach suggests three strategies for reducing drainage below the root zone: (1) improve irrigation scheduling to prevent overirrigation, (2) impose deficit irrigation on the crop, and (3) improve system uniformity.

7.6.1 IMPROVE IRRIGATION SCHEDULING

Irrigation scheduling can answer the questions of when to irrigate and how much water to apply. The answers to these questions can reduce drainage below the root zone by decreasing any overirrigation caused by excessive irrigation times and can also reduce surface runoff. Approaches to irrigation scheduling include estimating the crop evapotranspiration from climatic data and measuring or monitoring soil moisture content.

Many equations have been developed relating climatic data to a reference crop evapotranspiration.[47] The reference crop evapotranspiration is that of either alfalfa or grass, depending on the particular equation. The actual crop evapotranspiration is calculated by multiplying the reference crop evapotranspiration by a crop coefficient. Crop coefficients depend on crop type and stage of growth and can be found in Allen et al.[48] or in regional or state-wide material published by the Cooperative Extension Service of a particular state or the Natural Resources Conservation Service (USDA).

Measuring or monitoring soil moisture contents is recommended, even if the crop evapotranspiration is calculated from climatic data. Measuring soil moisture can help determine when to irrigate, how much water was used between irrigations, depth of wetting from an irrigation, and patterns of soil moisture extraction between irrigations. Instruments such as tensiometers and electrical resistance blocks measure the soil moisture tension, which can be used to determine when to irrigate. Guidelines are available for the maximum soil moisture tension that should occur before irrigating. These devices can also be used to determine depth of wetting from an irrigation, and extraction patterns between irrigations. They, however, do not directly measure soil moisture content. Calibration curves relating the reading of the instrument to soil moisture content are necessary to determine soil moisture depletions.

Measurements of soil moisture content can be made with devices such as the neutron moisture meter and dielectric soil moisture sensors. The neutron moisture meter has been used for decades to measure soil moisture content. It, however, uses a radioactive material, which means that the user must be licensed and trained by an appropriate agency. Thus, it is more appropriate for use by consultants, agency personnel, etc., than by growers. Many dielectric sensors are available for direct measurement of soil moisture content. Thus far, they have been used mainly by researchers. An evaluation of some of these devices conducted in California revealed that they may provide reasonably accurate measurements of soil moisture content in sandy soils with little soil salinity, but in finer-textured soils with moderate soil salinity, their built-in calibration curves may be inappropriate.[49]

A flowmeter is required to know the amount of applied water. The depth of the applied water is calculated using the following equation:

$$D = \frac{KQT}{A} \tag{7.2}$$

where Q = irrigation system flowrate; T = time required to irrigate the field; A = area irrigated; and K = 0.0022 where the units are gallons per minute for Q, hours for T, and acres for A, and K = 0.996 where the units are cubic meters per hour for Q, hours for T, and hectares for A.

Unfortunately, many irrigation systems lack flow meters. Based on an analysis of the data developed by mobile laboratories in California, flow meters were installed on 73% of microirrigation systems and on 24% of the sprinkler systems.[50] Few furrow and border irrigation systems appeared to have flow meters. Thus, a first step in improving irrigation water management is to install and use flow meters.

7.6.2 IMPOSE DEFICIT IRRIGATION

Irrigating at amounts of applied water less than that needed for maximum yield will reduce drainage below the root zone as shown in Figure 7.2 and in Pang et al.[45] At the same time, crop yield can be reduced (Figure 7.1). The effect on crop yield will depend on the amount of the deficit and on the tolerance of the crop to water stress.

Normally, deficit irrigation is discouraged because of its potential adverse effect on crop yield. For some crops, however, regulated deficit irrigation can result in less applied water with little or no effect on yield, and in some cases, can benefit crop quality.

Regulated deficit irrigation involves reducing the amount of applied water during periods of slow vegetative and reproductive growth. During other growth stages, amounts of water needed to maintain full crop evapotranspiration are applied. Tree growers have more potential to minimize adverse effects of deficit irrigation than do field and row crop growers because of the greater separation between vegetative and reproductive growth stages in trees compared with field and row crops. Research on prune,[51,52] peach,[53] pistachio,[54] olive,[55] and almond[56] showed regulated deficit irrigation to be an acceptable approach to reducing applied water yet maintaining crop yield. Research, however, showed that regulated deficit irrigation was not appropriate for walnut.[57] Regulated deficit irrigation may be particularly beneficial during drought conditions.

Opportunities for regulated deficit irrigation appear to be less for row crops. The few studies on this matter have shown that irrigation applications can be reduced or terminated before harvest earlier than normally practiced for sugar beet,[58] cotton,[59] cantaloupe,[60] and processing tomatoes[61] without substantial yield reductions. For many vegetable crops, however, deficit irrigation at any stage of growth can severely reduce yield.

7.6.3 IMPROVE SYSTEM UNIFORMITY

Options for improving irrigation system uniformity include upgrading existing systems or converting to a system with a potential for achieving a higher uniformity and irrigation efficiency.

7.6.3.1 Surface Irrigation

Improving the uniformity of surface irrigation requires reducing the variability in infiltration throughout the field. Strategies for improving uniformity include decreasing the time for water to reach the end of the field and reducing the infiltration rate. Measures commonly recommended for improving the uniformity of surface irrigation are as follows:

1. *Shorten the field length.* Shortening the length reduces differences in infiltration opportunity times down the furrow or border. This is the most effective measure for improving uniformity and reducing drainage below the root zone. Shortening the field length by one-half will generally reduce the drainage by at least 50%.[62] The DU may be increased by 10–15% points compared with the normal field length. This measure will be effective only if the irrigation set time is reduced because the time for water to reach the end of the shortened field generally will be 30–40% of the original time. The reduction in irrigation set time is equal to the difference between the original time to the field end and the new time. Failure to reduce the set time will greatly increase both drainage and surface runoff.

   A major problem with this measure is the potential for increased surface runoff. These studies indicate a potential increase of 2 to 4 times more runoff compared with the original field length. Cutback irrigation can alleviate this problem, provided the irrigation district will allow a decrease in the field inflow rate. Other measures for coping with this problem are to use tailwater recovery systems to recirculate the water back to the head of the field or to use the runoff on lower-lying fields. Reservoir storage is needed for both scenarios.
2. *Increase the unit inflow rate.* This commonly recommended measure reduces the time for water to reach the end of the field, thus decreasing differences in infiltration opportunity times along the field length.

However, this measure has a relatively small effect on both the uniformity and the drainage.[62] The higher furrow inflow rates increased the depth of flow in the furrow, which in turn increased the wetted area for infiltration of the furrow. Thus, the higher inflow rates caused higher infiltration rates, which offset the effect of the smaller time to the end of the field.

The infiltration rate under border irrigation would be only slightly affected by higher border inflow rates. Yet, field evaluations showed only a minor improvement in the performance of border irrigation under higher flow rates compared with lower flow rates.[63,64]

3. *Convert to surge irrigation.* Surge irrigation involves on-and-off cycling of the irrigation water. The water is first allowed to flow part way down the field and then is shut off. After the water applied by the first surge infiltrates the soil, the water is then applied again allowing water to advance an additional distance beyond that of the first surge. This surging is continued until the water reaches the end of the field.

The surging reduces the infiltration of coarse to medium-textured soils to values less than those under continuous-flow irrigation. Field evaluations have shown that the amount of water needed for water to reach the end of the field is about 30–40% less for surge irrigation compared with continuous-flow irrigation.[65]

Surge irrigation also appears to reduce the effect of soil variability on infiltration uniformity. Purkey and Wallender[66] found that surge irrigation not only reduced the average depth infiltrated by 31%, but also reduced infiltration differences caused by soil texture variation by 37%. Others found surge irrigation to reduce differences in infiltration rates between wheel and nonwheel furrows and to reduce seasonal differences in infiltration rates.[67,68]

Surge irrigation is most appropriate for furrow irrigation systems using gated pipe. Solar powered surge valves are available that control the surge times and also allow an adjustment in on/off times after water reaches the end of the field. Surge irrigation is difficult to apply to furrow irrigation systems using siphons and also to border or basin irrigation systems using alfalfa valves, ditch gates, and so forth.

4. *Other measures.* Other measures for improving the uniformity of infiltrated water include improving the slope uniformity through better land grading, and compacting the furrow surface using torpedoes (cylinder-shaped weights pulled in the bottom of the furrow) or tractor wheels. Field evaluations have shown these measures may have a minor effect on system performance.[69]

7.6.3.2   Sprinkler Irrigation

Recommended distribution uniformities under low-wind conditions range between 70 and 80% for periodic-move systems (hand-move, wheel-line) and solid-set sprinkler systems. Some measures for improving these systems are as follows:

1. Minimize pressure variation by using the proper combination of pipeline lengths and diameters. Limit field-wide pressure changes to less than 20% of the average pressure. Pipeline design procedures are given in Keller and Bliesner.[70]
2. Use flow control nozzles where the pressure variation exceeds 20%. These nozzles contain a flexible orifice that changes diameter as pressure changes.
3. Use appropriate sprinkler spacings.
4. Maintain appropriate sprinkler pressure. Low pressures cause a dough-nut-shaped pattern of applied water. Very high pressures cause much of the water to be applied very close to the sprinkler because of excessive spray breakup. Nozzles specially designed for low pressures are available, but field tests have revealed little difference in catch-can uniformity between those nozzles and the standard circular nozzles. Thus, uniformity problems caused by low pressure are not likely to be corrected by changing to low-pressure nozzles.
5. Offset lateral locations of periodic-move sprinkler systems such that the lateral positions of the succeeding irrigation are midway between those of the preceding irrigation. The distribution uniformity resulting from this measure is:

$$DU_o = 10\sqrt{DU}$$

where DU_o is the distribution uniformity of the offset moves and DU is the distribution uniformity of the normal system. The effect of this measure on yield is unknown.
6. Avoid mixing nozzle sizes, repair malfunctioning sprinklers and leaks, and maintain vertical risers.
7. Replace worn nozzles.

Distribution uniformities of center-pivot and linear-move sprinkler machines should be higher than those of the previously mentioned sprinkler systems. The more or less continuous movement of these machines reduces the effect of wind on uniformity. Recommended distributions uniformities of these machines are 80–90%.

7.6.3.3 Microirrigation

Microirrigation systems should be designed for a field-wide distribution or emission uniformity of at least 80%. This means that the design uniformity along the lateral must exceed 90% because the lateral uniformity is the largest contributor to the field-wide uniformity. Achieving this level of uniformity depends on the coefficient of manufacturing variation, emitter discharge rate, emitter spacing, tape or tubing diameter, slope, and lateral length. Design procedures are found in Keller and Bliesner,[70] Hanson et al.,[71] and Schwankl et al.[72]

Some measures for maintaining high uniformity of microirrigation systems are as follows:

1. Select emitters or microsprinklers with an excellent coefficient of manufacturing variation (CV). CVs less than 0.05 are excellent, CVs between 0.05 and 0.1 are acceptable, and CVs greater than 0.1 are marginal.
2. Use pressure-compensating emitters or microsprinklers where large pressure changes occur throughout the field. A minimum pressure is required for the pressure compensating features to operate properly.
3. Use proper filtration and chemical treatment of irrigation water to prevent or reduce clogging.
4. Flush laterals regularly to prevent clogging.
5. Maintain adequate pressure regulation.

7.7   REDUCING IMPACTS OF SURFACE RUNOFF

7.7.1   REDUCING FLOW EROSIVENESS

The erosiveness of furrow flows can be reduced by reducing flow rates. Reducing flow rate usually results in more time required to spread water across the field and thus lower irrigation water distribution uniformity. There is usually a tradeoff between reducing erosion and reducing irrigation uniformity, and thus between reducing surface runoff and drainage below the root zone. Infiltration-reducing management practices such as furrow packing and surge irrigation may counteract the impact of reduced flow rates on uniformity. Shortening furrow lengths by subdividing fields reduces required flow rates. However, as the number of shortened fields is increased, the amount of tailwater and sediment discharge may increase. Mid-field gated pipelines reduce run lengths without increasing field runoff.

Average furrow flow rates are set higher than necessary to ensure that all portions of all furrows are adequately irrigated. Reducing flow rate and allowing a small portion of the field to be inadequately irrigated may be a rational choice if erosion damage is a problem. Furrow application systems that facilitate uniform furrow flows allow reduced average flow rates. Reduced flow rate after stream advance is complete (cutback) will result in reduced runoff and erosion, although furrow erosion rates tend to decrease with time during an irrigation even with constant flow rates. Irrigation scheduling usually results in smaller total application amounts and times, and thus less erosion and runoff.

Flow velocity and thus erosiveness is also reduced by increasing furrow roughness. Furrow roughness can be increased by leaving or placing crop residue in the furrow.[73,74] A furrow straw-mulching machine is commercially available for this purpose. However, roughness also slows water advance and may reduce irrigation uniformity. Furrow residue is a good option for steep sections of furrows where erosion is greatest and water advance is rapid.[75] Straw mulching in combination with surge irrigation can reduce erosion and maintain irrigation uniformity.[76] No-till practices also resulted in lower infiltration during early-season irrigations so the remaining surface residue essentially eliminated erosion but irrigation uniformity was maintained.[73]

Erosion is reduced by reducing furrow slope, but changing field slopes is usually not practical. In some cases, the furrow direction can be oriented across the slope (contour furrows) to reduce effective furrow slope. This practice can result in severe concentrated flow erosion if water overtops and flows across beds.

On fields with a convex tail end, if the water flow in the tail ditch can be slowed, sediment deposition can fill in the depression. Carter and Berg[73] devised a buried pipe tailwater system that eliminates convex field ends. Eliminating tailwater ditches and planting close-growing crops on the convex end can slow the flow and reduce erosion and may result in sediment deposition on convex ends. Portable canvas dam checks across eroding tail ditches can reduce ditch erosion.

7.7.2 REDUCING SOIL ERODIBILITY

Our understanding of soil aggregate stability, cohesiveness, and erodibility is poor. Thus, few techniques are available to reduce erodibility. Erosion does tend to be higher after tillage. Thus, reducing the number and depth of tillage operations does reduce erosion.[73] Because sodium disperses clays and can increase erosion, decreasing the sodium adsorption ratio of the soil or using irrigation water lower in sodium or higher in calcium may reduce erosion.[77]

Polyacrylamide (PAM) applied in the irrigation water dramatically reduces furrow erosion. PAM has two effects—it acts as a soil stabilizer and reduces erodibility, and flocculates sediment particles, inducing them to deposit. When a low concentration of PAM (<10 mg/l) is applied with the irrigation water, erosion is reduced by over 90% in most cases.[78, 79] Material costs are about $5 and $10 per ha per application, and reapplication is recommended at least following every tillage operation. Although this application was developed recently, its use is growing rapidly in several states. PAM was used on over 200,000 ha in 1997.

7.7.3 REDUCING SEDIMENT DISCHARGE

If erosion cannot be adequately controlled on the field, off-field practices may be required to remove sediment from the runoff. These techniques are less desirable than on-field erosion control because they do not eliminate erosion damage to the field.

Sediment can be removed from water by slowing the flow to allow time for suspended sediment particles to settle out. Sediment basins with at least 2-hour residence time will settle out all of the sand-sized particles, most of the silt, and a portion of the clay.[80] For a tailwater flow of 30 l s^{-1} basin volume must be at least 220 m^3 for a 2-hour residence time. Sediment basin sizes vary from large ponds on major drains to small basins at the outflow point of a field. Sediment basins require the accumulated sediment to be periodically excavated and piled until it can be spread back onto the fields or other areas requiring topsoil fill. Basin size must account for expected sediment deposition amounts and desired cleanout intervals. An advantage of sediment basis is that they visually demonstrate to the farmer the amount of soil eroding from the field.

Sediment can be collected at the low end of fields by slowing the flow in the tailwater ditch with excavated pits or earthen surface checks. These "minibasins" are more efficient if water is directed from each basin into a ditch or buried tailwater collection system rather than allowing water to flow from basin to basin. Minibasins generally need to be rebuilt each year. Vegetative filter strips of small grains or permanent cover crops at the tail end of fields can also slow tailwater flows and accumulate sediment.

Sediment retention efficiency of adequately sized basins varies from 70 to 95%.[81] A weakness of sediment basins is that they least efficiently retain the small-sized sediment particles. Small soil particles have large specific surfaces compared with large particles and thus have more capacity to adsorb agricultural chemicals. Thus, a large proportion of the phosphorus and other chemicals that move with sediment is associated with the smallest sediment,[82] and sediment basin efficiency in containing phosphorus and other potential pollutants is lower than their sediment retention efficiency.

A portion of the agricultural chemicals such as phosphorus that are removed from fields with eroded sediment eventually come into solution in the runoff water. Research is currently being conducted to learn whether runoff flow through constructed wetlands will remove a portion of these dissolved materials as well as materials attached to clay particles that are not removed in sediment basins.[83] Questions about the eventual accumulation and recycling of these materials in the wetland are not yet answered.

7.7.4   Surface Runoff Containment and Reuse

Properly designed and used irrigation runoff reuse systems can contain all farm runoff and associated sediments and contaminants on the farm. These systems must have sufficient storage and pumping capacity to use the runoff water effectively.[84] With tailwater reuse, a portion of the sediment can be recycled back to the fields, reducing the required frequency for storage pond cleanout. Any soluble substances in the runoff are also contained on the farm. Of course, the farmer must be aware of potential problems with transporting pests or chemicals from one field to another, but it is preferable that a farmer deal with potential problems on the originating farm.

Nutrient and other farm chemical application in irrigation water is becoming a common practice. Nitrogen application in surface irrigation water is common in some areas. For surface irrigation with runoff, tailwater containment and reuse should be required when chemigating with materials that could be harmful to downstream farmers or ecosystems.

7.7.5   Conversion to Sprinkler Irrigation or
Microirrigation

Sprinkler irrigation and microirrigation produce little or no surface runoff. Converting from furrow irrigation to these irrigation methods will usually eliminate runoff and the associated water quality problems.

7.8   ECONOMIC CONSIDERATIONS IN REDUCING NONPOINT SOURCE POLLUTION

Which irrigation method is the best? The best irrigation method depends on one's perspective. For a farmer, the best irrigation method maximizes profits. For the environmentalist, the best method minimizes nonpoint source pollution by reducing drainage or surface runoff. An irrigation method that maximizes profit and minimizes nonpoint source pollution is the obvious choice. However, more efficient, less polluting irrigation methods are often more expensive, so some type of incentive may be needed to encourage improving irrigation efficiency where the existing irrigation system maximizes profit yet substantially contributes to nonpoint source pollution. Incentives for encouraging farmers to adopt measures to reduce nonpoint source pollution include improved farm-level economics as a result of improved irrigation water management, regulation, taxes, and subsidies.[85]

Several studies evaluated conditions that encourage the adoption of higher technology irrigation methods over surface irrigation.[86–88] They concluded that factors such as high water costs, marginal land quality, marginal weather conditions, and high cash value crops encourage the conversion from surface irrigation to sprinkler and drip irrigation. However, rotational or otherwise inconsistent surface water availability caused by irrigation district constraints tend to discourage conversions.

Irrigators of lower cash-value crops face a dilemma. Regardless of water costs, land quality, and so forth, adoption of sprinkler and drip irrigation may be uneconomical because of lower farm profits caused by increased irrigation costs.[89] An option for these irrigators is to provide subsidies to offset some of the costs of any improvements.

Table 7.4 lists yield and applied water from numerous field-scale comparisons of furrow and drip irrigation. Crops produced were cotton, tomato, and lettuce. These data show a broad range of results illustrating the difficulty in predicting the effect of converting from furrow to drip irrigation on crop yield and applied water. In some cases, drip irrigation produced higher yields with less water compared with furrow irrigation. Other cases showed similar yields but less applied water with drip irrigation. Still other cases showed similar yields but less applied water under furrow irrigation.

This range of responses reflect site-specific factors such as land quality (soil texture and variability), water quality, level of management of both irrigation methods, and factors such as nutrient levels and disease control. Some of these factors can be measured. Others cannot be measured with any reasonable degree of accuracy such as the uniformity of infiltrated water under surface irrigation as affected by soil variability and redistribution after an irrigation.

The economics of these various studies is also shown for cotton in Table 7.4. Production costs were not available for the lettuce and tomato crops. No tax or assessment on drainage was applied. As with crop yield, no trend clearly exists showing drip irrigation to be more profitable than furrow irrigation. For the lettuce and tomato crops, little difference in revenue would occur because of the similar yields between the furrow and drip systems. Less water was applied by the drip systems; however, the savings in water costs were insufficient to offset the cost of the drip systems.

TABLE 7.4
Comparison of Crop Yield, Applied Water, and Profit of Furrow and Drip Irrigation Systems.

Reference	Crop	Yield (Mg ha⁻¹)		Water (mm)		Profit ($ ha⁻¹)	
		Drip	Furrow	Drip	Furrow	Drip	Furrow
90	Cotton	1.582	1.419	556	612	504	990
91	Cotton	1.742	1.528	521	701	1,223	1,161
92, 93	Cotton	1.815	1.765	533	450	1,341	1,662
94							
1989	Cotton	1.714	1.214	584	749	1,149	610
1990	Cotton	1.449	1.431	610	500	689	1,079
1992	Cotton	1.758	1.533	599	500	1,218	1,252
1993	Cotton	1.645	1.454	455	643	1,087	1,060
95							
1990 (good)	Cotton	1.913	1.951	612	1,062	1,472	1,689
1991 (good)	Cotton	1.811	1.805	668	978	1,274	1,517
1990 (poor)	Cotton	1.838	1.622	581	1,166	1,358	1,131
1991 (poor)	Cotton	1.703	1.488	653	1,041	1,099	953
96							
1991	Lettuce	41.7	43.9	112	261	–	–
1992	Lettuce	40.9	41.0	229	335	–	–
97							
Variety 1	Tomato	114.7	112.7	686	970	–	–
Variety 2	Tomato	101.7	97.9	686	970	–	–

For those site-specific factors that result in higher profit and less applied water under drip irrigation compared with furrow irrigation, drip irrigation should be used instead of furrow irrigation. For conditions where profit is larger under surface irrigation, other incentives may be needed.

Several studies investigated the effect of various policy strategies for reducing drainage below the root zone in areas affected by saline, shallow groundwater. Dinar et al.[98] analyzed the policies of no regulation, direct fees on drainage discharges, and irrigation water pricing. The water pricing included flat fees on irrigation water use and a tiered pricing consisting of a base price until water use exceeded a chosen value, after which the water price increased. Results showed the unregulated policy to have a substantial cost to society for drainage water disposal. For the drainage fee policy, society net benefits were higher than for the unregulated case; however, net benefits decreased as the drainage fee increased. Most of the drainage reduction occurred for a fee increase from $300/ha-m to $794/ha-m. Further fee increases had a small effect of drainage reduction. Under this policy strategy, net benefits increased as the uniformity of the infiltrated water increased.

Under the policy of a flat fee on irrigation water, drainage disposal costs to growers were zero, but an additional charge was placed on the irrigation water. Results showed that substantial increases in irrigation water price were required to induce economically efficient water applications, which caused revenues to exceed drainage disposal costs. Under tiered water pricing, revenues were found to be less than the disposal costs.

Knapp et al.[99] investigated four policy strategies consisting of nonpoint incentives (tax on the estimated drainage discharges), nonpoint standards (specified maximum level of drainage discharge), management practice incentives (increased water price to induce source reduction), and management practice standards (specified level of irrigation water applications). For each policy strategy, the objective was to achieve economic efficiency. Results showed grower profits to decrease as either the price of irrigation water or drainage fees increased. Profits were significantly higher under the standard policies than the incentive policies. The incentive policies required substantially more transfer of information between regulators and growers compared with the standard policies.

Two other studies focused on drainage fees as a policy for inducing drainage reduction.[100,101] They assumed that reduced drainage from irrigation would occur because of changes in production practices (irrigation system, acreage allocation, and water applications) as drainage fees increased. Results showed the following:

1. Changes in irrigation systems occurred as drainage fees increased to maintain economic efficiency. The higher the cash value of a crop, the smaller the drainage fees at which a switch in irrigation system occurred.
2. Drainage fees could be increased up to a critical level with a minimal impact on net returns. Increases beyond that level greatly reduced net returns.

Although the studies reported different critical levels depending on the assumptions and methodology used for the economic models, they indicated that drainage reduction might be relatively easy in terms of costs and impact on net returns up to the critical level. Beyond that level, drainage reduction becomes relatively difficult.

7.9   OTHER CONSIDERATIONS

7.9.1   Physical Limitations

The feasibility of implementing these measures to reduce nonpoint source pollution can be affected by physical characteristics of a region. For example, an irrigation district might deliver water on a calendar basis only, and the duration of the delivery may also be a set time. This can greatly reduce the effectiveness of microirrigation, which is best suited for receiving water on demand. Irrigation districts using canals and ditches for water delivery may be unreceptive to a demand schedule because of the difficulty in regulating flows throughout the system in response to changes in water demand at the farm level.

Some measures may be inappropriate for a particular region. In the Salinas Valley of California, field sizes generally are small, ranging from about 4 to 16 ha in size. Opportunities to reduce furrow lengths are limited because the field lengths are already small. Linear-move and center-pivot sprinkler machines are not appropriate because of the small field sizes. The most appropriate measures include improved irrigation scheduling, converting to surge irrigation, and converting from furrow irrigation to drip irrigation for row crops.

In contrast, the west side of the San Joaquin Valley consists of large fields with lengths of 400 to 800 m and field sizes as large as 65 ha. Appropriate measures for this area include improving irrigation scheduling, reducing field lengths by one-half, converting to surge irrigation and drip irrigation, and converting to sprinkler irrigation including linear-move sprinkler machines. Center-pivot sprinkler machines are not appropriate because the relatively high application rates of these systems would create substantial surface runoff because of the low infiltration rates of the west side soils and natural slopes.

Along the east side of the San Joaquin Valley, an area experiencing nitrate and pesticide pollution of groundwater, tree crops are grown mostly on relatively small fields. The most appropriate measures are improving irrigation scheduling and converting to microirrigation or solid-set sprinklers. Reducing field lengths is impractical because of the tree crops and existing field lengths, and the tree crops and small field sizes restrict the use of linear-move and center-pivot machines.

7.9.2 SOIL SALINITY

In some arid areas, saline irrigation water may be used for irrigation, potentially causing adverse levels of soil salinity. Controlling soil salinity involves infiltrating an amount of water in excess of the soil moisture depletion to leach or transport salts below the root zone. This excess water is called the leaching fraction. The leaching fraction needed to prevent excessive soil salinity, called the leaching requirement, depends on the salinity of the irrigation water and the crop's tolerance to salt. Several sources for determining leaching requirements are available.[102,103]

As an example of the need for salinity control, irrigation water with salt concentrations ranging between 640 and 1,280 mg l^{-1} is used to irrigate vegetable crops in the coastal valleys of California, where nitrate pollution from nonpoint sources occurs. These crops are classified as salt-sensitive to moderately salt-sensitive and have a leaching requirement of about 14–26% (depending on crop type) to prevent yield reductions. This means that 14–26% of the infiltrated water must drain below the root zone for salinity control.

The need for salinity control means that a lower limit exists on the amount that drainage can be decreased for nonpoint source pollution reduction. Assuming that the leaching fraction is the amount of drainage occurring in the least watered part of the field, the total amount of drainage will be the leaching fraction plus the drainage from nonuniform infiltration. This suggests that, where moderately saline irrigation water is used to irrigate crops that are sensitive to moderately-sensitive to salt, substantial reductions in nonpoint source pollution of groundwater may not be possible.

7.9.3 SOLUTE TRAVEL TIMES

In areas where nonpoint source pollution of groundwater occurs, those involved in developing plans to address this problem must be aware of the time required for water

TABLE 7.5
Estimated Water Travel Rates in
Soil by Textural Class

Textural Class	Travel Rates (ft/year)
Sandy	6.2
Coarse-loamy	4.5
Fine-loamy	1.8
Fine	1.9

to travel through soil or aquifer material. Because of this travel time, many decades may pass before the impact of measures for reducing the pollution may be seen in well water. In some cases, the pollution may actually increase with time even if all leaching of contaminates below the root zone stops.

The time for nitrate to travel though an unsaturated soil depends on the water content of the soil, soil texture, soil profile depth, and drainage volume. Pratt et al.[104] estimated transit times of 12 to 47 years for nitrate to move through a 30-m unsaturated zone beneath citrus groves in southern California under a 40% leaching fraction. Table 7.5 lists estimated travel rates in the unsaturated zone for several soil textural classes as calculated by Ribble et al.[105]

7.10 SUMMARY

Nonpoint source pollution of groundwater from irrigated lands is caused by nonuniform water applications and excessive applications of water, both of which percolate water below the root zone. Pollution of surface water occurs from surface runoff of irrigated land, the result of water applications exceeding infiltration rates.

Measures for reducing nonpoint source pollution are discussed herein; however, planners and policymakers must be aware that some percolation below the root zone will occur for a properly managed irrigation system, regardless of the irrigation method. In some cases, percolation must occur to prevent adverse levels of soil salinity in the root zone. Some surface runoff may be necessary for furrow and border irrigation systems to irrigate the lower part of the field properly, but this runoff can be reduced or prevented for entering streams, rivers, and other channels.

All interested parties must be aware that even though the best of the measures are implemented, desired changes in groundwater quality may not be realized for many decades because of the slow movement of water in soil and aquifer material. Thus, unrealistic expectations and water quality requirements should be avoided. In some cases, treatment of municipal and domestic water supplies may be needed as an interim solution in coping with groundwater nonpoint source pollution.

REFERENCES

1. EPA, Agriculture and environment in California, Agriculture and Environment Fact Sheet, 1995.
2. EPA, Agriculture and environment in New Mexico, Agriculture and Environment Fact Sheet, 1995.
3. EPA, Agriculture and environment in Nebraska, Agriculture and Environment Fact Sheet, 1995.
4. EPA, Agriculture and environment in Texas, Agriculture and Environment Fact Sheet, 1995.
5. EPA, Agriculture and environment in Colorado, Agriculture and Environment Fact Sheet, 1995.
6. EPA, Agriculture and environment in Idaho, Agriculture and Environment Fact Sheet, 1995.
7. Coppock, R. and Meyer, R. D., Nitrate losses from irrigated cropland, Division of Agricultural Sciences, University of California, 1980, Leaflet 21136, 23 pg.
8. McNeal, B. L. and Pratt, P. F., Leaching of nitrate from soils, *Proceedings of National Conference on Management of Nitrogen in Irrigated Agriculture,* Sacramento, CA, 1978.
9. Schmidt, K. D. and Sherman, I., Effect of irrigation on groundwater quality in California, *ASCE Journal of Irrigation and Drainage,* 113(1):16, 1987.
10. Letey, J., Biggar, J. W., Stolzy, L. H., and Ayers, R. S., Effect of water management on nitrate leaching, *Proceedings of National Conference on Management of Nitrogen in Irrigated Agriculture,* Sacramento, CA. 1978.
11. Lund, L. J. and Wachtell, J. K., Denitrification potential of soils, in *Nitrate in Effluents from Irrigated Lands,* National Science Foundation, 1979, Chap. 18.
12. Pratt, P. F., Integration, discussion, and conclusions, in *Nitrate in Effluents from Irrigated Land,* Prepared for the National Science Foundation by the University of California, Riverside, pp. 719–758, 1979.
13. Monterey County Flood Control and Water Conservation District, Nitrates in Ground Water, Salinas Valley, CA., 1988.
14. Monterey County Water Resources Agency, Nitrates in Ground Water 1987–1993, Salinas Valley, CA., 1995.
15. Jury, W. A., Focht, D. D., and Farmer, W. J., Evaluation of pesticide groundwater pollution potential from standard indices of soil-chemical absorption and biodegradation, *Journal of Environmental Quality,* 16:42, 1987.
16. Spencer W. F., Volatization of pesticide residues, in *Fate of Pesticides in the Environment,* Publication 3320, Biggar, J. W. and Seiber, J. N., Agricultural Experiment Station, Division of Agriculture and Natural Resources, University of California, 1987.
17. Litwin, Y .J., Hantzsche, N. N., and George, N. A., Groundwater contamination by pesticides: a California assessment. Report prepared by RAMLIT Associates for the California State Water Resources Control Board, 208 pp., 1983.
18. Seiber, J. N., Solubility, partition coefficient, and bioconcentration factor, in *Fate of Pesticides in the Environment,* Publication 3320, Biggar, J. W. and Seiber, J. N., Agricultural Experiment Station, Division of Agriculture and Natural Resources, University of California, 1987.
19. Rao, P. S. C. and Hornsby, A. G., Behavior of pesticides in soils and water. Soil Science Fact Sheet, October 1989, Florida Cooperative Extension Service, Institute of Food and Agricultural Sciences, University of Florida, 1989.

20. Sacramento River Rice Water Quality Demonstration Project, Annual Progress Report, 1991.

21. Scardaci, S. C., Eke, A. U., Roberts, S. R., Hill, J. E., Hanson, B. R., and Anderson, W. J., Rice quality demonstration project: rice irrigation systems/pesticide monitoring study, 1991 Progress Report, 1991.

22. California State Department of Water Resources, San Joaquin Valley Drainage Monitoring Program, 1992, 53 pg., 1995.

23. San Joaquin Valley Drainage Program, 1990.

24. Koluvek, P. K., Tanji, K. K., and Trout, T. J., Overview of water erosion from irrigation, *ASCE Journal of Irrigation and Drainage Engineering* 119(6):929, 1993.

25. Brown, M. J., Carter, D. L., and Bondurant, J. A., Sediment in irrigation and drainage water and sediment inputs and outputs for two large tracts in southern Idaho, *Journal of Environmental Quality* 3(4):347, 1974.

26. Brockway, C. E. and Robison, C. W., Middle Snake River water quality study, Phase I, Final report, Idaho Water Resources Research Inst., Univ. of Idaho, Moscow, ID., 1992.

27. Berg, R.D. and Carter, C. L., Furrow erosion and sediment losses on irrigated cropland, *Journal of Soil and Water Conservation,* 35(6):267, 1980.

28. Tannahill, J. L., Crows Landing 319 demonstration project, West Stanislaus Resource Conservation Dist., Patterson, CA., 1995.

29. Carlile, B. L., Sediment control in Yakima Valley, *Proc. Nat. Conf. on Managing Irrigated Agri. To Imp. Water Quality,* Colo. St. Univ., Ft. Collins, CO. 77–82, 1972.

30. Idaho Dept. of Health and Welfare, Middle Snake River nutrient management plan, IDHW, Dept. Of Environmental Quality, SCIRO, Twin Falls, ID., 1995.

31. The Resources Agency, San Joaquin River Management Plan, California Dept. of Water Resources, Sacramento, CA., 1995.

32. Trout, T. J. and Neibling, W. H., Erosion and sedimentation processes on irrigated fields, *ASCE Journal of Irrigation and Drainage Engineering,* 199(6):947, 1993.

33. Trout, T. J., Furrow irrigation erosion and sedimentation—on field distribution. *Transactions ASAE,* Vol. 39(5):1717, 1996.

34. USDA-NRCS, Rock Creek Rural Clean Water Program—Ten-year Report, Twin Falls County USDA-NRCS office, Twin Falls, ID., 327 pp., 1991.

35. Kruse, E. G., Anderson, C. L., Bishop, A. A., Hotes, F., Keller, J., Merriam, J., Miller, A., Pinney, J., Jr., Smerdon, E., Winger, R. J., Jr. , Describing irrigation efficiency and uniformity, *ASCE Journal of the Irrigation and Drainage Division,* 104(IRl):35, 1978.

36. Warrick, A. W., Interrelationships of irrigation uniformity terms, *ASCE Journal of Irrigation and Drainage Engineering,* 109(3):317, 1983.

37. Burt, C. M., Walker, R. E., and Styles, S. W., Irrigation system evaluation manual, Department of Agricultural Engineering, California Polytechnic State University, San Luis Obispo, CA., 1985.

38. Hanson, B. and Schwankl, L., *Surface Irrigation,* Publication 3379, University of California Division of Natural and Agricultural Resources, 1995, 105.

39. Childs, J. L., Wallender, W. W., and Hopmans, J. W., Spatial and seasonal variation of furrow infiltration, *ASCE Journal of Irrigation and Drainage Engineering,* 119(1):74, 1993.

40. Tarboton, K. C., and Wallender, W. W., Field-wide furrow infiltration variability. *Transactions of the American Society of Agriculture Engineers,* 32(3):913, 1989.

41. Hanson, B. R., Practical potential irrigation efficiencies, in *Proceedings of the First International Conference on Water Resources Engineering,* San Antonio, TX. 1995.

42. EPA, Guidance specifying management measures for sources of nonpoint pollution in coastal waters, Department of Interior, 828 pp., 1993.

43. Grimes, D. W., Wiley, P. L., and Sheesley, W. R., Alfalfa yield and plant water relations with variable irrigation, *Crop Science,* 32:1381, 1992.

44. Pang, X. P., Letey, J., and Wu, L., Validity of the CERES-Maize model under semi-arid conditions, *Soil Science Society of America Journal,* 61:254, 1997a.

45. Pang, X. P., Letey, J., and Wu, L., Irrigation quantity and uniformity and nitrogen application effects on crop yield and nitrogen leaching, *Soil Science Society of America Journal,* 61:257, 1997b.

46. Tanji, K. K., Helfand, G., and Larson, D. M., BMP Assessment Model for Agricultural NPS Pollution, Final Report to the California State Water Resources Control Board, 1994.

47. Jensen, M. E., Burman, R. D., and Allen, R. D., *Evapotranspiration and Irrigation Water Requirements,* ASCE Manuals and Reports of Engineering No. 70, American Society of Civil Engineers, NY., 1990.

48. Allen, R. G., Pereira, L. S., Raes, D., and Smith, M., Crop Evapotranspiration: guidelines for computing crop water requirements, FAO Irrigation and Drainage Paper 56, Food and Agriculture Organization of the United Nations, Rome, 1998.

49. Hanson, B. R., and Peters, D., unpublished data, 1999.

50. Hanson, B. R., and Bowers, W., An analysis of mobile laboratory irrigation system evaluation data, Final report to the State Department of Water Resources (Division of Planning), 1994.

51. Goldhamer, D. A., Sibbett, G. S., Phene, R. C., and Katayama, D. G., Early irrigation cutoff has little effect on French prune production, *California Agriculture,* 48(4):13, 1994.

52. Lampinen, B. D., Shackel, K. A., Southwick, S. M., Olson, B., and Yeager, J. T., Sensitivity of yield and fruit quality of French prune to water deprivation and different fruit growth stages, *Journal of the American Society of Horticulture,* 120(2):139, 1995.

53. Goldlhamer, D. A., DeJong, T., Johnson, R. S., Girona, J., Mata, M., Handley, D., and Sanchez, M. R., Controlled deficit irrigation of early and late maturing peaches, Final Report to the California Department of Water Resources (Office of Water Conservation), Sacramento, Undated.

54. Goldhamer, D. A. and Beede, R. H., Regulated deficit irrigation (RDI) for pistachio orchards, *Purely Pistachio,* Undated.

55. Goldhamer, D. A., Regulated deficit irrigation for California olives, 1994 Crop Year Report, 1994.

56. Fulton, A. and Beede, R. H., Almond, Publication by University of California Cooperative Extension, Kings County, Undated.

57. Fulton, A. and Beede, R. H., Walnut, Publication by University of California Cooperative Extension, Kings County, Undated.

58. Howell, T. A., Ziska, L. H., McCormick, R. L., Vurtch, L. M., and Fischer, B. B., Response of sugarbeets to irrigation frequency and cutoff on a clay loam soil, *Irrigation Science,* 8:1, 1987.

59. Grimes, D. W. and Dickens, W. L., Dating termination of cotton irrigation from soil water-retention characteristics, *Agronomy Journal,* 66:403, 1974.

60. Hartz, T. K., Water management for drip irrigated melons, in *San Joaquin Valley Vegetable Crops Report,* 1:3, 1995.

61. May, D., unpublished data, 1998.

62. Hanson, B. R., Drainage reduction potential of furrow irrigation, *California Agriculture,* 43(1):6, 1989.

63. Howe, O. W., and Heerman, D. F., Efficient border irrigation design and operation, *Transactions of the American Society of Agricultural Engineers,* 13(1):126, 1970.

64. Schwankl, L., Improving irrigation efficiency of border irrigation, in *Proceedings of the 1990 California Plant and Soil Conference,* Fresno, CA., 1990.

65. Hanson, B., Schwankl, L., Bendixen, W., and Schulbach, K., Surge irrigation, University of California Division of Natural and Agricultural Resources Publication 3380, University of California, Davis, CA., 1994.

66. Purkey, D. R. and Wallender, W. W., Surge flow infiltration variability, *Transactions of the American Society of Agriculture Engineers,* 32(3):894, 1989.

67. Bishop, A. A., Walker, W. R., Allen, N. L., and Poole, G. J., Furrow advance rates under surge flow systems, *Journal of Irrigation and Drainage Division, American Society of Civil Engineers,* 107(IR3):257, 1981.

68. Iyuno, F. T., Podmore, T. H., and Duke, H. R., Infiltration under surge irrigation, *Transactions of the American Society of Agricultural Engineers,* 28(2):517, 1985.

69. Schwankl, L. J., Hanson, B. R., and Panoras, A., Furrow torpedoes improve irrigation water advance, *California Agriculture,* 46(6):15, 1992.

70. Keller, J. and Bliesner, R. D., Sprinkle and trickle irrigation, Van Nostrand and Reinhold, N.Y., 1990.

71. Hanson, B., Schwankl, L., Grattan, S., and Prichard, T., Drip irrigation for row crops, University of California Division of Natural and Agricultural Resources Publication 3376, University of California, Davis, CA., 1997.

72. Schwankl, L., Hanson, B., and Prichard, T., Micro-irrigation of trees and vines, University of California Division of Natural and Agricultural Resources Publication 3378, University of California, Davis, CA., 1995.

73. Carter, D. L. and Berg, R. D., Crop sequences and conservation tillage to control irrigation furrow erosion and increase farmer income, *Journal of Soil and Water Conservation,* 46(2):139, 1991.

74. Aarstad, J. S., and Miller, D. E., Effects of small amounts of residue on furrow erosion, *Soil Science Society of American Journal,* 45(1), 1981.

75. Brown, M. J., and Kemper, W. D., Using straw in steep furrows to reduce soil erosion and increase dry bean yields, *Journal of Soil and Water Conservation* 42(3):187, 1987.

76. Miller, D. E., Aarstad, J. S., and Evans, R. G., Control of furrow erosion with crop residues and surge flow irrigation, *Soil Science Society of American Journal,* 51:421, 1987.

77. Lentz, R. D., Sojka, R. E., and Carter, D. L., Furrow irrigation water-quality effects on soil loss and infiltration, *Soil Science Society of American Journal,* 60(1):238-245, 1996.

78. Lentz, R. D., Shainberg, R. I., Sojka, R. E., and Carter, D. L., Preventing irrigation furrow erosion with small application of polymers, *Soil Science Society of American Journal,* 56(6):1926, 1992.

79. Lentz, R. D. and Sojka, R. E., Field results using polyacrylamide to manage furrow erosion and infiltration, *Soil Science,* 158(4):274, 1994.

80. Carter, D. L., Controlling erosion and sediment loss on furrow-irrigated land, in *Soil Erosion and Conservation,* S. A. ElSwaity et al. (Ed.), Soil Conservation Soc. Am., Ankeny, IA. 1985.

81. Carter, D. L., Brockway, C. E., and Tanji, K. K., Controlling erosion and sediment loss in irrigated agriculture, *J. Irrig. Drain. Engr.* 199(6):975, 1993.

82. Agassi, M., Letey, J., Farmer, W. J., and Clark, P., Soil erosion contribution to pesticide transport by furrow irrigation, *Journal of Environmental Quality,* 24(5):892, 1995.

83. Smith, D. M., Flow through wetlands: a way to manage salts and trace elements, ASAE, ASAE Paper No. 94-2111, St. Joseph, Mich., 1994.

84. ASAE, Engineering Practice #EP-408.2: Irrigation Runoff Reuse, in *Standards,* ASAE, St. Joseph, MI., 1996.

85. Contant, C. K., Duffy, M. D., and Holub, M. A., Determining tradeoffs between water quality and profit ability in agricultural production: Implications for nonpoint source pollution policy, *Water Science Technology,* 28(3–5):27, 1993.

86. Caswell, M., and Zilberman, D., The choices of irrigation technologies in California, *American Journal of Agricultural Economics,* 67:224, 1985.

87. Caswell, M. and Zilberman, D., The effects of well depth and land quality on choice of irrigation technology, *American Journal of Agricultural Economics,* 68:798, 1986.

88. Dinar, A. and Zilberman, D., The economics of resource conservation, pollution-reduction technology selection—the case of irrigation water, *Resources and Energy,* 13:323, 1991.

89. Wichelns, D., Houston, L., and Cone, D., Economic analysis of sprinkler and siphon tube irrigation systems with implications for public policies, *Agricultural Water Management,* 32:259, 1997.

90. Fulton, A. W., Oster, J. D., Hanson, B. R., Phene, C. J., and Goldhamer, D. A., Reducing drainwater: furrow vs. subsurface drip irrigation, *California Agriculture,* 45(2):4, 1991.

91. Mateos, L., Berengena, J., Organg, F., Diz, J., and Fereres, E., A comparison between drip and furrow irrigation in cotton at two levels of water supply, *Agricultural Water Management,* 19:313, 1991.

92. Hodgeson, A. S., Constable, G. A., Duddy, G. R., and Daniells, I. G., A comparison of drip and furrow irrigated cotton on a cracking clay soil 2, Water use efficiency, waterlogging, root distribution and soil structure, *Irrigation Science,* 11:143, 1990.

93. Constable, G. A., and Hodgeson, A. S., A comparison of drip and furrow irrigated cotton on a cracking clay soil 3, Yield and quality of four cultivars, *Irrigation Science,* 11:149, 1990.

94. Boyle Engineering Corporation, Demonstration of Emerging IrrigationTechnologies, Final Report to the State of California Department of Water Resources, 1994.

95. Detar, W. R., Phene, C. J., and Clark, D. A., Full cotton production with 24 inches of water, ASAE Paper 92-2607, Presented at the 1992 International Winter Meeting, Nashville, TN, 1992.

96. Hanson, B. R., Schwankl, L. J., Schulbach, K. F., and Pettygrove, G . S., A comparison of furrow, surface drip, and subsurface drip irrigation on lettuce yield and applied water, *Agricultural Water Management,* 33:139, 1997.

97. Fulton, A. E., Subsurface drip irrigation: eastern San Joaquin Valley, Annual Report to the U.S. Salinity/Drainage Program and Prossier Trust, 1995.

98. Dinar, A., Knapp, K. C., and Letey, J., Irrigation water pricing policies to reduce and finance subsurface drainage disposal, *Agricultural Water Management,* 16:155, 1989.

99. Knapp, K. C., Dinar, A. S., and Nash, P., Economic policies for regulating agricultural drainage water, *Water Resources Bulletin, American Water Resources Association,* 26(2):289, 1990.

100. Posnikoff, J. F. and Knapp, K. C., Farm-level management of deep percolation emissions in irrigated agricultural, *Journal of the American Water Resources Association,* 33(2):375, 1997.

101. Knapp, K. C., Irrigation management and investment under saline, limited drainage conditions 3, policy analysis and extensions, *Water Resources Research,* 28(12):3099, 1997.

102. Hanson, B. R., Grattan, S. R., and Fulton, A., Agricultural Salinity and Drainage. University of California Division of Natural and Agricultural Resources Publication 3378, University of California, Davis, CA., 1995.
103. Agricultural Salinity Assessment and Management, American Society of Civil Engineers Manuals and Reports on Engineering Practice No. 71, K. K. Tanji, (Ed.).
104. Pratt, P. F., Jones, W. W., and Hunsaker, V. E., Nitrate in deep soil profiles in relation to fertilizer rates and leaching volume, *Journal of Environmental Quality,* 1:97, 1972.
105. Ribble, J. M., Pratt, P. F., Lund, L. J., and Holtzclaw, K. M., Nitrates in the unsaturated zone of freely drained soils, in *Nitrate in Effluents from Irrigated Land,* Prepared for the National Science Foundation by the University of California, Riverside, 1979, pg. 297.

8 Agricultural Drainage and Water Quality

William F. Ritter and Adel Shirmohammadi

CONTENTS

1-56670-222-4/01/$0.00+$.50
© 2001 by CRC Press LLC

8.1   INTRODUCTION

Water management for agricultural purposes can be traced to Mesopotamia about 9000 years ago.[1] Herodotus, a Greek historian of the fifth century B.C., wrote about a drainage works near the city of Memphis in Egypt.

Drainage has been part of American agriculture since colonial times. Without drainage, it is hard to imagine the U.S. Midwest as we know it in the 20th century, the epitome of agricultural production. Much of Ohio, Indiana, Illinois, and Iowa originally was swamp, or at least too wet to farm. Without drainage, irrigation development in the western United States would have failed because of waterlogging and salinity.

In the 1960s and 1970s, drainage was considered an honorable and viable soil and water conservation practice. Drainage technology developed rapidly during this era. In the 1990s, drainage is greeted with angry response in many quarters. Because of drainage, better than half the original wetlands in this country no longer excist. In addition, drainage has reduced the habitat for birds and wildlife and has had detrimental effects on water quality[2]. Today the design and operation of drainage systems must satisfy both agricultural and environmental objectives.

8.2   HISTORY OF DRAINAGE IN THE UNITED STATES

Early settlers brought European drainage methods with them to North America. These methods included small open ditches to drain wet spots in fields and to clean out small streams. In New York and New England, early settlers used subsurface drainage in addition to open ditches. Material used for buried drains prior to the use of clay-fired tile pipes included poles, logs, brush, lumber of all sorts, stones laid in various patterns, bricks, and straw.

In 1754, the Colony of South Carolina passed an act for draining the Cacaw Swamp.[3] The Dismal Swamp area of Virginia and North Carolina was surveyed by George Washington for reclamation in 1763, and in 1778 the Dismal Swamp Canal Company was chartered. A drainage outlet for the City of New Orleans was constructed around 1794.[4]

The first known colony-wide drainage law was enacted in New Jersey on September 26, 1772. Early drainage works were constructed in Delaware, Maryland, New Jersey, Massachusetts, South Carolina, and Georgia under the authority of colonial and state laws. The first organized drainage project in Maryland was authorized by the legislature for draining the Long Marsh in Queen Anne and Caroline Counties.[5] Similarly, legislation authorizing drainage projects in Delaware dates back to 1793.[6]

Drainage in the midwestern U.S. began after 1850, when the Swamp Land Act of 1849 and 1850 released large amounts of swamp and wetland still owned by the

Federal government. These lands were released for private development, with the funds from their sale used to build drains and levees. The Reclamation Act of 1902 established the Bureau of Agricultural Engineering within the U.S. Department of Agriculture, which was responsible for the design and construction of many of the major drainage ditches that were installed to create surface water outlets. Drainage districts began to be organized in the early 1900s. In its natural state, much of the fertile land in northwestern Ohio, northern Indiana, northcentral Illinois, northcentral Iowa, and southeastern Missouri was either swamp or frequently too wet to farm before drainage was installed. Drainage also permitted large areas in western Minnesota, the gulf plains of Texas, northeastern Arkansas, and the delta area of Mississippi and Louisiana to be cultivated.[7]

Drainage problems developed as a consequence of irrigation developed in the arid west. In the San Joaquin Valley of California, the Modesto Irrigation District drained more than 18,000 ha. In the Imperial Valley of California, over 81,000 ha of cropland had drainage problems by 1919. Today over 80% of the cropland in the Imperial Valley is drained. Bureau of Reclamation irrigation projects such as the Columbia Basin in Washington, the Grand Valley (Nebraska), Big Horn Basin (Montana and Wyoming), Oahe (South Dakota), Weber Basin (Utah), Garrison (North Dakota), and Big Thompson (Colorado) have required drainage as a consequence of irrigation.[3]

8.3   MATERIALS AND METHODS FOR SUBSURFACE DRAINAGE

The first use of clay tile for farm drainage is attributed to John Johnston, who lived in the Finger Lakes region of New York. Johnston imported patterns for horseshoe-type drain tile from Scotland in December 1835. Tiles were made from these patterns at the B.F. Whartenby pottery at Waterloo, N.Y. in 1835. They were made entirely by hand. A crude molding machine was installed in 1838 in the Whartenby factory that made the process cheaper and faster.[8] Sometime after 1851, John Dixon developed a much improved machine for making horseshoe tile. In the 1870s, another new method of tilemaking that used a rectangular slab of clay instead of a conventional mold was introduced.[8]

The first tilemaking machine, the "Scraggs," was brought to America in 1848 from England. The machine operated on the extrusion process.[8] Many locally manufactured tilemaking machines were patterned after the Scraggs machine; most of the early manufacturers were located in New York State.

Weaver[8] also discussed the early use of concrete tile for subsurface drainage. In 1862, David Ogden developed a machine for making drain tile from cement and sand. Until 1900, concrete drain tile was used primarily where good clay was not available.

In the 1940s, bituminized fiber pipe was used in the eastern States and early-generation plastic tubes were also introduced. By 1967, corrugated plastic tubing was manufactured commercially in the United States from polyvinyl and polyethylene resins. The agricultural market tubing was very light and flexible and greatly reduced handling and shipping costs. Tile alignment problems were avoided.[3] By 1983, 95%

of all agricultural subsurface drains installed annually in the U.S. and more than 80 percent in Canada consisted of corrugated plastic tubing.[9]

Subsurface drains were first installed in hand-dug trenches, followed by a combination of plowing and hand digging. The first trencher introduced in 1855 was the Pratt Ditch Digger revolving-wheel type that was horsedrawn.[3] The Hickok and the Rennie elevator ditchers were patented in 1869. Another early machine was the Johnston Tile Ditcher made in Ottawa, Illinois. All of the early machines required more than one pass over the trench to excavate it to the required depth. Singlepass machines powered by horses came next and included the Blickensderfer Tile Ditching Machine, the Heath's Ditching Machine, the Paul's Ditching Machine and the Fowler Drain Plow. In the early 1880s, steam-powered wheel trenches were introduced. The Bucheye steampowered trencher was introduced in 1882. In 1908, steam power was replaced with a gasoline engine on the Buckeye, which was the forerunner of today's high-speed trenchers and laser-controlled drain plows.

8.4 TYPES OF DRAINAGE SYSTEMS

8.4.1 SURFACE DRAINAGE

Surface drainage is used to remove water that collects on the land surface. Surface drainage is used primarily on flat or undulating land where slow infiltration, slow permeability, restricting layers in the soil profile, or shallowness of soil over rock or deep clays. A surface drainage system usually consists of an outlet channel, lateral ditches, and field ditches. Lateral ditches carry the water received from field ditches or from the field surface to the outlet channel.[10]

Surface drainage systems include land smoothing or grading, and field ditches. Land grading is the shaping of the land surface with scrapers and land planes to planned surface grades. Land smoothing removes small depressions and irregularities in the land surface.

Field ditches may be either random or parallel. The random ditch pattern is used in fields having depressional areas that are too large to be eliminated by land smoothing. Field ditches connect the low spots and remove excess water from them. When the topography is flat and regular, a parallel ditch pattern is used. The row direction should be perpendicular to the ditch. Drains do not have to be equally spaced and water may flow in only one direction. The drain should have a minimum depth of 0.23 m and have a minimum crosssectional area of 0.50 m^2. The sideslopes of the ditches should be 8:1 or flatter to allow machinery to cross.[11]

8.4.2 CONVENTIONAL SUBSURFACE DRAINAGE

Subsurface drains consist of underground pipe systems to collect excess water from the root zone and lower the water-table. Subsurface drainage falls into two classes: relief and interception drainage.[10] Relief drainage is used to lower a high water-table that is generally flat or of very low gradient. Interception drainage is to intercept, reduce the flow, and lower the flowline of the water in the problem area. Relief drains normally consist of a system of parallel collection drains connected to a main drain located on the low side of a field or along a low waterway in the field. The main drain transports the collected water to the outlet. An interceptor drain often consists of a single drain which intercepts lateral flow of groundwater caused by canal seepage, reservoir seepage, or levee-protected areas.

8.4.3   WATER-TABLE MANAGEMENT

The trend in the humid areas of the United States is to develop a total water management system. Water-table management strategies can be grouped into three types: subsurface drainage, controlled drainage, controlled drainage–subsurface irrigation.[12] Subsurface drainage alone lowers the water table during wet periods and is governed by drainage system depth. Controlled drainage is achieved by placing a control structure, such as a flashboard riser in the outlet ditch or a subsurface drain outlet, to control the rate of subsurface drainage. Controlled drainage-subirrigation is similar to the controlled drainage system, except that supplemental water is pumped into the system to maintain the water table at a current level during drought periods. Drainage is provided during wet periods by allowing excess water to flow over the control structure, which may be adjusted in elevation depending upon the rainfall (Figure 8.1). The practice has been used for years in peat and muck soils with high permeability and an impervious layer below the drains or with a naturally high water table.[13]

The system can be applied in both the field and watershed scale using various water control structures and operational procedures.[12,14] Water-table management offers more possibilities for flood control, improved water conservation, and improved water quality than conventional drainage systems[15]. The greatest potential for water-table management systems is on relatively large flat land areas where high water tables persist for long periods during the year. There have been a number of papers in recent years dealing with the design, economics, and environmental impacts of controlled drainage systems.[16,17]

8.5   WATER TABLE MANAGEMENT DESIGN

Shirmohammadi et al.[12] outlined five tasks that must be performed to design a successful and efficient water-table management system. These tasks include preliminary evaluation and feasibility of the site, detailed field investigation, design computations, system layout and installation, and operation and management. Each of these tasks is discussed by Evans and Skaggs[18] in detail. ASAE[19] has also developed a design, installation, and operation standard for water table management systems.

8.5.1   PRELIMINARY EVALUATION AND FEASIBILITY OF SITE

Six site characteristics should be considered for successful performance of water-table management systems:

8.5.1.1   Drainage Characteristics

The site must require improved subsurface drainage to remove excess water that otherwise would restrict farm operations and crop growth. Soils classified as

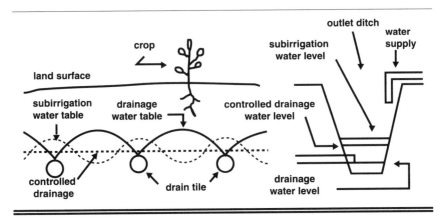

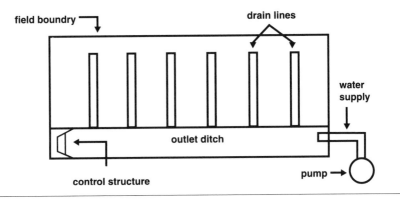

FIGURE 8.1   Schematic of a water-table management system.

"somewhat poorly drained," "poorly drained," and "very poorly drained" are prime candidates for water-table management. Natural Resources Conservation Service soil survey manuals provide soil maps and classifications for each state within the Atlantic Coastal Plain.

8.5.1.2 Topography

Surface slopes should not exceed 1% for the system to be economically feasible. As the slope increases, more control structures are required to maintain a uniform water table.

8.5.1.3 Barrier

A shallow natural water table or shallow impermeable layer within 1.8 to 6.1 m of the soil surface should exist for controlled drainage or controlled drainage–subirrigation systems to perform satisfactorily. The deeper the barrier, the larger the volume of water required to fill the soil profile and raise the water table during irrigation.

8.5.1.4 Hydraulic Conductivity

Moderate to high soil hydraulic conductivity values (about $K_s > 1.9$ cm/hr) are required for efficient system performance and timely water table response, especially in the subirrigation mode. Soils with low hydraulic conductivity values require closer tile spacings, which will increase system cost and reduce its cost effectiveness. Hydraulic conductivity values reported in the SCS Soils 5 form for individual series may be sufficient for preliminary planning. A detailed measured hydraulic conductivity value is required to compute the system design, however.

8.5.1.5 Drainage Outlet

A good gravity or pumped drainage outlet is needed to provide adequate flow capacity for expected peak discharges. For gravity flow systems, the drainage outlet should be at least 1.2 m below the average land surface. A sump equipped with an appropriate pump can be constructed to collect the surface and subsurface drainage flow where an adequate natural drainage outlet is not present.

8.5.1.6 Water Supply

An adequate water supply must be available for the subirrigation mode. Location, quantity, and quality of the water must be taken into consideration during the planning stage.

8.5.2 DETAILED FIELD INVESTIGATIONS

For efficient design, soil type and arrangement of soil horizons, soil hydraulic properties, crops, water supply, and various climatological and topographical parameters must be considered. Soil type, arrangement of soil horizons, soil hydraulic properties, and hydraulic conductivity (lateral conductivity values and soil water characteristic data) determine drain line depth and spacing. The crop and its rooting depth may also influence system design.

An accurate topographic map is required to evaluate the slope of the land and its adequacy for any type of water-table management system. A general guideline is to install the drain lines perpendicular to the slope, but this guideline can be modified, depending upon site conditions.

Climatological data, such as rainfall, temperature, and solar radiation, are important parameters. Knowledge of climatological data can provide a good understanding of crop water use and periods of peak water requirement. Crop water requirement information is required for a controlled drainage–subirrigation system to determine the external water supply size, pumping plant size, and overall management strategy. Design criteria also should be evaluated for each site based on economic and environmental quality considerations.

8.5.3 DESIGN COMPUTATIONS

Data collected from the field investigation enables the design engineer to compute proper drain depth, drain spacing, drain grades, number and size of control structures needed to maintain a uniform water table, and a proper pump capacity required for the water supply and the drainage outlet if a sump is used at the outlet. Soil horizon arrangement data, topography, and crop-rooting characteristics will help to determine the proper drain depth, which generally ranges from 1 to 2 m, depending upon site conditions.[18] Soil hydraulic conductivity values and depth to the impermeable layer will enable the engineer to evaluate the drain spacings, using the Hooghoudt's steady state drainage rate method for drainage conditions. However, other procedures must be used to evaluate the drain spacings if subirrigation is a part of the overall plan.[18]

DRAINMOD, a water table management model for shallow water table conditions, is probably the most comprehensive model available for design of subsurface drainage, controlled drainage, and controlled drainage–subirrigation systems, provided the required input data are available.[20]

8.5.4 SYSTEM LAYOUT AND INSTALLATION

Using the information obtained during the first three steps, the design engineer needs to prepare a map showing the field, location of laterals and mains, and location and number of control structures. Appropriate grades for drains must also be specified using the design standards and site information. The type of water table management system should also be specified.

A contour map prepared during the second phase of planning must be used to identify the location and grade of the drain lines and the control structures. Locations of the control structures are selected so that they provide the most uniform water table elevations possible. Water table fluctuations of 0.30 to 0.45 m and 0.15 to 0.20 m may be tolerated for grain crops and shallow-rooted vegetable crops.[18]

Once the system layout is completed on a well prepared map, the size, spacing, and grade of drain lines and the size and capacity of the control structures are specified. A contractor then can initiate the installation according to specifications. Autolevel, laser-controlled plows and trenchers that provide accurate and fast installation of the system are currently available. However, caution is necessary regarding the hand installation of laterals and main to the drain in a closed system to ensure that none will be left unattached.

8.5.5 OPERATIONS AND SYSTEM MANAGEMENT

This task is one of the most important aspects of the overall effort; traditionally, it has been performed by the producer and most usually on a trial-and-error basis. Selecting the proper weir elevation, maintenance of the system, and timing of the subirrigation and drainage phases are part of the operation and management of the system. On large-scale fields (40.5 ha), there may be high spots and depressions that were not considered in designing the depth and spacings of the drain lines because of the economics of the system. During the operation mode, however, a producer may adjust the control structure setting so that neither drought in high spots nor excess water in depressions will harm the crop. Similarly, knowing when to reverse from the drainage mode to the subirrigation mode in a controlled drainage-subirrigation system requires experience as well as soil moisture measurement, using such devices as tensiometers. Tensiometers indicate the soil-water potential from which one may judge the timing of subirrigation. Weather forecasts can be used to evaluate the time for lowering the water table to provide proper storage for incoming rain.

Manual adjustment of the control structure setting is laborious; consequently, it is often not adjusted because of the farmer's conflicting schedule. Research developments have enabled linking weather forecast data to the control structures through computers, modems, and telephone lines.[21] In the future this type of system will probably be used in commercial systems.

8.6 SOIL AND CROP MANAGEMENT ASPECTS OF WATER-TABLE MANAGEMENT

The Southeast and Mid-Atlantic Coastal Plain have variable rainfall during the growing season. This, combined with sandy soils with low water holding capacity, can cause drought conditions.[22] These conditions are worse in soils with shallow root zones caused by subsurface hardpans that could be controlled by deep chisel plowing. Water-table management by controlled drainage–subirrigation can ameliorate variability of water supply.[22, 23]

Intense rains in some regions are possible during the growing season.[22] As a result of such rainfall, the shallow water tables that result from controlled drainage–subirrigation leave fields vulnerable to flooding. To prevent this, systems have been designed to link controlled drainage–subirrigation to weather predictions. Fouss and Cooper[21] stopped subirrigation when a 55% or greater rainfall probability is predicted. They also recommended free drainage of the soil in advance of a predicted storm. If free drainage is used, precautions must be taken not to drop the water table so much that reestablishment of the desired level would be difficult.[23,24]

For controlled drainage–subirrigation systems to be successful, the depth of the water table must be low enough to prevent aeration problems and high enough to permit capillary rise into the root zone for plant uptake. The capillary water contribution to root uptake is negligible for water table depths 76 cm below the bottom of the root zone in sandy soils or 92 cm in clay soils.[23] Doty[26] found the best water-table depth for corn on sands or sandy loam in the Coastal Plain was 76–89 cm. The recommended depth of the water table is 92–153 cm for clay soils.[27] The crop type and climate in addition to soils determine where, within these ranges, the water table should be set.

If the ratio of deep percolation to infiltration is greater than 1:10, a water table will not perch adequately and the site is unsuitable for controlled drainage–subirrigation.[27] Other soil factors that affect water-table management are poor surface drainage, organic soils that subside, and soil strength. Poor surface drainage may affect trafficable conditions and soil aeration.[22] Shih et al.[28] recommended different water table depths for different crops and different times of the year on organic soils to provide irrigation and reduce subsidence. Deep tillage combined with controlled water table depth can eliminate hard-pan problems that limit root growth depth.[29]

8.7 WATER QUALITY IMPACTS

8.7.1 HYDROLOGY

8.7.1.1 Conventional Drainage

Land development using conventional drainage generally increases total annual outflows from fields and peak outflow rates. Studies in North Carolina have shown that annual outflows increased 5% for surface drainage and 20% for subsurface drainage[30, 31] when compared with natural undrained conditions. Peak flow rates typically increased up to four times with surface drainage compared with natural conditions. Subsurface drainage peak flow rates doubled compared with natural systems. Peak outflow rates varied greatly depending upon storm intensity, antecedent moisture, and drainage intensity. The natural areas used for comparison were unmanaged forested areas without drainage improvement, flat (0.01 slope or less), and broad (exceeding km^2).

Bengston et al.[32] measured surface runoff and outflow from four plots in Louisiana on Commerce clay loam soil from 1982 to 1991. Two of the plots had both surface and subsurface drainage and two of the plots had surface drainage only. The average annual surface drainage was 402 mm from the surface and subsurface-drained plots and 614 mm from plots only with surface drainage. The annual runoff from surface and subsurface-drained versus only surface drained plots ranged from a high of 775 and 1085 mm in 1989 to a low of 150 and 208 mm in 1984, respectively. Subsurface drainage reduced surface runoff by an average of 35%, but the total drainage flow from surface and subsurface drain plots (i.e., runoff plus subsurface drain outflow) was about 35% more than for the plots with only surface drainage.

8.7.1.2 Controlled Drainage

Evans et al.[14] reported controlled drainage may reduce total outflow by approximately 30% when managed all year compared with conventional drainage. The effect of controlled drainage on outflows varies with soil type, rainfall, type of drainage system, and management intensity. In wet years, controlled drainage may have little or no effect on total outflow. During dry years, flow may be eliminated in some cases. Much of the outflow reductions occurs during the winter and early spring. If controlled drainage is used only during the growing season, typical outflows are lower by less than 15% compared with conventional drainage.

8.7.2 NUTRIENTS

8.7.2.1 Conventional Drainage

The earliest research on tile drainage water quality was reported by Willrich.[33] Willrich collected water samples twice a month from 10 subsurface drainage outlets draining 2.4–148 ha in Iowa. The median values for chemical properties of the drainage water ranged as follows: total N = 12 to 27 mg/L, ortho P = 0.1 to 0.3 mg/L; K = 0.2 to 0.8 mg/L; hardness = 350 to 440 mg/L as $CaCO_3$, alkalinity = 260 to 330 mg/L, and pH from 7.4 to 7.8. The N was mostly in the NO_3 form.

Bolton et al.[34] were the first to study the effect of agricultural drainage on water quality in Ontario. They measured nutrient losses in tile drainage on a Brookston clay soil in continuous corn, continuous bluegrass, and a four-year rotation of corn, oats, alfalfa, and alfalfa. No fertilization was compared with fertilizer application rates of 17 kg/ha of N and 67 kg/ha P for all crops except first- and second-year alfalfa in the rotation. The corn received an additional 112 kg/ha of N. The average annual N and P losses are presented in Table 8.1. Nitrogen losses increased with fertilizer applications in four of the six cropping seasons. Nitrate concentrations in the tile outflow were above 10 mg/L for fertilized rotation corn and second-year alfalfa. Cropping systems had little effect on P concentrations. Fertilizer application caused a small increase in P losses.

Baker and Johnson,[35] in a summary paper of several studies, concluded that concentrations of NO_3-N were greater in subsurface drainage than in surface runoff; NH_3 concentrations in runoff were usually greater than in subsurface drainage and P concentrations in subsurface drainage were usually less than in runoff. Baker and Johnson based their conclusions on a number of studies in different locations and represent general conditions that exist for runoff and subsurface drainage water quality. Other studies have also shown that N losses in tile drainage increase with fertilizer application. Logan and Schwab[36] monitored subsurface drainage water quality from three field-sized areas on glacial till soils in Union County, Ohio. They found seasonal N losses varied from 0.1 to 45.6 kg/ha. The highest loss was on a site where 224 kg/ha of N was applied preplant to corn. In 1972, only 22 kg/ha of N fertilizer was applied, but the seasonal N loss was still 36.4 kg/ha. On the site where continuous alfalfa was grown, the seasonal N losses were 0.1 and 0.9 kg/ha in 1972 and 1973.

TABLE 8.1
Average Annual N & P Losses in Tile Drains[34]

	Nitrogen		Phosphorus	
Crop	No fertilizer (kg/ha)	Fertilizer (kg/ha)	No fertilizer (kg/ha)	Fertilizer (kg/ha)
(a) Rotation				
Corn	8.5	14.0	0.13	0.24
Oats and alfalfa	6.4	8.5	0.13	0.13
Alfalfa-first year	6.3	5.8	0.13	0.15
Alfalfa-second year	9.3	10.1	0.08	0.22
(b) Continuous				
Corn	4.4	8.9	0.26	0.24
Bluegrass	3.5	1.1	0.01	0.12

No fertilizer was applied to the alfalfa, and the tile discharge was much lower than from the other two sites where corn was grown.

Baker and Johnson[37] compared differential nitrogen fertilization rates and tile NO_3-N discharge rates on a Webster slit loam soil in Iowa. The 5-year average annual NO_3-N loss from an area receiving an average of 56 kg/ha of N fertilizer was 26 kg/ha. The high fertilization rate area had an average annual NO_3-N loss of 48 kg/ha and received an average of 116 kg/ha/yr of N fertilizer. The average annual flow volume from the tile lines was 132 mm, which represents a significant contribution to stream flows in central Iowa.

In another study on a Webster clay loam soil in southern Minnesota, Gast et al.[38] measured NO_3-N losses from tile lines for annual N applications of 20, 112, 224, and 448 kg/ha to continuous corn. Each treatment was replicated three times on plots 13.7 by 15.3 m. Nitrate losses and tile flow volumes are summarized in Table 8.2. Water flow through the tile lines occurred annually for approximately 6 weeks in the period from mid-April through early July and constituted an equivalent flow from 7 to 22% of the annual precipitation during the 3-year study. Nitrate losses from the tile lines after fertilizer applications for 3 years (1975) were 19, 25, 59, and 120 kg/ha/yr for the 20, 112, 224, and 448 kg/ha N application rates. Application of the recommended 112 kg/yr resulted in only slight increases in NO_3-N concentrations in the tile water or total losses from the tile lines compared with the 20 kg/ha treatment.

Tillage also has an effect on the amount and timing of NO_3-N and total N in subsurface drainage waters. Gold and Loudon[39] compared P and N losses from conservation tillage (chisel plow) and conventional tillage (moldboard plow) from two 4-ha watersheds in the Saginaw Bay area of Michigan. Total P and soluble P concentrations were higher in tile flow from conservation tillage than conventional tillage. The greater losses of P in surface runoff for conventional tillage more than offset the larger losses in P in tile flow for conservation tillage. Nitrate concentrations were similar in the tile flow from both tillage systems (11.7 and 10.5 mg/L) but were higher than in the surface runoff. Kjeldahl N concentrations were higher in surface runoff than in tile flow.

TABLE 8.2
Average Tile Line Flow and Nitrogen Losses as Influenced by Nitrogen Fertilizer Application[38]

Treatment (kg N/ha)	Tile Flow			Nitrate Losses		
	1973	1974 (cm)	1975	1973	1974 (kg N/ha)	1975
20	3.5	9.6	10.3	5 (0.6)[a]	17 (1.0)	19 (2.6)
112	3.5	9.1	12.0	6 (0.1)	22 (1.6)	25 (4.0)
224	2.8	8.4	13.3	4 (0.8)	20 (2.9)	59 (8.9)
448	5.0	9.9	15.1	6 (0.1)	54 (6.7)	120 (26)

[a] Means of three replications with standard errors of the means indicated in parenthesis.

Kanwar et al.[40] studied the effects of no-tillage and conventional tillage, and single N and split applications of N fertilizer on tile water quality in a Nicollet loam soil in Iowa. Tillage did not have a significant effect on tile drainage NO_3-N concentrations during the first year, but by the third year the average NO_3-N concentrations in drainage from conventional tillage was significantly higher than from no-tillage for a single N application of 175 kg/ha. Nitrate concentrations in drainage from conventional tillage the third year ranged from 16.3 to 34.7 mg/L with an overall average of 23.2 mg/L. For the same year, the average NO_3-N concentrations in drainage water from no-tillage ranged from 9.7 to 18.4 mg/L with an overall average of 14.7 mg/L. The effect of three split N applications totaling 125 kg/ha compared with a single application of 175 kg/ha was investigated only under no-tillage. In the third year, NO_3-N concentrations in the tile drainage were significantly lower from the split N applications than the single application. Overall average NO_3-N concentrations in drainage under split and single applications were 11.4 and 14.7 mg/L, respectively.

Several researchers[41, 42] also studied the effect of tillage on NO_3-N in groundwater and tile outflow in eastern Ontario. Nitrate loads over a 2-year period ranged from 20.0 kg/ha/yr for no-tillage to 29.0 kg/ha/yr for conventional tillage. Nitrate loads and concentrations were higher in conventional tillage than in no-tillage. The NO_3-N loads were not significantly different between tillage systems, but the NO_3-N concentrations were significantly different in 1991. Groundwater was sampled at depths of 1.2, 1.8, 3.0, and 4.8 m.

Nitrate concentrations exceeded the drinking water standard of 10 mg/L in 93% of the samples collected at 1.2 m, 80% at 1.8 m, 76% at 3.0 m, and only 15% at 4.6 m. Average NO_3-N concentrations under no-tillage and conventional tillage, respectively, were 29.4 and 35.6 mg/L at 1.2 m, 19.6, and 26.5 mg/L at 1.8, 18.5, and 13.9 mg/L at 3.0 m, and 2.4 and 4.5 mg/L at 4.6 m. The difference between tillage systems was only significant only at the 4.6 m depth. More data are needed to determine the long-term effect of tillage on groundwater and tile-drain-water quality.

In another study in southern Ontario, Kachanoski and Rudra[43] found there was no significant difference in the total drainage water between the no-tillage (NT) and moldboard-tillage (MB) treatments. However, NT had a significantly higher average concentration and flow-weighted concentration of NO_3-N in the tile outflow during spring and early fall periods than MB. The opposite trend was observed for late-fall and early-winter periods, when MB had significantly higher NO_3-N concentrations than NT. Yearly flow-weighted concentrations were similar for both treatments, and the average groundwater NO_3-N concentrations between 1 m and 5 m depth were similar. Tracer experiments revealed more preferential flow occurred in the MB tillage treatment. Overall bulk average velocity was higher in the case of the NT treatment. Tile water quality has also been investigated in areas other than the Midwest and Ontario. Madramootoo et al.[44] measured N, P, and K losses in subsurface drainage from two potato fields. Nitrogen concentrations in the tile effluent ranged from 1.70 to 40.02 mg/L. Phosphorus concentrations ranged from 0.020 to 0.052 mg/L. Potassium concentrations ranged from 2.98 to 21.4 mg/L. The total N loads in subsurface drainage during the growing season (April–November) from the two fields were 14 and 70 kg/ha in 1990. Phosphorus loads were less than 0.02 kg/ha.

In a 2-year study involving five farm sites in New Brunswick, flow-weighted average NO_3-N concentrations of the subdrain discharge (April–December) were greater than 10 mg/L for established potato rotation sites, both in the year with potatoes and in the subsequent nonpotato year when the rotation crop received little or no fertilizer.[45] Corresponding average NO_3-N concentrations at low input, nonpotato rotation sites were approximately 3 mg/L. The total mass of NO_3-N removed in the drainage water are summarized in Table 8.3. The annual NO_3-N load varied from 1 kg/ha in a hay, hay, potato, winter wheat, and hay five-year rotation to 33 kg/ha in a potato, potato, oats, hay, and potato rotation.

Bengston et al.[32] measured nutrient losses from research plots with surface drainage only and from plots with both surface and subsurface drainage from 1982 to 1991 in Louisiana. The plots were located on an alluvial Commence clay loam soil. Average rainfall for the period was 156.8 cm. The average annual surface drainage was 40.2 cm from the surface and subsurface-drained plots and 61.4 cm from the only surface-drained plots. The average annual P loss was 7.1 kg/ha from the surface and subsurface-drained plots and 10.2 kg/ha from only the surface-drained plots. The average annual N loss was 8.2 kg/ha from only the surface-drained plots and 6.8 kg/ha from the surface- and subsurface-drained plots. From 1982 to 1987, corn was grown on the plots and from 1988 to 1992, soybeans were grown. Corn received 109 and 38 kg/ha of N and P fertilizer and the soybeans received 40 kg/ha of P and no N.

Evans et al.[46] found a threefold and sixfold increase in total N transported at the field edge in surface and subsurface drainage, respectively, compared with natural conditions in North Carolina. Total N transported from subsurface drainage was 31.1 kg/ha/yr. Phosphorus transported by surface drainage was doubled compared with undeveloped (0.48 versus 0.20). Subsurface drainage had little effect on P transport compared with undeveloped sites but decreased P transport by 40–50% compared with surface drainage. Evans et al.[46] concluded the increase in N and P transport in drainage outflow is caused primarily by the addition of fertilizer, which results from

TABLE 8.3
Nitrates Removed by Tile Drainage for Different Cropping Rotations[45]

Site No.	Crops		N Applied		NO₃-N Removed	
	1987 (kg/ha)	1988 (kg/ha)	1987 (kg/ha)	1988 (kg/ha)	1987 (kg/ha)	1988 (kg/ha)
		(a) Established Potato Rotation Sites				
1	potato	barley[a]	110	45	16	28
2	potato	barley	150	35	33	25
3	fall rye	fall rye, peas	0	60	11	10
		(b) Nonpotato Rotation Sites				
4	hay	potato	0	200	1	5
5	potato	peas	165	50	11	7

[a] Underseeded to clover-grass mixture.

the change in land use following drainage instead of from mere installation of drainage.

Applying liquid manure to fields with tile drainage may have an increased impact on tile effluent water quality. Dean and Foran[47] found higher concentrations of bacteria and N and P in tile drainage discharge when rainfall occurred shortly before or shortly after manure spreading. McLellan et al.,[48] in a study in southwestern Ontario on a Brookston clay loam soil, found tile discharge NH_4-N concentrations increased from 0.2 to 0.3 mg/L before spreading to a peak of 53 mg/L shortly after manure was spread. Land application of liquid manure did not increase NO_3-N concentrations in the tile effluent but significantly increased fecal coliform bacteria. Blocking the drains to simulate controlled drainage decreased NH_4-N and bacteria concentrations.

In a 3-year study in southern Ontario, Fleming[49] found no significant relationship between NO_3-N levels and either time of year or number of weeks after spreading of manure. He sampled 14 tile lines on a weekly basis and six stream sites. Only five of the sites had NO_3-N levels above 10 mg/L. Total P concentrations in the tile water were significantly higher at sites receiving regular applications of manure compared with sites receiving only occasional manure applications or none at all. Sites where manure was spread regularly had higher fecal coliform concentrations in the tile effluent, but the results were not significantly different. Fecal coliform concentrations were higher in six stream sites than in the tile water, but NO_3-N and total P concentrations were lower. The stream flow consisted of tile discharge, surface runoff, and groundwater.

Geohring[50] discussed control methods to reduce the environmental impacts of tile drainage effluent from manure spreading. He discussed controlled drainage, time and rate of manure application, and tillage as viable control methods. When tiles are flowing, liquid manure application should be avoided or low applications of 0.3 to 0.8 cm should be applied. Tillage before application of liquid manures will reduce and delay the opportunity for preferential flow, minimizing the incidence of high concentrations of bacteria and NH_4-N entering the drains.

8.7.2.2. Controlled Drainage

In recent years, controlled drainage has been recognized as a best-management practice for reducing nutrient outflow from drained land. Evans et al.,[46] in evaluating 10 studies, found controlled drainage has shown significant reductions in N and P transport at the field edge. Total P concentrations in drainage outflow have been similar in controlled drainage and conventional drainage, but there was a reduction in outflow volume with controlled drainage that reduced the total mass of N and P. Controlled drainage reduced the annual transport of total N leaving the edge of the field by 45% and total P in surface runoff by 40%. Controlled drainage had little effect on P in subsurface flow.

Iziuno et al.[51] recommended improved drainage practices that reduce outflows, but also maintain flood control and crop protection as one method to reduce P loads from the Florida Everglades Agricultural Area (EAA). They investigated P concentrations in drainage water from muck soils of the EAA to identify critical P loss problems for the development and implementation of BMPs. The cropping systems during the study included sugarcane, radish, cabbage, rice, drained fallow, and flooded fallow. Total dissolved P loading rates from the overall cropping system represented from 50 to 80% of the total P loading rates. In some cases, under less-fertilized crops, the P concentrations in drainage water were lower compared with the drained fallow fields.

In another study, Izuno and Bottcher[52] evaluated the effects of slow versus fast drainage on N and P losses, along with crop management alternatives. Their results indicated that basin-wide implementation of BMPs could potentially reduce P loadings by 20–40%.[53] The most significant P loading reductions were attributed to altering farm drainage practices to slow drainage release.

Research in the Corn Belt with controlled drainage has been very site- and management-specific. However, research indicates that properly designed and operated controlled drainage systems provide both water quality and economic benefits.

Michigan researchers monitored N and P concentrations in subsurface drainage at sites near Bannister and Unionville.[54] At the Bannister site, dissolved NO_3-N concentrations were reduced from 9.0 mg/L for subsurface drainage to 5.7 mg/L with controlled drainage. The mass of NO_3-N was reduced 64% by controlled drainage. Controlled drainage had little effect on the dissolved ortho P loads delivered to the drainage ditch. At the Unionville site, for two growing seasons (May through October), a 58% reduction in NO_3-N and a 16% reduction in ortho P were observed with controlled drainage compared with only subsurface drainage. Average NO_3-N concentrations were reduced from 41.3 to 13.3 mg/L in 1990, and 18.2 to 9.9 mg/L in 1991. Corn was grown on both sites.

Kalita et al.[55] conducted a study in Iowa using variable water table depths for subirrigation. Average water-table depths were maintained at 0.3 (shallow), 0.6 (medium) and 1.0 (deep) m. Nitrate concentrations in the groundwater under shallow water-table depths were always less than those with medium and deep water-table depths. Nitrate concentrations in the groundwater decreased with increasing soil depth under all three water table conditions. When the water table was maintained at depths of 0.3 to 0.6 m, NO_3-N concentrations were reduced to below 10 mg/L.

Drury et al.[56] evaluated controlled drainage for reducing NO_3-N on a Brookston clay loam soil in Ontario planted to corn. Over a 2-year period, controlled drainage reduced NO_3-N concentrations by 25% and effectively reduced NO_3-N loss in the tile drainage water by 41% compared with conventional drainage. The flow-weighted mean NO_3-N concentrations were above 10 mg/L for conventional drainage but were less than 10 mg/L for the controlled drainage. This research, along with other results, indicated controlled drainage has the potential to reduce NO_3-N concentrations below the EPA drinking water standard of 10 mg/L.

8.7.3 PESTICIDES

8.7.3.1 Conventional Drainage

Pesticides have been measured in tile drainage in a number of locations in North America. Steenhuis et al.[57] measured pesticide concentrations in suction lysimeters, and groundwater and tile outflow under conventional tillage and conservation tillage on Rhinebeck sandy clay loam and variant clay loam soils. Low concentrations of atrazine (0.2–0.4 μg/L) and alachlor (0.1 μ/L) were detected in the groundwater 1 month after application. Only atrazine was detected in the conventional tillage in groundwater in low concentrations (0.4 μ/L) in November. They concluded that pesticide leaching to the groundwater was by macropore flow.

A project in the eastern region of Ontario studied the effect of tillage on the pesticides atrazine and metolachlor in groundwater and tile outflow.[41,42] During the first 2 years, concentrations and loadings of atrazine and deethylatrazine were higher for no-tillage than for conventional tillage. Cumulative loading rates and average concentrations of atrazine, deethylatrazine, and metolachlor in the tile outflow are summarized in Table 8.4. The loading rate of atrazine was significantly different between the conventional tillage and no-tillage, whereas for deethylatrazine the loading rate was not significantly different between the two tillage systems. Atrazine and deethylatrazine concentrations were significantly different for the two tillage systems in 1991 but not in 1992. Metolachlor was detected only for a short period during the winter of the second year. Groundwater was sampled at depths of 1.2, 1.8, 3.0, and 4.8 m.[42] Atrazine was detected in 71% of the samples. Average concentrations decreased with depth. Concentrations were significantly higher under no-tillage than conventional tillage at the 3.0 m and 4.8 m depths. The Environmental Protection Agency (EPA) drinking water standard of 3 μg/L was exceeded in only 7 of 418 samples. Deethylatrazine was detected in 85% of the samples. Average deethylatrazine concentrations were higher than average atrazine concentrations at all depths. There was a significant difference at all depths between tillage systems, with the no-tillage having the higher deethylatrazine concentrations. Metolachlor was detected in only 4% of the samples. All concentrations were below the EPA health advisory limit of 10 μg/L.

Bastien et al.[58] detected metribuzen in the tile flow at concentrations up to 3.47 μg/L in the two potato fields where Madramootoo et al.[44] measured nutrient losses. Concentrations in surface runoff samples were much higher (33.6–47.1 μg/L). Aldicarb, fenvalerate, and phorate were not detected in the drainage waters.

The influence of drainage systems design and pesticide fate and transport have not been clearly documented. Kladivako et al.[59] evaluated the effect of drain spacing

TABLE 8.4
Herbicides in Tile Effluent[41]

	1991		1992	
	Conventional tillage (g/ha)	No-tillage (g/ha)	Conventional tillage (g/ha)	No-tillage (g/ha)
Atrazine	0.90	1.82	0.58	1.48
Deethylatrazine	1.55	2.05	0.06	1.20
Metolachlor	0.00	0.00	0.04	0.49

on subsurface drainage water quality in Indiana. The amount of water and pesticides that moved offsite were greater with narrow (6 m) than with wider (12 m and 2 4 m) drain spacing. Most pesticide removal occurred within 2 months after application. Annual carbofuran losses in subsurface drainflow ranged from 0.79 to 14.1 g/ha. Atrazine, alachlor, and cyanazine losses ranged from 0.10 to 0.69 g/ha, 0.04 to 0.19 g/ha, and 0.05 to 0.83 g/ha, respectively.

Concentrations of most pesticides studied have been several times higher on surface drainage than in subsurface drainage. Bengston et al.[60] found that losses of atrazine and metolachlor were less than one-half in subsurface drainage plots than surface drained plots (22.8 g/ha versus 57.6 g/ha for atrazine and 23.1 g/ha versus 52.7 g/ha for metolachlor).

Recently, subsurface drainage systems have been examined for their possible contribution of pesticide pollution to surface water. It is believed that some of the agricultural chemicals that leach beyond the crop root zone into the shallow groundwater migrate with the drain water to the local streams, rivers, and lakes as part of drain effluent. Masse et al.[61] reported that atrazine and its dealkylated-N metabolites were found in the shallow groundwater zone of a corn field on a clay loam soil in Quebec. Many times, the concentrations were found to be higher than the 3-μg/L advisory limit of EPA. Muir and Baker[62] observed atrazine concentrations in tile-drain water in the range of 0.20–3.85 μg/L in Quebec corn fields. In eastern Ontario, Patni et al.[63] detected atrazine and deethylatrazine in 75% and metolachlor in 32% of the tile-drain water samples from a clay loam soil where corn was being grown under conventional tillage.

Most research shows pesticide occurrence in subsurface drainage water can be related to pesticide solubility, sorption coefficients, and soil persistence characteristics.[64]

8.7.3.2   Controlled Drainage

Several field-scale studies have been initiated in the last few years to investigate the role of water-table management systems in reducing pesticide discharges from subsurface-drained farmlands. One of the hypotheses driving these investigations is that the drain effluent will become less toxic if the water can be held within the farm boundaries for extended periods of time, a typical phenomenon-controlled drainage system. Most pesticides have a field half-life of a few weeks to a few months under aerobic conditions; therefore, the tile effluent would contain a lower concentration of pesticides if the drainage water is prevented from escaping the farm boundaries for an extended period of time. With controlled drainage systems, it is possible to maintain favorable moisture content levels in the soil profile which, in turn, can lead to higher adsorption and microbial degradation rates of pesticides in such fields.

Arjoon et al.[67] found that the leaching of prometryn herbicide in water table-managed plots was slower than in subsurface-drainage plots in an organic soil in Quebec. Similar results were obtained by Aubin and Prasher[65] for the herbicide metributzen in a potato field in Quebec. However, Arjoon and Prasher[67] found there was no difference in the leaching of metolachlor in controlled drainage and regular subsurface drainage in a loamy sand soil.

Ng et al.[68] found total atrazine and metolachlor losses did not differ between controlled and noncontrolled drainage in a Brookston clay loam in southwestern Ontario. The controlled drainage increased the amount of surface runoff compared with the uncontrolled drainage. For the controlled drainage, 23% of the rainfall was lost as surface runoff, whereas 12% of the rainfall was lost as surface runoff with the uncontrolled drainage.

Kalita et al.[55] found atrazine and alachlor concentrations in groundwater were decreased by maintaining shallow water table depths of less than 1m in the field. Atrazine concentrations were reduced from 67 to 0 µg/L by maintaining shallow water-table control.

8.8   IMPACT OF DRAINAGE OF SURFACE WATER QUALITY

Drainage outflows, whether from surface or subsurface, eventually enter surface water systems. The scientific link between drainage and the health of receiving streams is not fully understood. Nutrients from drainage outflows can cause eutrophication and make receiving bodies more susceptible to undesirable blooms of blue-green algae. The salinity of estuary headwaters could be reduced by periodic high outflow rates from artificial drainage which might change the ecosystem of the estuary.[69,70]

Lakshminarayana et al.[71] investigated the impact of subdrainage discharge containing atrazine on planktonic drift of the receiving natural stream. Maximum measured atrazine concentrations were 13.9 µg/L in the subdrain discharge and 1.89 µg/L in the stream. No negative impacts on plankton populations were evident beyond 50 m downstream from the drainage outlet. A section 20 m downstream was affected during low-flow conditions. Ambient environmental conditions and atrazine were thought to be contributing to the measured results.

Fausey et al.[72] concluded well-planned and well-managed drainage systems change the hydrologic relationships on the land where applied. Erosion can be reduced with surface drainage. Subsurface drainage can reduce the amount of runoff and the peak rate of discharge, thereby further reducing erosion and the associated off-site impacts of erosion.

8.9   INSTITUTIONAL AND SOCIAL CONSTRAINTS

Improved drainage of agricultural land purposes is increasingly viewed as being against the public's best interest. The pendulum has swung away from development in the last 20 years as a balance has been sought between development, reclamation, and drainage on the one hand and preservation of environmental values on the other. The U.S. National Environmental Policy Act of 1969, the Clean Water Act as amended in 1977, and the Food Security Act of 1985 have had an effect on agricultural drainage development. The Food Security Act of 1985 and 1990 Farm Bill deny price support and other farm program benefits to producers who grow crops on converted or drained wetlands. Also, the elimination of investment tax credits and restrictions on expending farm conservation investment under the Tax Reform Act of 1986 are further disincentives to bring new lands into production through drainage.

The Upper Choptank River Watershed, covering 40,713 ha in Kent County, Delaware, and Caroline and Queen Anne Counties, Maryland, was initiated in 1965. This project called for the reduction of flooding and drainage problems to cropland. The conflict between environmental interests and drainage problems on cropland forced the Upper Choptank River Watershed to be put on hold as a major construction project. Construction of the Maryland portion occurred during the late 1970s through the early 1980s. After construction had begun, the project required an environmental impact statement. Although the project is still actively addressing nonpoint source pollution control, federal assistance for maintaining the drainage infrastructures was lost.

Increased public concern about negative impacts of drainage on water quality brought about the failure in implementing the Upper Chester River Project, which was proposed by local sponsors with assistance of the Natural Resources Conservation Service in the state of Maryland in 1982. The failure of this project has increased institutional barriers and social constraints in implementing drainage research in most of the Mid-Atlantic states.[6]

In the midwestern U.S., many soils have problems with excess soil water in the spring and fall, which leads to excessive runoff and erosion, which in turn can impair surface-water quality. Excess soil water also poses a problem for timely planting and harvesting of crops and tillage operations. To alleviate these problems, both quantity and quality of water must be considered when assessing water management practices. The problem is that only water quality has received public concern and attention in recent years. Wise management of our water resources is important in developing sustainable agricultural production systems.

8.10 SUMMARY

Early settlers brought European drainage methods with them to North America. The first use of clay tile for agricultural drainage occurred in the Finger Lakes region of New York in 1835. Clay tile was the main material used for agricultural drainage until the early 1970s, when corrugated plastics tubing became popular. The drainage trenching machine was introduced in 1855.

Conventional drainage systems generally will increase total annual outflows from fields and peak outflow rates compared with naturally drained land. The earliest reports on tile drainage water quality was reported by Willrich.[33] Following this study, many studies have been reported in the literature. Most of these studies have shown that concentrations of NO_3-N are greater in subsurface drainage than in surface runoff, and that NH_4-N and P concentrations are greater in surface runoff than in subsurface drainage. Tillage also has an effect on the amount and timing of NO_3-N in subsurface drainage. Applying liquid manure to fields with subsurface drainage may increase N, bacteria, and P concentrations in drainage outflows. Atrazine and its degradation products and other pesticides have been detected in tile drainage waters in a number of studies. Pesticide occurrence in subsurface drainage can be related to pesticide solubility, sorption coefficients, and persistence characteristics.

Since the 1980s, the trend in the humid areas of the U.S. has been to develop a total water management system. Water-table management strategies can be grouped into three types: subsurface drainage, controlled drained, and controlled drainage–subsurface irrigation. There has been extensive research in North Carolina on water-table management. Controlled drainage may reduce N loads to streams by over 40% and it has the potential for reducing P loads under certain soil and geological conditions.

Although drainage has been part of agriculture since colonial times, in the 1990s drainage is greeted with angry responses in many quarters. Environmental concerns with drainage have stopped the implementation of several drainage projects. Today, both environmental and agricultural production concerns must be addressed in the design and operation of drainage systems.

Although there has been considerable research done on drainage and water quality, a number of needs must be addressed in future research. These research needs include the following:

1. Evaluate the impact of controlled drainage on pesticide transport.
2. Evaluate the overall economic benefits of water-table management systems to reduce water-quality degradation and improve crop yields.
3. Quantify the impacts of controlled and uncontrolled drainage on water quality with land application of animal wastes.
4. Evaluate the effect of drainage and water-table water management on on-site and off—site water quality in the Mid-Atlantic states.

REFERENCES

1. van Schilfgaarde, J., Drainage yesterday, today and tomorrow, in *Proc. of the ASAE Nat. Drain. Symp.,* ASAE, St. Joseph, MI, 1971, 1.
2. Skaggs, R. W., Drainage and water management modeling technology, in *Proc. 6th Int. Drain.* Symp., ASAE, St. Joseph, MI, 1992, 1.
3. Beauchamp, K. H., A history of drainage and drainage methods, in *Farm Drainage in the United States: History, Status and Prospects,* Pavelis, G. A., Ed., Mis. Pub. 1455, ERS, USDA, Washington, DC, 1987, Chap. 2.
4. Gain, E. W. and Patronsky, R. J., Historical sketches on channel modification, Paper No. 73-2537, ASAE, St. Joseph, MI, 1973.
5. Green, R. L. and Merrick, C. P., The drainage law of Maryland, Extension Bull. 196, Cooperative Extension Service, University of Maryland, College Park, MD, 1962.
6. Smith, R. T. and Sprague, L. A., Change and accommodations of environmental issues in drainage projects; a missing documentation, Paper No. 88-2604, ASAE, St. Joseph, MI, 1988.
7. Wooten, H. H. and Jones, L. A., The history of our drainage enterprises, in *Yearbook of Agriculture,* USDA, Washington, DC, 1955, 478.
8. Weaver, M. M. History of tile drainage, M. M. Weaver, Waterloo, NY, 1964.

9. Schwab, G. O. and Fouss, J. L., Plastic drain tubing: successor to shale tile, *Agric. Eng.,* 65(7), 23, 1985.

10. U. S. Department of Agriculture, Soil Conservation Service, Drainage of Agricultural Land, Water Information Center, Inc., Port Washington, NY, 1973, Chap. 3.

11. Schwab, G. O., Fangmeier, D. D., and Elliott, W. J., *Soil and Water Management Systems,* 4th ed., John Wiley & Sons, Inc., New York, NY, 1996, Chap. 12.

12. Shirmohammadi, A., Camp, C. R., and Thomas, D. L., Water-table management for field-sized areas in the Atlantic Coastal Plain, *J. Soil Water Cons.,* 47(1), 52, 1992.

13. Schwab, G. O., Fangmeier, D. D., Elliott, W. J., and Frevert, R. K., *Soil and Water Conservation 4th ed.,* John Wiley & Sons, Inc., New York, NY, 1993.

14. Evans, R. O., Gilliam, J. W., and Skaggs, R. W., Controlled drainage management guidelines for improving water quality, Publ. AG-443, North Carolina Agr. Ext. Serv., Raleigh, NC, 1990.

15. Thomas, D. L., Hunt, P. G., and Gilliam, J. W., Water-table management for water quality improvement, *J. Soil Water Cons.,* 47(1), 65, 1992.

16. Belcher, H. W. and D'Itri, F. M. *Subirrigation and Controlled Drainage,* Lewis Publishers, Ann Arbor, MI, 1995.

17. American Society of Agricultural Engineers, Drainage and water-table control, in *Proc. 6th Int. Drain. Symp.,* ASAE Pub. 13-92, St. Joseph, MI, 1992.

18. Evans, R. O., and Skaggs, R. W., Design guidelines for water-table management systems on Coastal Plain soils, *J. Applied Eng. Agric.,* 5, 82, 1989.

19. American Society of Agricultural Engineers, Design, construction, and operation of water-table management systems for subirrigation/controlled drainage in humid regions, EP479 in *ASAE Standards 1993,* ASAE, St. Joseph, MI, 1993, 744.

20. Skaggs, R. W., A water-table management model for shallow water-table soils, Rpt. No. 134, Water Resources Res. Inst., Univ. North Carolina, Raleigh, NC, 1978.

21. Fouss, J. L., and Cooper, J. R.,Weather forecasts as control input for water-table management in coastal areas, *Trans. ASAE,* 31, 61, 1988.

22. Buscher, W. J., Sadler, E. J., and Wright, F. S., Soil and crop management aspects of water-table control practices, *J. Soil Water Cons.,* 47(1), 71, 1992.

23. Doty, C. W., Cain, K. R., and Fanner, L. J., Design, operation, and maintenance of controlled drainage/subirrigation (CD-DI) systems in humid areas, *J. Applied Eng. Agric.,* 2, 114, 1986.

24. Evans, R. E., and Skaggs, R. W., Operating controlled drainage and subirrigation systems, Publ. AG-356, North Carolina Agr. Ext. Serv., Raleigh, NC, 1985.

25. Verhoeven, B., Over de zout en vochthurshouding in gemundeerdegronden, M.S. Thesis, Netherlands Agr. College, Wageningen, The Netherlands, 1953.

26. Doty, C. W., Crop water supplied by controlled and reversible drainage, *Trans. ASAE,* 22, 1122, 1987.

27. Williamson, R. E., and Kriz, G. I., Response of agricultural crops to flooding, depth of water-table and soil gaseous composition, *Trans. ASAE,* 13, 216, 1970.

28. Shih, S. F., Vandergrift, D. E., Myhre, D. L., Rahi, G. S. and Harrison, D. S., The effect of land forming on subsidence, in the Florida Everglades organic soil, *Soil Sci. Soc. Am. J.,* 45, 1206,1981.

29. Reicosky, D. C., Campbell, R. B., and Doty, C. W., Corn plant water stress as influenced by chiseling, irrigation, and water-table depth, *Agron. J.,* 68, 499, 1976.

30. Gregory, J. D., Skaggs, R. W., Broadhead, R. G., Culbreath, R. H., Bailey, J. R., and Foutz, T. L., Hydrologic and water quality impacts of peat mining in North Carolina, Rep. No. 214, North Carolina Water Resour. Res. Inst., Raleigh, NC, 1984.

31. Skaggs, R. W., Gilliam, J. W., Sheets, T. J., and Bames, J. S., Effect of agricultural land development on drainage waters in the North Carolina Tidewater Region, Rep. No. 159, North Carolina Water Resour. Res. Inst., Raleigh, NC, 1980.

32. Bengston, R. L., Carter, C. E., Fouss, J. L., Southwick, L. M., and Willis, G. H., Agricultural drainage and water quality in Mississippi delta, *J. Irrig. Drain. Eng.*, 121, 292, 1995.

33. Willrich, T. L., Properties of tile drainage water, Completion Rep., Project A-013-1-A, Iowa State Water Resour. Res. Inst., Ames, IA, 1969.

34. Bolton, E. F., Aylesworth, J. W., and Hore, F. R., Nutrient losses through tile drains under three cropping systems and two fertility levels on a Brookston clay soil, *Can. J. Soil Sci.*, 50, 275, 1970.

35. Baker, J. L., and Johnson, H. P., Impact of subsurface drainage on water quality, in *Proc. ASAE 3rd Nat. Drain. Symp.*, ASAE, St. Joseph, MI, 1977, 91.

36. Logan, T. J., and Schwab, G. O., Nutrient and sediment characteristics of tile effluent in Ohio, *J. Soil Water Cons.*, 31(1), 24, 1976.

37. Baker, J. L., and Johnson, H. P., Nitrate-nitrogen in tile dainage as affected by fertilization, *J. Environ. Qual.*, 10, 1981, 519.

38. Gast, R. G., Nelson, W. W., and Randall, G. W., Nitrate accumulation in soils and loss in tile lines following nitrogen applications to continuous com, *J. Environ. Qual.*, 7, 1978, 258.

39. Gold, A. J. and Loudon, T. L., Tillage effects on subsurface runoff water quality from artificially drained cropland, *Trans. ASAE*, 32, 1989, 1329.

40. Kanwar, R. S. Baker, J. L., and Baker, D. G., Tillage and split N-fertilization effects on subsurface drainage water quality and corn yield, *Trans. ASAE*, 31, 1988, 453.

41. Patni, N. K., Masse, L., Clenz, H. S., and Jui, P., Tillage effect on tile effluent quality and loading, Paper No. 87-2627, ASAE, St. Joseph, MI, 1992.

42. Masse, L., Patni, N. K., Clegg, S., and Jui, P., Tillage effects on groundwater quality, Paper No. 92-2615, ASAE, St. Joseph, MI, 1992.

43. Kachanoski, R. G., and Rudra, R. P., Effect of tillage on the quality and quantity of surface and subsurface drainage waters, Final Rep., Technology and Development Sub-Program, SWEEP, Univ. of Guelph, Guelph, Ont., Canada, 1991.

44. Madramootoo, C. A., Wiyo, K. A., and Enright, P., Nutrient losses through tile drains from two potato fields, *J. Applied Eng. Agric.* 8, 1992, 639.

45. Milburn, P., Gartley, C., Richards, J., and O'Neill, M., Effects of potato production in groundwater quality: Observations in New Brunswick Canada, Paper No. NABEC 90-302, ASAE, St. Joseph, MI, 1990.

46. Evans, R. O., Skaggs, R. W., and Gilliam, J. W., A field experiment to evaluate the water quality impacts of agricultural drainage and production practices, in *Proc. Nat. Conf. on Irrig. and Drain. Eng.*, ASCE, New York, NY, 1991, 213.

47. Dean, D. M. and Foran, M. E., The effect of farm liquid waste application on receiving water quality, Project Rep. No. 512G, Ontario Ministry of Environment, Research Management Office, Toronto, Ont., Canada, 1990.

48. McLellan, J. E., Fleming, R. J., and Bradshaw, S. H., Reducing manure output to streams from subsurface drainage systems, Paper No. 93-2010, ASAE, St. Joseph, MI, 1993.

49. Fleming, R. J., Impact of agricultural practices on water quality, Paper No. 90-2028, ASAE, St. Joseph, MI, 1990.

50. Geohring, L. D., Controlling environmental impact in tile-drained fields, in *Proc. of Liquid Manure Application Systems Conf.*, NRAES-89, Cornell Univ., Ithaca, NY, 1994, 194.

51. Izuno, F. T., Sanchez, C. A., Coale, F. J., Bottcher, A. B., and Jones, D. B., Phosphorus concentrations in drainage water in the Everglades Agricultural Area, *J. Environ. Qual.,* 20, 1991, 608.

52. Izuno, F. T., and Boucher, A. B., The effects of on-farm agricultural practices in the organic soils of the EAA on nitrogen and phosphorus transport: screening BMPs for phosphorus loadings and concentration reductions, Phase II Final Rep., South Florida Water Mgmt. Dist., West Palm Beach, FL., 1987.

53. Izuno, F. T., and Bottcher, A. B., The effects of on-farm agricultural practices in the organic soils of the EAA on nitrogen and phosphorus transport: screening BMPs for phosphorus loadings and concentration reductions, Final Rep., South Florida Water Mgmt. Dist., West Palm Beach, FL., 1991.

54. Fogiel, A. C., and Belcher, H. W., Water quality impacts of water-table management systems, Paper No. 91-2596, ASAE, St. Joseph, MI, 1991.

55. Kalita, P. K., Kanwar, R. S., and Melvin, S. W., Subirrigation and controlled drainage: management tools for reducing environmental impact of non-point source pollution, in *Proc., 6th Int. Drain. Symp.,* ASAE Pub. 13-92, St. Joseph, MI, 1992, 129.

56. Drury, C. F., Tan, C. S., Gaynor, J. D., Oloya, T. O., and Welacky, T. W., Influence of controlled drainage/subirrigation on nitrate loss from Brookston clay loam soil, Paper No. 94-2068, ASAE, St. Joseph, MI, 1994.

57. Steenhuis, T., Paulsen, P., Richard, T., Staubitz, W., Andreini, M., and Surface, J., Pesticide and nitrate movement under conservation and conventional tilled plots, in *Proc., ASCE Irrig. and Drain. Div. Conf.,* ASCE, New York, NY, D. R. Hay, ed., 1988, 587.

58. Bastien, C., Madramootoo, C. A., Enright, P., and Caux, P. Y., Pesticide movement on agricultural land in Quebec, Paper No. 90-2513, ASAE, St. Joseph, MI, 1990.

59. Kladivko, E. J., Van Scoyoc, G. E., Monke, E. J., Oates, K. M. and Pask, W., Pesticide and nutrient movement into subsurface tile drains on a silt loam soil in Indiana, *J. Envir. Qual.,* 20,1991,264.

60. Bengtson, R. L., Southwick, L. M., Willis, G. H., and Carter, C. E., The influence of subsurface drainage practices on herbicide losses, *Trans. ASAE,* 33, 1990, 415.

61. Masse, L., Prasher, S. O., and Khan, S. U., Transport of metolachlor, atrazine and atrazine metabolites to groundwater, in *Proc. Annu. Conf. and 1st Biennial Envir. Spec. Conf.* Can. Soc. for Civil Eng., Toronto, Ont., Canada, 1990, 925.

62. Muir, D. C. and Baker, B. E., Detection of triazine herbicides and their degradation products in tile-drain water from fields under intensive corn (maize) production, *J. Agric. Food Chem.,* 24, 1976, 122.

63. Patni, N. K., Frank, R., and Clegg, S., Pesticide persistence and movement under farm conditions, Paper No. 87-2627, ASAE, St. Joseph, MI, 1987.

64. Evans, R. O., Skaggs, R. W., and Gilliam, J. W., Controlled versus conventional drainage effects on water quality, *J. Irrig. Drain. Eng.,* 121, 1995, 271.

65. Arjoon, D. and Prasher, S. O., Reducing water pollution from a mineral soil, in *Proc., 1993 Joint CSCE-ASCE Nat. Conf. on Envir. Eng.,* ASCE, New York, NY, 1993, 589.

66. Aubin, E. and Prasher, S. O., Impact of water table on metribuzen leaching, in *Proc., 1993 Joint CSCE-ASCE Nat. Conf. on Envir. Eng.,* ASCE, New York, NY, 1993, 557.

67. Arjoon, D., Prasher, S. O., and Gallichand, J., Reducing water pollution from an organic soil, in *Proc., 1993 Joint CSCE-ASCE Nat. Conf. on Environ. Eng.,* ASCE, New York, NY, 1993, 573

68. Ng, H. Y. F., Gaynor, J. D., Tan, C. S., and Drury, C. F., Atrazine and metolachlor transport in well-drained and poorly drained soils, ASAE-NABEC Meeting, Tech. Paper, ASAE, St. Joseph, MI, 1994.

69. Hobbie, J. E., Copeland, B. J., and Harrison, W. G.,Nutrients in the Pamlico River estuary, NC, 1969–1971, Rep. No. 76, North Carolina Water Resour. Res. Inst., Raleigh, NC, 1972.
70. Pate, P. P. and Jones, R., Effects of upland drainage on estuarine nursery areas of Pamlico Sound, North Carolina, Working Paper No. 81-10, UNC Sea Grant, UNC, Raleigh, NC, 1981.
71. Lakshminarayana, J. S. S., O'Neill, A J., Jonnovithula, S. D., Leger, D. A., and Milbum, P., Impact of atrazine-bearing agricultural tile draiange discharge on planktonic drift of a natural stream, *Environ. Poll.* 76, 1992, 201.
72. Fausey, N. R., Brown, L. C., Belcher, L. C., and Kanwar, R. S., Drainage and water quality in Great Lakes and cornbelt states, *J. Irrig. Drain Eng,* 115, 1995, 283.

9 Water Quality Models

*Adel Shirmohammadi, Hubert J. Montas, Lars Bergstrom,
and Walter G. Knisel, Jr.*

CONTENTS

9.1 INTRODUCTION

The quality of our water resources has been of both national and global concern for decades. Similarly, the manmade environmental problems of freshwater and marine eutrophication and contamination of groundwater have increased over the last few decades. The potential negative impact of agricultural chemicals on the quality of both surface and groundwater resources has been a major concern of scientists and engineers worldwide as well. Such adverse effects include deteriorating surface water and groundwater quality by plant nutrients and pesticides[1-6] and accumulation of agrochemicals in the soil to toxic levels (Torstensson and Stenstrom[7]).

Agricultural chemicals can contaminate water resources by one or more of the following pathways (Shirmohammadi and Knisel[8]): (1) surface runoff to streams and lakes, (2) lateral movement of chemicals through unsaturated or saturated soil media to bodies of surface water, or (3) vertical percolation of chemicals through unsaturated or saturated soil media to underlying groundwater.

Climate, soils, geology, land use, and agricultural management practices influence the quantity of water and chemicals that move through each of the aforementioned pathways. Because of the complex nature of nonpoint source (NPS) pollution, the development of detection and abatement techniques is not a simple process. Only two methods for tracking the environmental fate of chemicals and assessing the effectiveness of NPS management techniques in preventing water quality deterioration exist: (1) actual field monitoring, and (2) computer modeling (Shoemaker et al.,[9] Shirmohammadi and Knisel[8]). Field monitoring imposes many limitations, considering the variable nature of soils, geology, cropping and cultural systems, and, more importantly, climate. Collection of statistically sound data on the environmental fate of chemicals under varying physiographic and climatic conditions may be very costly and would require several years of field monitoring. Thus, computer models are viable alternatives in examining the environmental fate of chemicals under different physiographic, climatic, and management scenarios.[10-15] Process models can also be linked with economic models to determine the economic feasibility of environmentally sound agricultural management scenarios (Roka et al.[16]). The Geographic Information System (GIS) has also been used to evaluate the critical areas regarding NPS pollution of surface and groundwater.[17-18]

This chapter intends to provide the governing philosophy behind model development, types of water quality models and their intended uses, role of GIS in conjunction with the water quality models, and associated limitations and misuses of water quality models. The overall goal of the chapter is to provide a state-of-the-art review of the status of water quality models, thus assisting scientists and engineers in using the existing models and creating a platform for future research and developments in the area of water quality modeling.

9.2 CONCEPT OF MODELING

Models are used for better understanding and explanation of natural phenomena, and, under some conditions they may provide predictions in a deterministic or probabilistic sense (Woolhiser and Brakensiek[19]). To understand an event in our natural environment, we may need to provide a scientific explanation of it, as was described by Hempel.[20] "Scientific explanation" of an event, E, can be inferred from a set of general laws or theoretical principles $(L_1, L_2 \ldots \ldots L_n)$ and a set of statements of empirical circumstances $(C_1, C_2 \ldots \ldots C_n)$ (Woolhiser and Brakensiek[19]). Such an explanation can be represented by the following equation:

$$E = f(L_1, L_2 \ldots \ldots L_n) + g(C_1, C_2 \ldots \ldots C_n) \qquad (9.1)$$

where f and g represent subfunctions, combination of which describe the event of our interest, E. Equation 9.1 indicates that formal models (empirical and theoretical) are required for scientific explanation of a natural event. However, one should be aware of the limitations that each type of formal model may impose in trying to describe an event. For example, an empirical model is generally derived from a set of observed data under specific conditions; thus, application of such models to the conditions other than the ones under which they have been developed may pose a significant error in our predictions. Most of the hydrologic and water quality models are formal and generally include both empirical and theoretical principles.

9.3   MODEL PHILOSOPHY

To understand the "role of models," it may be appropriate to have an understanding about the term model and the philosophy behind model development. The model may have different interpretations based on its discipline of use. In hydrology, water quality, and in engineering, models are used to explain natural phenomena and, under some conditions, to make deterministic or probabilistic predictions (Woolhiser and Brakensiek[19]). In other words, a modeler tries to use the established laws or circumstantial evidence to represent the real-life scenario, which is called "model." Although each modeler tries to represent the real system, the strengths and weaknesses of their models depend on the modeler's background, the application conditions, and scale of application. One should note that Aristotle and his idea that "inaccessible is more challenging to explore than the accessible in the everyday world" seem to have had a guiding influence on the development of water quality models. Additionally, the "particle theory" of Einstein that "universe has a grain structure and each grain is in a relative state with respect to the others," has formed the basis for describing interrelationships between different components of water quality models. For instance, a natural scientist is concerned about the interrelationships governing the state of a given environment and tries to understand such relationship using experimental procedures and biological principles. The products of such studies are generally a set of factual data and possibly some empirical models describing such relationships. A physicist and an engineer, on the other hand, try to use physical laws and mechanistic approaches to describe interrelationships governing the state of an event and produce deterministic and mechanistic models. Such models are not complete until they have been calibrated, validated, and tested against experimental data.

To address the interaction between human life and the surrounding environment in the landscape, the "peep-hole" principle has mostly been used (Hagerstrand[21]). The result is that the landscape mantle is understood to a limited degree only, mainly as related to biological systems and to components of economic importance related to the use of natural resources. Recent needs for sustainability has encouraged scientists to evaluate the multicause problems of the environment in relation to human life under diverse conditions (Falkenmark and Mikulski[22]). Efforts to respond to the issue of sustainability have produced multicomponent water quality models describing hydrologic and water quality responses of the landscape under diverse climatic and managerial conditions. And in most cases, these models have used the systems approach in describing a natural event rather than looking at each event as an isolated phenomenon.

9.4 MODEL CLASSIFICATION

A model, an abstraction of the real system, may be represented by a "black box" concept where it produces output in response to a set of inputs (Novotny and Olem[23]). To describe the interrelationship between the outputs, different approaches have been used to create several types of models. Figure 9.1 shows the type of classification that was introduced by Woolhiser and Brankensiek[19] in describing hydrologic models.

Although each of the above forms of models tries to represent the real system, all have their own strengths and weaknesses depending upon the application conditions, and scale of application. For example, an empirical model is derived from a set of measured data for specific site conditions and therefore its application to other sites may create a real concern. A regression model relating a dependent variable such as nitrogen concentration at a watershed outlet to an independent variable such as fertilizer application rates is an example of the empirical model.

Theoretical models, as opposed to empirical models, use certain physical laws governing the behavior of the real system, and thus have a more generic application. Such models are composed of both variables describing the physical system (system parameters) and those describing the state of the system (state variables). The physical characteristics of the watershed such as soils, slope, and surface conditions may be considered as the system parameters. Climatic factors such as temperature and solar radiation coupled with management factors such as tillage and vegetation cover may be considered as the state variables or "driving variables." A thorough knowledge of both system parameters and state variables is essential to the model accuracy. Relationships (equations) are proposed for the observed processes based on the understanding of basic physical, biological, chemical, and mathematical principles (Piedrahita et al.[24]). Because they are based on general principles and not on specific site data, physical models tend to be applicable to a wider variety of situations, but, as a result, tend to be less accurate predictors than empirical models. However, a major asset of physical models is their usefulness in gaining insight on how a particular system or process works, and on being able to identify how a system or process might perform under conditions different from those for which data are available (Piedrahita et al.[24]).

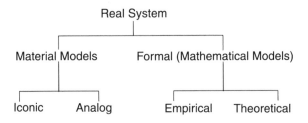

FIGURE 9.1 Representation of real systems by different models, Woolhiser and Brankensiek.[19]

Novotny and Olem[23] used Chow's concept of model classification (Chow[25]) and divided the diffuse-pollution models into three basic groups as follows: (1) simple statistical routines and screening models, (2) deterministic hydrologic models, and (3) stochastic models.

The first category of models in Novotny's classification are analogous to the empirical models described in the Woolhiser and Brakensiek[19] classification. They are simply regression models of different forms relating a dependent variable to the independent variable with a certain accuracy level described by the correlation coefficient, and are derived from observed data. A deterministic model, on the other hand, provides only one set of outputs for a given single set of inputs (Jarvis et al.[26]). No matter how many times the model is run for the given input, the output will always be the same. The third category of models—stochastic models—considers the output to be uncertain and uses mean and probabilistic ranges to describe the output.[27–28] Stochastic models are usually used where a great deal of variability and uncertainty is expected in both input parameters and outputs. For example, soil physical and hydraulic properties are known to be both spatially and temporally variable, thus causing uncertainty in the predicted leaching and groundwater loading of water and chemicals. In certain instances, deterministic models can be used in a stochastic or probabilistic way. For example, incorporating the deterministic models into a shell program to run Monte Carlo simulations constitutes such a marriage between deterministic and stochastic models.[29–33]

Unlike stochastic models, deterministic models ignore the input of random perturbations and variations of system parameters and state variables. The two approaches used in constructing a deterministic model are lumped parameter and distributed parameter, and accordingly, they are referred to as "lumped parameter models" and "distributed parameter models." Lumped parameter models are the more common of the two approaches and are characterized by treating the watershed hydrologic system, or a significant portion of it, as one unit. Using the lumped parameter approach, the watershed characteristics are lumped together in an empirical equation, and the final form and magnitude of the parameters are simplified as a uniform system (Novotny and Olem[23]). Lumped parameter models require calibration of coefficients and system parameters by comparing the response of the model with field data. Additionally, lumped parameter models may be both deterministic and stochastic. Because hydrologic systems possess dynamic fluctuations caused by meteorological events or basin physical characteristics, and deterministic models ignore these random fluctuations, using statistical routines to estimate probabilistic characteristics by a deterministic model may provide erroneous information of the modeled phenomenon, Novotny and Olem.[23] An example of a lumped parameter model is the HSPF model (Donigian et al.[34]) where the model uses lump-sum parameters for the physical processes in the watershed.

The distributed parameter approach involves dividing the watershed into smaller homogenous units with uniform characteristics. Each areal unit is described as a set of differential mass-balance equations. When the model is run, the mass balance for the entire system is solved simultaneously. Distributed parameter files may provide information from each subunit, therefore allowing the consideration of the effects of changes in the watershed in the model. The drawback with distributed parameter files is that they require a lot of computer storage space and an extensive detailed description of system parameters from each areal unit. A benefit of these models is that they are more suitable to be included in the geographic information systems (GIS) and computer-aided design (CAD) environments, which makes the models more robust in a spatial sense (Montas et al.[35–36]). Moreover, a routing algorithm may be necessary to route the output from one subunit to the next and finally to the outlet of the watershed. Models such as SWAT (Arnold et al.,[37] Chu et al.[38]) and ANSWERS-2000 (Bouraoui and Dillaha[39]) are examples of distributed parameter models.

As stated above, the failure of deterministic models, especially for complex hydrologic systems, is their inability to represent the variability of data. Additionally, deterministic steady-state models are unable to detect nondeterministic variation in the output. Because hydrologic responses vary according to state variables, stochastic models are more appropriate for analyzing time series (Coyne et al.[27]). Stochastic models possess both the deterministic and the stochastic nature of the underlying processes, enabling them to differentiate between deterministic relationships and noise (Novotny and Olem[23]). Although they are more crude, incorporating only a few input and system parameters and requiring data over an uninterrupted time series, stochastic models are a good, unbiased tool for prediction and control.

9.5   TYPES OF WATER QUALITY MODELS

Numerous models have been developed and are in use either as research, management, or regulatory tools. Table 9.1 shows selected water quality models that range from profile scale to watershed scale models. Ghadiri and Rose[40] provide a comprehensive review of these models. Water quality models range in complexity from detailed research tools to relatively simple planning tools and index-based models. Research models usually incorporate the state-of-the-art understanding of the processes being modeled and are aimed at improving our understanding of the complex processes governing the hydrologic and water quality response of a system, identifying gaps in our knowledge of these processes, and generating new researchable issues and hypotheses (Jarvis et al.[26]). On the other hand, management models use physical or empirical relationships to represent the natural system and provide guidance regarding the wise use of the agricultural and natural resources. These models can be developed directly, or through the simplification of more detailed mechanistic models. For example, GLEAMS (Knisel and Davis[41]) is a nonpoint source pollution management model where it is capable of simulating the relative impacts of different agricultural management systems on water quality over a long duration. It uses both physical-based as well as empirical functions to describe the flow of water and contaminants on the land surface and through the vadose zone.

Research models have generally been more deterministic, thus considering detailed processes. However, recent modeling efforts have attempted to develop research models with an ultimate goal of using them to answer management questions. For example, MACRO (Jarvis et al.,[26] Larsson and Jarvis[42]) and LEACHP

TABLE 9.1
Selected Water Quality Models and their Practical Attributes

Model	Type	Scale	Purpose	Validation Level	Documentation On (User's Manual)
PLM (Nichols and Hall)[45]	Process-based profile model	Unit area process model	Predicts water and pesticide leaching using 3-domain (slow, medium, fast) flow pathways in the soil column	Fair	Fair
TRANSMIT (Hutson and Wagenet)[46]	Process-based profile model	Unit area process model	Predicts movement of water and chemicals through soil profile	Fair	Fair
GLEAMS (Knisel and Davis)[41]	Unit management model	Field	Predicts surface and root zone hydrologic and water quality response	Well validated	Excellent
PRZM-3 (Carsel et al.)[52]	Unit management model	Field	Predicts pesticide and nitrogen fate in surface and crop root zone	Reasonable	Excellent
EPIC (Williams et al.)[49]	Unit management model	Field	Predicts surface and root zone hydrologic and water quality response	Reasonable	Good
ANSWERS-2000 (Bouraoui and Dillaha)[39]	Distributed parameter model	Watershed	Predicts surface and root zone hydrologic and water quality response—stream routing for hydrology	Intermediate	Poor
SWAT (Arnold et al.)[37]	Distributed parameter model	Watershed	Predicts surface and subsurface hydrologic and water quality response—with stream routing	Fair	Good
SWRRB (Arnold et al.)[55]	Distributed (up to 10 subwatersheds)	Watershed	Predicts surface and root zone hydrology and sediment yield—has sediment routing but has no flood routing	Fair	Fair
AGNPS (Young et al.)[54] and AnnAGNPS (Cronshey et al.)[117]	Distributed/lumped	Watershed	Predicts surface hydrologic and water quality response—with stream routing	Fair	Fair
HSPF	Lumped parameter	Watershed	Predicts the hydrologic and water quality response of the watersheds	Fair	Good

(Hutson and Wagenet[43]) use mechanistic relationships to simulate pesticide movement through the soil profile while attempting to consider the impact of different management scenarios. Some models such as the pesticide root zone model, PRZM (Carsel et al.[29]), and PRZM2 (Mullins et al.[44]) use a simple capacitance-type water flow model and a physical-based solute transport model to simulate the movement of water and contaminants through the soil profile under diverse management scenarios.

Water quality models have also been developed to consider the issue of scale. Most of the process-oriented and mechanistic models such as PLM (Nichols and Hall[45]), TRANSMIT (Hutson and Wagenett[46]), SOIL (Jansson[47]), and SOILN (Johnsson et al.[48]) are one-dimensional or two-dimensional column-based models. They are generally used to predict transport and chemical distribution profiles in the vadose zone and are limited in their ability to examine the water quality impacts of different agricultural management systems. On the other hand, field scale models such as CREAMS by Knisel,[10] GLEAMS by Knisel and Davis,[41] EPIC by Williams et al.,[49] ADAPT by Chung et al.[50] and Gowda et al.,[51] PRZM-2 by Mullins et al.,[44] and PRZM-3 by Carsel et al.[52] are unit-management models and are used as research, management, and regulatory tools to evaluate the impact of different agricultural management systems on water quality. These models are generally physically based but use many empirical equations to describe many of the processes within the model. Most of these models use the familiar SCS-Curve Number Method (Shirmohammadi et al.[53]) as a basis for hydrologic predictions. It is also important to note most of these field-scale models use daily climatic data as opposed to many of the process-based models that use event climatic data.

Watershed scale nonpoint source pollution models use the principles used in the field-scale models and extend them to mixed land use scenarios. For example, AGNPS by Young et al.,[54] SWRRB by Arnold et al.,[55] and SWAT by Arnold et al.[37] all are built upon the strength of the USDA's CREAMS model (Knisel[10]). They all are continuous simulation models with daily time steps. Some watershed models such as ANSWERS-2000 (Bouraoui and Dillaha[39]) are event-based, thus requiring more detailed climatic data. Watershed scale models such as SWAT and ANSWERS-2000 are distributive parameter models, thus enabling the user to consider the diversities in land use, soils, topography, and management alternatives within the watershed. These models generally contain routing algorithms that consider the attenuation of sediment and chemicals through the upland areas as well as the stream system. The distributive parameter nature of these models make them more viable to be used in conjunction with GIS environments.

The most extensively used water quality model is HSPF (Donigian et al.[34]), which extends the field-scale ARM model (Donigian and Crawford[56]) to basin-size areas. Its hydrology is simulated using modification of the famous Stanford Watershed Model, based on the infiltration concept. This model is generally used for large basins such as the Chesapeake Bay Basin on the eastern coast of the United States. The limitation of the HSPF is its requirement of large amounts of input data and a considerable amount of computer storage. BASINS (Lahlou et al.[57]), a recently developed basin-scale model, uses HSPF model in the GIS environment and helps to reduce some of the difficulties in preparing input data by using an electronically available GIS data base.

Index-based approaches to evaluate the nonpoint source pollution impacts of different land uses under varying climatic, soils, and management scenarios have also been paving their way into the literature. Aller et al.[58] developed a model called DRASTIC, which is a standardized system to evaluate the vulnerability of any hydrogeologic setting to groundwater pollution in the United States. The application of DRASTIC provides mappable results that can be used as a quick reference of relative pollution potential of different areas within a region or a watershed. Similar concepts have recently been developed within the GIS environment whereby layering of different data sets influencing the quality of water within a region or a watershed enables identification of critical pollution areas within a watershed (Hamllet,[17] Shirmohammadi et al.[59]). For example, Shirmohammadi et al.[59] used the GIS system and indexing approach to identify the critical pollution areas within an agricultural watershed and then used the GLEAMS model to prescribe a management system for the polluted areas of the watershed.

9.6 MODEL DEVELOPMENT

Model development may consist of (1) problem identification and algorithm development, (2) data base compilation, (3) model calibration and sensitivity analysis, (4) model validation and verification, (5) model documentation, and (6) model support and maintenance. Renard[60] listed nine steps for model development that are generally comparable to those listed and discussed in this section.

9.6.1 PROBLEM IDENTIFICATION AND ALGORITHM DEVELOPMENT

9.6.1.1 Problem Definition

It is essential to clearly identify the problem and the purpose of the modeling effort. For instance, assessing hydrologic and water quality response of an agricultural watershed may be the problem for which one desires to develop a model. Responses to the following questions may assist one in determining the type and level of modeling effort needed:

(1) Is the model to be constructed for prediction, system interpretation, or a generic modeling exercise? Is it a research or a management model?

(2) What do we want to learn from the model? What questions do we want the model to answer?

(3) Is a modeling exercise the best way to answer the questions?

(4) What is the scale of the model? As the scale increases, the uncertainty increases in the model. Therefore, a decision about the desired level of confidence in the output should be made.

9.6.1.2 Algorithm Development

The problem should first be well defined. The goal of the modeling exercise is to simulate information that can be used to make predictions for the real systems. The first approach in algorithm development may involve the development of a conceptual framework (Sargent[61]). For a mathematical model, the governing equations should be identified for each component and process involved in the model. The key processes involved in modeling a system should be considered, thus proper input parameters to get the desired output may be identified. For example, a desired output, Y, may be related to a set of input parameters as:

$$Y = f(X_1, X_2, X_3, \ldots X_n) \tag{9.2}$$

where $X_1 \ldots X_n$ represents the input variables and system parameters. Once the governing equation is identified, then the boundary and initial conditions for the problem should be identified. The solution (e.g., exact or numerical) to the equation should be detailed, including the relevant assumptions. Solving the equation with the help of the initial and boundary conditions will lead us to obtaining the particular solution of interest. It is in this step of the model development that one needs to identify programming language and strategy to handle the computations necessary for solving governing equations Renard[60]).

9.6.2 DATABASE REQUIREMENT

Data collection is a compromise between precision and expenditure. There may be many input data needed for running the model. Some may need to be highly precise; others do not make a difference. Sometimes data over a long period may be needed. The period of data collection for statistical viability is another major concern (Haan[62]). It is the modeler's dream to have access to a database that is already available. It not only helps the process to be faster but also eliminates the expense involved in the collection of such data. Therefore, the databases that act as a common record from which modelers can pull out information is essential and important. Collection or the existence of standard databases can be an immense help in model calibration and testing (Bergstrom and Jarvis[63]). However, collection and compilation of databases for modeling purposes have generally been use-oriented; thus, the databases do not render themselves into generic use. One should note that, on macro scale, certain databases such as weather data collected by U.S. National Oceanic and Atmospheric Administration, flow quantity and quality data collected by U.S. Geological Survey for different river basins, and soils data collected by the USDA Natural Resource Conservation Service (NRCS) have generic use and may be very useful in model testing and evaluation

The input parameters that are both site- and model specific have to be collected by the model developer. Some databases such as the natural resources data obtained by the U.S. Environmental Protection Agency may even contain calibration and verification surveys for runoff modeling (Huber et al.[64]). Some default parameter values can be obtained through a user's manual or front-end electronic database for some models such as GLEAMS (Knisel and Davis[41]).

9.6.3 SENSITIVITY ANALYSIS

Sensitivity analysis refers to the evaluation of model sensitivity to uncertainty in estimated parameter values. It depends on the quantity considered and on the parameter values in the standard calculation to which all sensitivity results are compared. Sensitivity analysis helps determine which of the parameters can be estimated and which should be measured with high accuracy. It involves a calibration step. Calibration means varying the coefficients of the designed model within the acceptable range until a satisfactory agreement between measured and computed output values is achieved. The variable to which the model is most sensitive should be calibrated first. The values of the input variables are needed for calibration. The data obtained from a standard database, that collected by different agencies, or the data measured in the field will be used at this stage.

Once the model is calibrated, it should be verified. Verification is done by running the model with the coefficients established during calibration and with input corresponding to another standard database. Calibration and verification need to be done during the design process itself. For example, Boesten[65] used a standard value of 0.9 for Freundlich exponent (l/n) in a model exercise for pesticide leaching to groundwater. The results showed that the exponent increased with increasing value of a coefficient (K_{om}) that represents the sorptivity. Further analysis revealed that the exponent is highly sensitive to pesticides that are sorbed. Therefore, the steps in estimating the Freundlich exponent should be attempted carefully, and it also means that the sorption properties of different soil layers need to be measured with high accuracy. Similarly, Wei et al.[66] performed a comprehensive sensitivity analysis of the MACRO model and identified both physical and chemical parameters to which the model was most sensitive. Caution must be exercised in making a sensitivity analysis because it may be site-specific. For example, the land surface slope and slope shape may be highly sensitive. If a plot or field has a concave or complex slope, the overland flow parameters are not sensitive in the calculation of sediment yield because the system is transport-limited. On the other hand, if the slope shape is convex, the overland flow parameters will be highly sensitive. Also, if a concentrated flow (channel) occurs in the field/basin, the overland flow parameters will not be sensitive because it generally has a flatter slope than the overland flow, and is transport limited. Basins or watersheds generally include channels that dominate the sensitivity of overland rill and interrill erosion.

9.6.4 MODEL VALIDATION AND VERIFICATION

Model validation is the assessment of accuracy and precision, and a thorough test of whether a previously calibrated parameter set is generally valid. In other words, validation in a strict sense requires that no input parameters should be obtained via calibration. It involves both operational and scientific examination. The scientific component should assess the consistency of the predicted results with the prevailing scientific theory. It may not be perfect in the case of empirical models. The evaluation should be done through statistical analyses of observed and predicted data. The model performance is accepted if there is no significant difference between the observed and predicted data. Under- or over-prediction by the model may be characterized through many factors of analysis like the modeling efficiency (EF) (Wright et al.[67]). If EF is less than zero, it means that the model predictions are worse than the observed mean, and refinement of the model may be necessary. Graphical displays can also be used to test the model performance because they will show the trend, type of errors, and distribution patterns. For example, the nutrient component of the GLEAMS model was validated with readily available published data over a range of soils, climate, and management scenarios (Knisel and Davis[41]).

Bergstrom and Jarvis[63] provided results of a comprehensive evaluation of pesticide leaching models in a special issue of the *Journal of Environmental Sciences and Health*. Models evaluated included CALF by Nichols,[68] PRZM by Mueller,[69] GLEAMS by Shirmohammadi and Knisel,[8] PELMO by Klein,[70] PLM by Hall,[71] PESTLA by Boesten,[72] and MACRO by Jarvis et al.[26] All these models used a single set of bentazon and dichlorprop pesticide leaching data to calibrate the models and then used another set of data on the same pesticides to validate the models. Measured leaching data used during the validation phase was not made available for the users before the simulations were complete. This model evaluation exercise indicated that both caution with input parameter values and careful interpretation of the output results are needed for each of the models tested in this study. It also indicated that models should not be used beyond the conditions for which they are developed. Thomas et al.[73] provided a comprehensive discussion on the use and application of nonpoint source pollution models, including their evaluation and validation.

9.6.5 DOCUMENTATION

A good documentation report is essential to the effective completion of a modeling study. Because of many changes in parameter values, boundary conditions, and even modeling strategies between the start and finish of the model development, documentation becomes very crucial. It becomes almost impossible for another modeler to reconstruct the original modeler's ideas without proper documentation. Therefore, a good documentation of the various steps in the model development is essential. It should list chronologically the purpose of each model run, the changes in the input file, the rationale for the changes, and the effect of changes on the results. Maclay and Land[74] showed that the report should contain the following materials and any related extra information: (1) purpose, (2) formulation, (3) assumptions, (4) governing equations, (5) boundary and initial conditions, (6) parameters, (7) grid of the numerical model, (8) calibration results, (9) sensitivity analysis, (10) results, and (11) references. The modeler should also provide sufficient data so that the reader can understand and reproduce the results. Table 9.1 indicates our assessment of the quality documentation for some selected models.

9.6.6   MODEL SUPPORT AND MAINTENANCE

Managing the models over a long period needs continuous support. Constant monitoring of data may be necessary for long-term estimation by modeling. Managing the water quality is done by assessing the existing or future uses of a water body. This will detect the long-term trends or changes in the water quality, and also may provide background data for future purposes. Recently developed models may contain concepts and parameters that require new data not available from earlier data collection projects. The new data also helps in checking if the model predictions are agreeable. To monitor the parameters continuously over time, the means of measuring the parameters need to be maintained. It involves several monitoring stations with several instruments for recording the data, timely retrieval of the data, and periodic checking. If the model is supported by several users, then the model may even become refined over time. Support provided by USDA-ARS to maintain the GLEAMS model and the U.S. Environmental Protection Agency support of the PRZM-2 model are examples of model support and maintenance.

9.7   WATER QUALITY MODELS AND THE ROLE OF GIS

Geographic Information Systems (GIS) are DataBase Management Systems (DBMS) for georeferenced spatial data. These systems were originally developed for automated map production (Monmonier[75]) but have since been applied to a variety of spatial analysis problems in the areas of ecology, epidemiology, and the environment (Moilanen and Hanski,[76] Matthew,[77] Goodchild et al.[78]). GIS have been applied to the analysis of water quality (WQ) problems since the early 1980s (Logan et. al.[79]) and their use in this area has steadily increased since.

GIS can be viewed as extensions of standard DBMS that provide tools for storage, processing, and visualization of spatially distributed data. The spatial data stored in a GIS are georeferenced, their positions are specified in relation to an earth-centered coordinate system (Wolf and Brinker[80]). These data are typically stored in one of two formats—vector or raster—where, in the former, the positions of feature boundaries are specified explicitly as lists of coordinates whereas, in the latter, positions are specified implicitly using a grid of square pixels (Samet[81]). Vector format is often judged best for cartography, whereas raster format is considered best for modeling because it directly provides the spatial discretization required by numerical solution techniques (Vieux and Gauer,[82] Montas et. al.[35]). Data stored in a GIS are further characterized by their map scale which specifies their accuracy (Wolf and Brinker[80]). Small-scale data (e.g., 1:250,000) cover large areas with positional accuracies of the order of 100 m or less, whereas large-scale data (e.g., 1:24,000) typically cover smaller areas with accuracies of the order of 10 m or better. These data may come from a variety of sources including ground surveys, remote sensing, and hard-copy or digital maps. Remote sensing is particularly well suited to data acquisition for GIS-based WQ analysis, because it provides high-resolution and up-to-date data (Lillesand and Kiefer[83]). Current commercial earth-orbiting satellites that can be used for this purpose include IKONOS, IRS, SPOT-4, and the Landsat Thematic

Mapper (TM), with spatial resolutions of 1m to 25 m and 1 to 8 bands of data. Digital maps are also being increasingly used as data sources for GIS analysis. In the U.S., many such digital data products are made available to the public by the USGS, USDA, and EPA, on the Internet (e.g., ⟨at mcmcweb.er.usgs.gov, edcwww. cr.usgs. gov, ftw.nrcs.usda.gov⟩ and ⟨epa.gov/oppe/spatial.html⟩).

GIS is being increasingly used to store, process, and visualize the spatial and non-spatial (attribute) data used for WQ modeling (Goodchild et al.[78]). They have been applied at field, watershed, and regional scales with quantitative analysis tools ranging from WQ indices to detailed, physicallybased process models. Four levels of GIS-model linkages have been used: no direct linkage, nongraphical file-transfer interfaces, Graphical User Interfaces (GUI), and integration of the model inside the GIS. The scale of analysis, type of quantitative tool, and linkage level are generally interrelated. For example, index-based techniques are often used over large areas (e.g., region or river basin) and implemented within the GIS using its data overlay facilities (Johnes,[84] Navulur and Engel,[85] Secunda et. al.[86]). Conversely, detailed models are typically applied over small areas (e.g., a single field) and have either no direct linkage or a nongraphical interface with the GIS (Searing et. al.,[18] Wu et al.[87]). Intermediate scale WQ modeling of nonpoint source (NPS) pollution over watersheds is often performed with models of intermediate descriptiveness and GIS linkage levels that range from nongraphical interfaces to full integration.

Although the original application of GIS in WQ modeling was on a regional level (Logan et. al.[79]), they are being increasingly used to perform field-level WQ analyses. Searing et. al.,[18] for example, used a GIS to derive appropriate input parameters for GLEAMS that they then used to evaluate the effectiveness of BMPs at the field level. A WQ index had been previously integrated in the GIS (ERDAS Inc. IMAGINE) and used, at the watershed level, to identify fields with high pollution potential (critical areas) on which GLEAMS was then run (Searing and Shirmohammadi,[88] Shirmohammadi et. al.[59]). Another example is Wu et. al.,[87] who used a GIS (ESRI Inc. Arc/Info) to separate a heterogeneous 30-ha plot into 34 homogeneous zones and then applied GLEAMS to each of these units in a stochastic framework to evaluate the effects of heterogeneity on nitrate leaching. In both cases, the GIS was used to support field-level analysis but there was no direct linkage between GIS and model. Foster et al.[89] developed interfaces between GLEAMS and the USA CERL GRASS GIS (U.S. Army Construction Engineering Research Lab Geographical Resources Analysis Support System). They applied the GIS and model in a two-scale approach similar to that of Searing and Shirmohammadi[88] where critical areas are identified first at the watershed level and GLEAMS is then used to evaluate BMPs. Field level WQ applications of GIS that explicitly consider spatial variability are also being developed to support precision farming activities. Mulla et al.,[90] for example, integrated WQ index calculations in a farm-scale GIS to precisely identify zones of high pesticide leaching potential within this small area. Verma et al.[91] used GIS-calculated indices to identify minimal spray zones associated with active subsurface drains in east-central Illinois in support of variable rate application of agrichemicals. Field and farm-level combinations of GIS and modeling are expected to become more prominent in the future because they have the potential to conjunctively promote crop yield and WQ.

GIS and WQ modeling are often combined in watershed scale analysis of NPS pollution. The reason is probably that distributed parameter hydrologic models used in this application require extensive data sets that are tedious to prepare without appropriate data management tools. Several interfaces have hence been developed between GIS and WQ models. The AGNPS model, for example, has been interfaced with GRASS by Line et. al.,[92] Arc/Info by Haddock and Jankowski,[93] Liao and Tim,[94] and Generation 5 Technology Inc. Geo/SQL by Yoon.[95] Similarly, ANSWERS has been interfaced with GRASS (Rewerts and Engel[96]) and with GIS developed in-house (Montas and Madramootoo,[97] DeRoo[98]). The updated version of SWRRB—SWAT—has also been interfaced to both GRASS (Srinivasan and Arnold[99]) and Arc/Info (Bian et al.,[100] Ersoy et. al[101]). In all of these examples, the WQ model and GIS retain their distinct identities and are developed independently by different groups of individuals. The GIS model interface itself is often developed by a third group. The interface generally provides significant support for preparing input files, running the model, and visualizing its results. However, the fact that the model, interface, and GIS are of different origins may cause compatibility problems between each upgraded version of individual components, not to mention operating system and CPU type (Bekdash et al.[102]). One way of avoiding such problems, and the development of external interfaces altogether, is to integrate the model in the GIS. For example, Vieux and Gauer[82] integrated a finite element surface flow model in GRASS using the C language (McKinney and Tsai[103]), and Montas et al.[35] developed subsurface and surface flow and transport models, respectively, directly inside of a GIS using its high-level scripting language. In these cases, the models have direct access to GIS data and do not require file-formatting interfaces. They are run from within the GIS, using its native user interface, but cannot be used independently. The major advantage of the approach is in portability because the models are expected to run, without modification, on any platform where the GIS is installed.

Regional WQ modeling analyses have benefited from GIS in much the same way as larger-scale analyses. The GIS typically stores the spatial data required for the analysis and permits visualization of spatially distributed results. Because regional analyses are most often performed with WQ indices, the GIS also performs the required processing of spatial data. Shuckla et al.[104] used this approach with an Attenuation Factor (AF) to classify Louisa County, VA, into zones having unlikely high potential for pesticide contamination of groundwater. Navulur and Engel[85] implemented the SEEPAGE and DRASTIC WQ indices in a GIS and used them to determine groundwater vulnerability to nitrate pollution over the state of Indiana. Zhang et al.[105] and Secunda et al.[86] implemented modified DRASTIC indices in GIS and used them to evaluate groundwater vulnerability to NPS pollution in Goshen County, Wyoming, and the Sharon coastal region of Israel, respectively. A similar technique was used by Fraser et al.[106] to determine the potential for pathogen loading from livestock in a tributary of the Hudson River. Regional WQ analyses are also starting to be performed using physically based models rather than indices. The HUMUS project, for example, integrates GRASS and SWAT to perform WQ modeling at scales that can exceed the conterminous U.S. (Srinivasan et. al.[107]). One can certainly expect that the application of GIS-driven physically based models at regional scales will increase in the future.

As linkages between GIS and WQ models reach maturity, new research avenues for GIS model interaction emerge. One important avenue of research is the addition of graphical, statistical, and qualitative analysis tools to the model GIS to form Decision Support Systems (DSS). The additional tools are meant as aids for decision-making processes that use WQ modeling results. The US EPA has recently developed such a DSS that links HSPF and other models and indices with Arc/View GIS of ESRI (Lahlou et. al.[57]). The DSS incorporates several graphical and statistical analysis and reporting tools. Similarly, USDA researchers developed a DSS for nutrient management on beef-ranch operations that integrates a GIS, WQ model, and economic analysis tools (Fraisse and Campbell[108]). Advanced DSSs that incorporate Artificial Intelligence (AI) to aid in the selection of BMPs based on simulation results and GIS data are also being developed by researchers (Montas and Madramootoo,[97] Foster et. al.,[89] Montas et. al.[36]). Another emerging research area is the Internet delivery or operation of GIS-driven WQ models. Internet delivery permits remote access to GIS data, WQ models, and analysis tools, possibly through hand-held devices in the field, and significantly decreases the likelihood of compatibility problems between WQ analysis tools (e.g., model, GIS, and interface). Examples of Internet-oriented systems are quite scarce at present (Srinivasan et. al.,[107] Line et. al.,[92] Lee et. al.[109]), but their number is expected to increase rapidly in the future. A third research area is in the expansion of GIS dimensionality. Because of their origins in cartography, most GIS are overwhelmingly two-dimensional and static in nature. Most spatial data used in WQ analyses are, however, three-dimensional and often time-dependent. Research is needed to develop and apply 3-D data structures and processing techniques to improve the capabilities of current GIS-WQ-modeling systems (Lee et. al.,[109] Tempfli,[110] Lin and Calkins[111]). Finally, results of WQ analyses performed with GIS and models are typically interpreted deterministically, suggesting that both data and process equations are known with infinite precision. Spatial data used in WQ analyses are, however, often highly variable over a wide range of scales and hence best characterized statistically using, at least, a mean and variance. This suggests that stochastic approaches to data storage and process modeling will play an increasing role in future GIS-based WQ modeling analyses (Bonta (112) Fisher.[43]

9.8 USE AND MISUSES OF WATER QUALITY MODELS

Models, whether index-based such as DRASTIC or process-based and management models such as PRZM and GLEAMS, and research-oriented models such as MACRO can be used in one or all of the following ways:

(1) Models can be used to evaluate the potential loadings of agricultural chemicals such as nutrients and pesticides to surface water and groundwater systems based on the soil, geology, culture and, climatic characteristics of any given physiographic region.

(2) Models can be used to identify the impact of climatic variations on chemical loadings to groundwater.

(3) Models can also be used to identify the critical areas regarding the chemical loading to the groundwater, which can assist in selecting the field monitoring site.

(4) Models can help to evaluate the timing and frequency of sampling for a field monitoring project such that the sampling time will coincide with the recharge periods.

(5) Models can help to identify the degree of vulnerability of each aquifer system based on its hydrogeologic setting and other relevant physical and hydrologic characteristics.

(6) Models can be used to evaluate the relative impacts on different agricultural (BMPs) on nutrient and pesticide loadings to groundwater.

(7) Models can be used to evaluate the environmental and economic feasibility of system of BMPs under variable conditions.

(8) Models can provide an in-depth understanding of the pathyways through which chemicals move. This can help to implement BMPs in a proper manner to remediate the pollution problem.

(9) Models can also help to evaluate the significance of processes such as macropore flow on groundwater loading of chemicals.

Recognizing the model classifications and using them within the frame of their capability is an extremely vital principle and is most often a violated one. A common error made by model users is that they tend to consider the simulation results as true and absolute for unknown conditions. Output of a model may be affected by input errors as wells as algorithm errors (Scheid,[114] Loague and Green[115]). Model errors may be caused by incorrect or undue simplification of representing process in the model (Russel et al.[116]). Novotny and Olem[23] indicated that errors in nonpoint source pollution increase with the size of the watershed for which the model is being applied. They also reported lower confidence on model simulations for biological constituents such as bacteria than chemicals, sediments, and hydrology. Therefore, it should be kept in mind that nonpoint source pollution models try to represent complexities of the natural environment with all its associated heterogeneities, thus they seldom are perfect. Following may be possible guidelines to follow in using models:

(1) Perform a sensitivity analysis on model parameters using a reliable set of measured data and identify the most sensitive parameters in the model.

(2) Calibrate the model by the same set of data used to perform the sensitivity analysis.

(3) Validate the applicability of the model using a set of measured data other than the set that was used in steps 1 and 2 above.

(4) Apply the model to any area or condition of interest and interpret the output within the range of the capabilities of the model. For instance, models built to simulate the relative impacts of different agricultural

practices on hydrologic and water quality response of watersheds should not be used as the absolute predictors.

(5) Keep in mind the uncertainties in the model simulations and apply the results with caution.

REFERENCES

1. Foster, S. S. D., A. K. Geake, A. R. Lawrence, and J. N. Parker, *Memories of the 18th Congress of the International Association of Hydrogeologists,* Cambridge, 168, 1985.
2. Angle, S. J., V. A. Bandel, D. B. Beegle, D. R. Bouldin, H. L. Brodie, G. W. Hawkins, L. E. Lanyon, J. R. Miller, W. S. Reid, W. F. Ritter, C. B. Sperow, and R. W. Weismiller, Extension Service, *Chesapeake Basin Bull.,* 308, 1986.
3. Burt, T. P., B. P. Arkell, S. T. Trudgill, and D. E. Walling, *Hydrol. Proc.,* 2, 267, 1988.
4. Cohen, S. Z., C. Eiden, and M. N. Lorber, Monitoring groundwater for pesticides in the USA, in: Evaluation of Pesticides in groundwater, W. Y. Garner, R. C. Honeycutt, and H. N. Nigg, ACS Symposium Series, No. 315, Am. Chem. Soc., Washington, D.C., 170, 1986.
5. National Research Council, Pesticide and groundwater quality: issues and problems in four states. National Academy Press, Washington, D.C., 1986
6. Shirmohammadi, A. and W. G. Knisel, Irrigated agriculture and water quality in the south. *J. Irrig. Drainage Eng.,* 115(5), 791, 1989.
7. Torstensson, L. and J. Stenstrom, Persistence of herbicides in forest nursery soils. *Scand. J. For. Res.,* 5, 457, 1990.
8. Shirmohammadi, A. and W. G. Knisel, Evaluation of GLEAMS model for pesticide leaching. *J. Environ. Sci. Health—Part A, Environ. Sci. Eng.,* A29 (6), 1167, 1994.
9. Shoemaker, L. L., W. L. Magette, and A. Shirmohammadi, Modeling management practice effects on pesticide movement to groundwater. *Ground Water Mon. Rev.,* X(1), 109, 1990.
10. Knisel, W. G. (Ed.), A field scale model for chemical, runoff, and erosion from agricultural management systems. Conservation Service Report 26, U.S. Department of Agriculture, Washington, DC, 1980.
11. Pacenka, S. and T. Steenhuis, User's Guide for MOUSE computer program. Agricultural Engineering Department, Cornell University, Ithaca, NY, 1984.
12. Baker, D. B., Regional water quality impacts of intensive row crop agriculture: A Lake Erie Basin Case Study. *J. Soil Water Conserv.* 40(1), 125, 1985.
13. Donigian, A. S., Jr., and R. F. Carsel, Overview of terrestrial processes and modeling, in *Vadose Zone Modeling of Organic Pollutants,* S. C. Hern and S. M. Melancon (Eds.), Lewis Publishers, Chelsea, MI, 1986.
14. Leonard, R. A., W. G. Knisel, and D. A. Still, GLEAMS: groundwater Loading Effects of Agricultural Management Systems. *Trans. ASAE* 31(3), 776, 1987.
15. Shirmohammadi, A., L. L. Shoemaker, and W. L. Magette, Model simulation and regional pollution reduction strategies. *J. Environ. Sci. Health—Part A, Environ. Sci. Eng.,* A27(8), 2319, 1992.
16. Roka, F. M., B. V. Lessley, and W. L. Magette, Economic effects of soil conditions on farm strategies to reduce agricultural pollution. *Water Res. Bull.,* 25(4), 821, 1989.
17. Hamlett, J. M., D. A. Miller, R. L. Day, G. W. Peterson, G. M. Baumer, and J. Russo, Statewide GIS-Based Ranking of Watersheds for Agricultural Pollution Prevention. *J. Soil Water Conserv.,* 47(5), 339, 1992.

18. Searing, M. L., A. Shirmohammadi, and W. L. Magette, Utilizing GLEAMS model to prescribe best management practices for critical areas of a watershed identified using GIS. ASAE Paper No. 95-3248, ASAE, St. Joseph, MI 49085, 1985.

19. Woolhiser, D. A. and D. L. Brakensiek, Hydrologic System Synthesis, in Haan, C. T., H. P. Johnson, and D. L. Brakensiek (eds.) *Hydrologic Modeling of Small Watersheds.* ASAE Nomograph Number 5, published by Am. Soc. of Ag. Eng, 3, 1982.

20. Hempel, C. G., Explanation and prediction by covering laws, in B. Baumrin (Ed.), *Philosophy of Science,* The Delaware Seminar, Interscience. John Wiley and Sons, NY and London, 107, 1963.

21. Hagerstrand, T, Landskapsmanteln. Input to a dialogue on "Understanding landscape changes." Swedish Council for Planning and Coordination of Research, Friiberghs Herrgard, 8, 1992.

22. Falkenmark, M. and Z. Milulski, The key role of water in the landscape system—conceptualization to address growing human landscape pressure. *GeoJournal,* 33.4, 55, 1994.

23. Novotny, V. and H. Olem, *Water Quality: Prevention, Identification, and Management of Diffuse Pollution.* Van Nostrand Reinhold. New York, 1054, 1994.

24. Piedrahita, R. H., et al., Computer Applications in Pond Aquaculture: Modeling and Decision Support Systems, Unpublished paper., 1994.

25. Chow, V. T, Hydrologic Modeling. *J. Boston Soc. Civ. Eng.,* 60,1, 1972.

26. Jarvis, N. J., L. F. Bergstrom, and C. D. Brown, Pesticide leaching models and their use for management purposes, in T. R. Roberts and P. C. Kearney (Eds.), *Environmental Behavior of Agrochemicals,* 185, 1995.

27. Coyne, K. J., A. Shirmohammadi, H. J. Motnas, and T. J. Gish, Prediction of Pesticide transport Through the Vadose Zone Using Stochastic Modeling. ASAE Paper No. 992066, ASAE, St. Joseph, MI 49085, 1999.

28. Jury, W. A, Simulation of Solute Transport Using a Transfer Function Model. *Water Resources Res.* 18, 363, 1982.

29. Carsel, R. F., R. L. Jones, J. L. Hansen, R. L. Lamb, and M. P. Anderson, A simulation procedure for groundwater quality assessments of pesticides. *J. Contam. Hydrol.,* 2, 125, 1988a.

30. Carsel, R. F., R. S. Parrish, R. L. Jones, J. L. Hansen, and R. L. Lamb, Characterizing the uncertainty of pesticide leaching in agricultural soils. *J. Contam. Hydrol.,* 2, 111, 1988b.

31. Petach, M. C., R. J. Wagenet, and S. D. DeGloria, Regional water flow and pesticide leaching using simulations with spatially distributed data. *Geoderma,* 48, 245, 1991.

32. Zhang, H., C. T. Haan, and D. L. Nofziger, An approach to estimating uncertainties in modeling transport of solutes through soils., *J. Contam. Hydrol.,* 12, 35, 1993.

33. Nofziger, D. L., S. S. Chen, and C. T. Haan, Evaluating the chemical movement in layered soil model as a tool for assessing risk of pesticide leaching to groundwater, *J. Environ. Sci. Health,* A29, 1133, 1994.

34. Donigian, A. S., J. C. Imhoff, B. R. Bricknell, and J. L. Kittle, Application Guide for Hydrological Simulation Program FORTRAN (HSPF). Environmental Research Laboratory, U.S. Environmental Protection Agency, Athens, GA, 1993.

35. Montas, H. J., A. Shirmohammadi, P. Okelo, A. M. Sexton, J. S. Butler, and T.-W. Chu, Targeting Agrichemical Export Hot Spots in Maryland using Hydromod and GIS. Paper No. 99-3123, ASAE, St. Joseph, MI 49085, 1999a.

36. Montas, H. J., A. Shirmohammadi, J. S. Butler, T.-W. Chu, P. Okelo, and A. M. Sexton, Decision Support for Precise BMP Selection in Maryland. Paper No. 99-3049, ASAE, St. Joseph, MI 49085, 1999b.

37. Arnold, J. G., J. R. Williams, R. Srinivasan, and K. W. King, SWAT-Soil and Water Assessment Tool. USDA-ARS, Temple, TX, 1996.

38. Chu, T. W., A. Shirmohammadi, and H. J. Montas, Validation of SWAT Model's Hydrology Component on Piedmont Physiographic Region. ASAE Paper No. 992105, ASAE, St. Joseph, MI 49085, 1999.

39. Bouraoui, F. and T. A. Dillaha, ANSWERS-2000: Runoff and sediment transport model. *J. Environ. Eng.*, ASCE 122(6), 1996.

40. Ghadiri, H. and C. W. Rose, *Modeling Chemical Transport in Soils: Natural and Applied Contaminants.* Lewis Publishers, Boca Raton, Florida, 217, 1992.

41. Knisel, W. G. and F. M. Davis, GLEAMS: Groundwater Loading Effects of Agricultural Management Systems, Version 3.0, Users Manual. USDA- Agricultural Research Service, Southeast Watershed Research Laboratory, Tifton, GA—SEWRL-WGK/FMD-050199, 182, 1999.

42. Larsson, M. H. and N. J. Jarvis, Evaluation of a Dual Porosity Model to Predict Field-Scale Solute Transport in a Macroporous Soil. *J. Hydrol.* 215, 1999.

43. Hutson, J. L., and R. J. Wagenet, A Pragmatic Field-Scale Approach for Modeling Pesticides. *J. Environ. Qual.,* 22, 494, 1993.

44. Mullins, J., R. Carsel, J. Scarbough, and A. Ivery, PRZM-2, A Model for Predicting Pesticide Fate in the Crop Root and Unsaturated Soil Zones: User's Manual for Release 2.0; EAP/600/R-93/046; Athens, GA, U.S. Environmental Protection Agency, 1993.

45. Nichols, P. H. and D. G. M. Hall, Use of the pesticide leaching model (PLM) to simulate pesticide movement through macroporous soils, in *Proceedings of BCPC, Pesticide Movement to Water,* (eds. A. Walker, R. Allen, S. W. Bailey, A. M. Blair, C. D. Brown, P. Gunther, C. R. Leake, and P. H. Nicholls), Warwick, U.K., Monograph No. 62, 187, 1995.

46. Hutson, J. L. and R. J. Wagenet, Multi-Region Water Flow and Chemical Transport in Heterogeneous Soils: Theory and Applications. p. 171–180, in A. Walker et al. (ed.) Pesticide Movement to Water. Proc. BCPC Symposium, Monograph No. 62, Warwick Univ., Coventry, UK, 1995.

47. Jansson, P.-E, Simulation model for soil water and heat conduction. Description of the SOIL model, Rep. No. 165, Dept. Of Soil Sci., Division of Agric. Hydrotechnique, SLU, Uppsala, Sweden, 1991.

48. Johnsson, H., L. Bergstrom, P.-E. Jansson, and K. Paustian, Simulated nitrogen dynamics and losses in a layered agricultural soil. *Agric. Ecosyst. Environ,* 18, 333, 1987.

49. Williams, J. R., P. T. Dyke, and C. A. Jones, EPIC: a model for assessing the effects of erosion on soil productivity, in *Analysis of Ecological Systems: State of the Art in Ecological Modeling,* W. K. Lauenroth et al. (Eds.), Elsevier, Amesterdam, 553, 1983.

50. Chung, S. O., A. D. Ward, and C. W. Schalk, Evaluation of the ADAPT Water Table Management Model. *Trans. ASAE,* Vol. 35(2), 571, 1992.

51. Gowda, P., A. Ward, D. White, J. Lyon, and E. Desmond, The Sensitivity of ADAPT Model Predictions of Streamflow to Parameters Used to Define Hydrologic Response Unit. *Trans. ASAE,* Vol. 42(2), 381, 1999.

52. Carsel, R. F., J. C. Imhoff, R. R. Hummel, J. M. Cheplick, and A. S. Donigian., PRZM-3, A Model for Predicting Pesticide and Nitrogen Fate in the Crop Root and Unsaturated Soil Zones: Users Manual for Release 3.0, USEPA-Athen, GA, 1998.

53. Shirmohammadi, A., K. S. Yoon, J. W. Rawls, and O. H. Smith, Evaluation of Curve Number Procedures to Predict Runoff in GLEAMS. *J. An. Water Res. Assoc.,* 33(5), 1069, 1997.

54. Young, R. A., C. A. Onstad, D. D. Bosch, and W. P. Anderson, ANGPS: a nonpoint source pollution model for evaluating agricultural watersheds. *J. Soil Water Conserv.* 44(2), 168, 1989.

55. Arnold, J. G., Williams, J. R., Nicks, A. D., and Sammons, N. B, *SWRRB, A Basin Scale Simulation Model for Soil and Water Resources Management.* Texas A&M University Press, College Station, Texas, 143, 1990.

56. Donigian, A. S., and N. H. Crawford, Modeling pesticides and nutrients on agricultural lands. USEPA, Environmental Protection Technology Series, EPA-600/2-76-043, Washington, D.C., 1976.

57. Lahlou, M., L. Shoemaker, S. Choudhury, R. Elmer, A. Hu, H. Manguerra, A. Parker, Better Assessment Science Integrating Point and Nonpoint Sources, BASINS Verson 2.0 User's Manual. EPA-823-B-98-006, Washington, D.C., 1998

58. Aller, L., T. Bennett, J. H. Lehr, and R. Petty, DRASTIC: A system to evaluate the pollution potential of hydrogeolgic settings by pesticides. *Proceedings of the American Chemical Society Symposium,* Washington D.C., February 19–20, 141, 1986.

59. Shirmohammadi, A., H. J. Montas, M. L. Searing, and A. Gustafson, *Proceedings of Work Sciences, CIGR-CIOSTA, XXII,* June 14–17, 1999, Horsen, Denmark, 418, 1999.

60. Renard, K. G, Introduction to models, in Belsley D. B., W. G. Knisel, and A. P. Rice (eds.), *Proceedings of the GLEAMS/CREAMS Symposium,* Sept. 27–29, 1989, Athens, GA, Published by Univ. of Georgia—Coastal Plain Experiment Station, Tifton, GA, 3, 1989.

61. Sargent, R., Simulation Model Validation, Simulation and Model-Based Methodologies: An Integrative View, NATO ASI Series, Vol. F10, 1984.

62. Haan, C. T, Parameter uncertainty in hydrologic modeling. *Trans. ASAE,* 32(1), 137, 1989.

63. Bergstrom, L. and N. Jarvis, editors, Special Issue on the Evaluation and Comparison of Pesticide Leaching. *J. Environ. Sci. Health, Part A—Environ. Eng.* A29(6), 1061, 1994.

64. Huber, W. C., et al., Urban Rainfall-Runoff-Quality Database, EPA 600/S2-81-238, U.S. Environmental Protection Agency, Cincinnati. O.H., in Novonty, V. and H. Olem (eds.), 1994. *Water Quality Prevention, Identification, and Management of Diffuse Pollution.* Van Nostrand Reinhold, New York, 507, 1982.

65. Boesten, J. J. T. I, Sensitivity analysis of a mathematical model for pesticide leaching to groundwater. *Pestic. Sci.* 31, 375, 1991.

66. Wei, S., A. Shirmohammadi, A. Sadeghi, and W. J. Rawls, Predicting Atrazine Leaching Under Field Conditions Using MACRO Model. ASAE Paper No. 992067, ASAE, St. Joseph, MI 49085, 1999.

67. Wright, J. A., A. Shirmohammadi, W.L. Magette, J.L. Fouss, R.L. Bengston, and John E. Parsons, Combined WTM and PMP effects on water quality. *Trans. ASAE,* 35(3), 823, 1992.

68. Nichols, P. H., Simulation of the movement of bentazon in soils using the CALF and PRZM models. *J. Environ. Sci. Health,* A29(6), 1157, 1994.

69. Mueller, T. C., Comparison of PRZM computer model predictions with field lysimeter data for dichlorporp and bentazon. *J. Environ. Sci. Health,* A29(6),1183, 1994.

70. Klein, M., Evaluation of comparison of pesticide leaching models for registration purposes. Results of performed by Pesticide Leaching Model. *J. Environ. Sci. Health,* A29(6), 1197, 1994.

71. Hall, D. G., Simulation of dichlorprop leaching in three texturally distinct soils using the pesticide leaching model. *J. Environ. Sci. Health,* A29(6), 1211, 1994.

72. Boesten, J. J. T. I., Simulation of bentazon leaching in sandy loam soil from Melby (Sweden) with the PESTLA model. *J. Environ. Sci. Health,* A29(6), 1231, 1994.

73. Thomas, D. L., R. O. Evans, A. Shirmohammadi, and B. A. Engel, Agricultural nonpoint source water quality models: their use and applications. ASAE Paper #98-2193, ASAE, St. Joseph, MI 49085, 1998

74. Maclay, R. W. and L. F. Land, Simulation of flow in the Edwards aquifer, San Antonio region, Texas, and refinement of storage and flow concepts, USGS, Water Supply Paper 2336-A, pp. A1-A48, in Anderson, P.M. and W.W. Woessner. 1992. *Applied groundwater Modeling.* Academic Press, Inc., San Diego, 275, 1988.

75. Monmonier, M. S., *Computer-Assisted Cartography: Principles and Prospects.* Prentice-Hall, Inc. Englewood Cliffs, NJ, 1982.

76. Moilanen, A. and I. Hanski, Metapopulation Dynamics: Effects of Habitat Quality and Landscape Structure. *Ecology,* 79(7), 2503, 1998.

77. Matthews, S. A., Epidemiology using a GIS. The Need for Caution. *Comput. Environ. Urban Sys.,* 14(3), 213, 1990.

78. Goodchild, M. F., L. T. Steyaert, B.O. Parks, C. Johnston, D. Maidment, M. Crane and S. Glendinning (Eds.), *GIS and Environmental Modeling: Progress and Research Issues.* GIS World Books, Boulder, CO, 1996.

79. Logan, T. J., D. R. Urban, J. R. Adams, and S. M. Yaksich, Erosion Control Potential with Conservation Tillage in the Lake Erie Basin: Estimates using the Universal Soil Loss Equation and the Land Resource Information System (LRIS). *J. Soil Water Conserv.,* 31(1), 50, 1982.

80. Wolf, P. R. and R. C. Brinker, *Elementary Surveying, 9th ed.* Harper Collins Pub., New York, NY, 1994

81. Samet, H., *The Design and Analysis of Spatial Data Structures,* Addison Wesley Inc., New York, NY, 1990.

82. Vieux, B. E. and N. Gauer, Finite-Element Modeling of Storm Water Runoff Using GRASS GIS. *Microcomp. Civil Eng.,* 9, 263, 1994.

83. Lillesand, T. M. and R. W. Kiefer, *Remote Sensing and Image Interpretation, 2nd ed.* John Wiley and Sons, Inc., New York, NY, 1987.

84. Johnes, P. J., Evaluation and Management of the Impact of Land Use Change on the Nitrogen and Phosphorus Load Delivered to Surface Waters: The Export Coefficient Modeling Approach. *J. Hydrol.,* 183(3–4), 323, 1996.

85. Navulur, K. C. S. and B. A. Engel, Groundwater Vulnerability Assessment to Non-Point Source Nitrate Pollution on a Regional Scale using GIS. *Trans. Am. Soc. Agric. Eng.,* 41(6), 1671, 1998.

86. Secunda, S., M. L. Collins, and A. J. Melloul, Groundwater Vulnerability Assessment using a Composite Model Combining DRASTIC with Extensive Agricultural Land Use in Israel's Sharon Region. *J. Environ. Manage.,* 54(1), 39, 1998.

87. Wu, Q. J., A. D. Ward, and S. R. Workman, Using GIS in Simulation of Nitrate Leaching from Heterogeneous Unsaturated Soils. *J. Environ. Qual.,* 25(3), 526, 1996.

88. Searing, M. L. and A. Shirmohammadi, The Design, Construction and Analysis of a GIS Database for use in Reducing Nonpoint Source Pollution on an Agricultural Watershed. ASAE Paper No. 94-3551, ASAE, St. Joseph, MI 49085, 1994

89. Foster, M. A., P. D. Robillard, R. Zhao, and L. E. Low, An Expert GIS/Modeling System for Water Quality Control Practices. GIS/LIS '94 Proceedings. Bethesda: ACSM-ASPRS-AAG-URISA-AM/FM, 1994, 1, 331, 1994.

90. Mulla, D. J., C. A. Perillo, and C. G. Cogger, A Site-Specific Farm-Scale GIS Approach for Reducing Groundwater Contamination by Pesticides. *J. Environ. Qual.,* 25(3), 419, 1996.

91. Verma, A. K., R. A. Cooke, M. C. Hirschi, and J. K. Mitchell, GIS and GPS Assisted Variable Rate Application (VRA) of Agri-Chemicals. *J. Geogr. Info. Syst. Dec. Anal.*, 2(1), 17, 1998.

92. Line, D. E., S. W. Coffey, and D. L. Osmond, WATERSHEDSS GRASS-AGNPS Model Tool. *Trans. Am. Soc. Agric. Eng.*, 40, 971, 1997.

93. Haddock, G. and P. Jankowski, Integrating Nonpoint Source Pollution Modeling with a Geographic Information System. *Comput. Environ. Urban Sys.*, 17, 437, 1993.

94. Liao, H. H. and U. S. Tim, An Interactive Environment for Non-Point Source Pollution Control. *J. Am. Water Res. Assoc.*, 33(3), 591, 1997.

95. Yoon, J., Watershed-Scale Nonpoint Source Pollution Modeling and Decision Support System Based on a Model-GIS-RDBMS Linkage. Presented at the AWRA Symposium on GIS and Water Resources, Sept. 22–26, 1996, Ft. Lauderdale, FL, 1996

96. Rewerts, C. C. and B. A. Engel, ANSWERS on GRASS: Integrating a Watershed Simulation Model with a GIS. ASAE Paper No. 91-2621. ASAE, St. Joseph, MI 49085, 1991

97. Montas, H. J. and C. A. Madramootoo, A Decision Support System for Soil Conservation Planning. *Comput. Elect. Agric.*, 7, 187, 1992.

98. DeRoo, A. P. J., Modelling surface runoff and soil erosion in catchments using geographical information systems: validity and applicability of the "ANSWERS' model in two catchments in the loess area of South Limburg (the Netherlands) and one in Devon (UK). *Nederlandse Geograf. Stud.* 157, 295, 1993.

99. Srinivasan, R. and J.G. Arnold, Integration of a basin scale water quality model with GIS. *Water Resource Bull.*, 30(3), 453, 1994.

100. Bian, L., H. Sun, C. Blodgett, S. Egbert, W. Li, L. Ran, and A. Koussis. An Integrated Interface System to Couple the Swat Model and Arc/Info. Third International NCGIA Conference/Workshop on Integrating GIS and Environmental Modeling, Santa Fe, New Mexico, January 21–25, 1996.

101. Ersoy, Y. Y., C. J. Skonard, J. Arumi, D. L. Martin, and D. G. Watts, Evaluation of Best Management Practices using an Integrated GIS and SWAT Model for Field Sized Areas. American Society of Agricultural Engineers, August 10–14, 1997, Papers v.2, ASAE 9, 0145, 1997.

102. Bekdash, F. A., A. Shirmohammadi, and T. H. Ifft, Three pixel window for the delineation of spacial features in a DEM/GIS. ASAE Paper No. 93-3549, American Society of Agricultural Engineers, St. Joseph, MI 49085, 1993.

103. McKinney, D. C. and H.-L. Tsai, Multigrid Methods in GIS Grid-Cell-Based Modeling Environment. *J. Comput. Civil Eng.*, 10(1), 25, 1996.

104. Shukla, S., S. Mostaghimi, V. O. Shanholtz, and M. C. Collins, A GIS-based modeling approach for evaluating groundwater vulnerability to pesticides. *J. Am. Water Res. Assoc.*, 34(6),1275, 1998.

105. Zhang, R., J. D. Hamerlinck, S. P. Gloss, and L. Munn, Determination of nonpoint-source pollution using GIS and numerical models. *J. Environ. Qual.*, 25(3), 411, 1996.

106. Fraser, R. H., P. K. Barten, and D. A. K. Pinney, Predicting stream pathogen loading from livestock using a geographical information system-based delivery model. *J. Environ. Qual.*, 27(4), 935, 1998.

107. Srinivasan, R., J. G. Arnold, and C. A. Jones, Hydrologic modeling of the United States with the soil and water assessment tool. *Int. J. Water Res. Dev.*, 14(3), 315, 1998.

108. Fraisse, C. W. and K. L. Campbell, BRADSS: A Decision Support System for Nutrient Management in Beef Ranch Operations. American Society of Agricultural Engineers, August 10–14, 1997, Papers v.1, ASAE 17, 0145, 1997.

109. Lee, H.-G., K.-H. Kim, and K. Lee, Development of a 3-Dimensional GIS Running on Internet. Int. Geosci. and Remote Sens. Symp. (IGARSS) Sponsored by IEEE, July 6–10, 1998. 2, 1046, 1998.

110. Tempfli, K., 3D Topographic mapping for urban GIS. *ITC J.*, 3–4, 181, 1998.

111. Lin, H. and H. W. Calkins, Rationale for Spatiotemporal Intersection. ACSM-ASPRS Annual Convention Technical Papers, ACSM Pub., v.2, 204, 1991.

112. Bonta, J. V., Spatial variability of runoff and soil properties on small watersheds in similar soil-map units. *Trans. Am. Soc. Agric. Eng.*, 41(3), 575, 1998.

113. Fisher, P., Pixel: A snare and a delusion. *Int. J. Remote Sens.*, 18(3), 679, 1997.

114. Scheid, F., *Schaum's Outline of Theory and Problems of Numerical Analysis.* Schaum's Outline Series, McGraw-Hill Book Company, 442, 1968.

115. Loague, K. M. and R. E. Green, Statistical and graphical methods of evaluating solute transport models: overview and application. *J. Contaminant Hydrol.*, 7, 51, 1991.

116. Russel, M. H., R. J. Layton, and P. M. Tillotson, The use of pesticide models in a regulatory setting: an industrial perspective. *J. Environ. Sci. Health,* A29(6), 1105, 1994.

117. Cronshey, R.G. and F.D. Theurer, AnnAGNPS – Non-point pollutant loading model. *Proceedings of the First Federal Interagency Hydrologic Modeling Conference.* Las Vegas, Nevada. April 19–23, 1998, 1–9, 1998.

10 Best Management Practices for Nonpoint Source Pollution Control: Selection and Assessment

Saied Mostaghimi, Kevin M. Brannan, Theo A. Dillaha III, and Adriana C. Bruggeman

CONTENTS

10.1 INTRODUCTION

Activities associated with modern agricultural practices could potentially degrade our water resources. During the 1960s, people became skeptical of the environmental benignity of agricultural chemicals on the environment, culminating in the publication of Rachel Carson's book *Silent Spring*. Other past events brought on by human activities or natural events, such as the Dust Bowl of the 1930s, demonstrated how agriculture may influence the environment. Out of disasters like the Dust Bowl, conservation programs at all levels of government evolved. These conservation programs were mainly focused on soil erosion with the goal of increasing on-farm production. Since the 1960s, the focus of conservation programs has shifted from on-farm productivity to off-farm impacts on the environment.[1] Examples of off-farm impacts include pesticide leaching to groundwater and nutrient enrichment of surface waters bodies caused by the transport of excess fertilizers and manure by agricultural runoff. The approach commonly used to minimize the off-site impacts is to implement management practices that reduce the mass of pollutants exiting the agricultural system while maintaining the system's economic viability.

Before the development of modern agrochemicals and mechanization, agriculture was commonly considered a struggle pitting farmers against nature. These farmers fed their families and the world while facing blight, locusts, and other catastrophic events. However, this depiction of an adversarial relationship between farmers and nature is not entirely true. Many ancient agricultural practices took advantage of natural processes and cycles to produce food. For example, the ancient Egyptians developed an irrigation system that utilized the flood cycles of the Nile River and grew enough crops on the edge of a desert to support a vast population. Other examples of ancient farming practices include the development of terracing and cropping systems. Terracing, which has been used throughout the world, demonstrates the farmers' intuitive understanding of the basic mechanics of soil erosion control and water conservation. Examples of cropping systems include the sabbatical year of Judea and the three sisters of the Iroquois. In ancient Judea, the land was given a rest (left in fallow) every seventh year. The three sisters of the Iroquois nation of North

America included maize, beans, and squash.[2] The Iroquois use of these three crops formed a symbiotic system for producing food. In all of these examples except terracing, farmers worked within the environmental constraints to grow crops. These environmental constraints also presented challenges to farmers who needed to produce more food for growing populations.

Modern farming practices have reduced many of the food production obstacles faced by farmers in the past. Examples of these obstacles are short-term drought, low soil fertility, pests, and weeds. Generally, modern approaches have resulted in increased yields along with new environmental problems. In most cases, these new problems are directly related to the practices and technologies that allowed farmers to overcome earlier obstacles. Current societal concerns focus on the environmental consequences of modern agricultural practices. Runoff and leachate from agricultural areas transport pollutants, such as chemicals and sediment, downstream to water bodies. These pollutants could degrade downstream water resources. Examples of these repercussions are depletion of ground water resources from excessive pumping for irrigation, eutrophication of surface water bodies by excessive use of fertilizers, and health risks related to pesticide use.

The main approach used to minimize pollution resulting from agricultural activities is implementation of Best Management Practices (BMPs). The basic paradigm of the BMP approach is to implement an economically feasible practice or combination of practices that will address a particular water quality problem. Although cost-share incentives and some regulations are used, current nonpoint pollution abatement programs rely mostly on voluntary implementation of management practices. Consequently, practices with prohibitive costs will not be accepted or implemented by landowners and may create opposition to pollution abatement programs. Therefore, when selecting BMPs, one must consider not only whether the practices will provide pollutant reductions that will achieve water quality goals, but also whether implementation of the practices is economically feasible for the parties involved. After BMPs are implemented, their effectiveness in achieving the goals of the pollution abatement program needs to be assessed. In the following sections, various BMPs are discussed with respect to pollution reductions and economic impacts along with procedures to assess their effectiveness in reducing pollutant losses.

10.2   AGRICULTURAL BEST MANAGEMENT PRACTICES

10.2.1   GENERAL CONSIDERATIONS

Before proceeding with descriptions of specific practices, a general discussion of BMPs is necessary. There is no universally accepted definition available for BMP. The Soil and Water Conservation Society (SWCS) defines a BMP as "a practice or combination of practices that are determined by a state or designated area wide planning agency to be the most effective and practicable (including technological, economic, and institutional considerations) means of controlling point and nonpoint source pollutants at levels compatible with environmental quality goals."[3] An alternative definition presented by Novotny and Olem[4] states that "BMPs are methods and practices or combination of practices for preventing or reducing nonpoint source pollution to a level compatible with water quality goals." The two definitions given here both state that the purpose of BMPs is to reduce pollutant levels to achieve water quality goals. However, the SWCS definition is more comprehensive because it also states that the practices are to be practicable. Most pollution abatement programs currently rely on voluntary compliance; therefore, the pollution control practices must be feasible if landowners are to adopt them.

In the following sections, classification of BMPs and some general characteristics are discussed. For each BMP, the discussion contains four components. The first component is the definition of the BMP, which explains the important characteristics of the practice. These characteristics relate to farm management issues and the impact of the practice on physical, chemical, and biological processes that control the generation and transport of pollutants. In the definition, the practice is also categorized either as a source reduction, transport interruption, or a combination of the two. Moreover, the BMP is classified as either a managerial or structural BMP. In the second component of the BMP classification, the situations and pollutants for which the BMP is appropriate are discussed. The discussion of these situations involves the consideration of hydrologic, topographic, economic, soils, and farm management information. The third component discusses the possible negative effects of the BMP, if any, and limitations that it may have. In the discussion of the negative effects, both environmental as well as economic aspects of the BMPs are considered. Finally, the potential combinations of practices that may increase the overall effectiveness of the BMP are discussed. In addition, the practice code used by the Natural Resource Conservation Service (NRCS) of the U.S. Department of Agriculture (USDA) is also provided. The NRCS practices codes can be used to obtain detailed descriptions of the BMPs from the National Handbook of Conservation Practices (NHCP).[5] Although many variations of BMPs can be found among different state and local agencies, the NHCP provides a description of the basic components common to many of the most frequently used BMPs. Table 10.1 provides a summary of the BMPs discussed in the following sections.

When selecting a BMP, all the physical, chemical, and biological processes affected by the practice should be considered. Some BMPs protect both surface-water and groundwater resources simultaneously. Other BMPs protect one resource at the expense of the other. The selection of BMPs depends not only on the physical and managerial characteristics of the farm, but also on the objectives and priorities of the parties involved.

The generation and transport of agricultural chemicals by surface runoff is the cause of much of the pollution of streams, rivers, lakes, and other water bodies in the U.S. Over 35% and 25% of river miles in the U.S. are impacted by sediment and nutrients, respectively.[6] These pollutants are normally associated with surface runoff. Surface water processes are usually driven by meteorological events, such as rainfall and snowmelt. These meteorological events are highly episodic, resulting in the random behavior of surface water transport processes. The main pollutants associated with surface runoff are sediment, nutrients,

TABLE 10.1
Description and Classifications of BMPs

BMP	Pollutants Treated	Type	NRCS Code(s)[5]	Major Concerns
Conservation tillage	Sediment, sediment-bound pollutants	Source reduction; managerial	329A to 329C, 344	Increased potential of groundwater pollution. Accumulation of nutrients on the soil surface.
Contour farming	Sediment, sediment-bound pollutants	Source reduction; managerial	330	Not effective on steep slopes Potential for increased erosion during highly-intense storms
Contour strip cropping	Sediment, sediment-bound pollutants	Source reduction; managerial	585	Cropland taken out of production
Field strip cropping	Sediment, sediment-bound pollutants	Source reduction; managerial	586	Cropland taken out of production
Filter strips	Sediment, sediment-bound, biological and some soluble pollutants	Transport interruption; structural	393A	Cropland taken out of production. Long-term maintenance necessary. Occurrence of concentrated flow within the strip.
Riparian buffers	Sediment, sediment-bound, biological and some soluble pollutants	Transport interruption; structural	391A	Cropland taken out of production. Nitrate retention
Cover crop	Sediment, sediment-bound and soluble pollutants	Source reduction; managerial	340	Increased use of herbicides
Conservation crop rotation	Sediment, sediment-bound and soluble pollutants	Source reduction; managerial	328	Economic risk due to fluctuating commodity prices
Nutrient management	Sediment, sediment-bound, biological and soluble pollutants	Source reduction; managerial	590	Costs associated with equipment and increased labor.
Manure storage facilities	Sediment, sediment-bound, biological and soluble pollutants	Source reduction; structural	313	Costs associated with construction. Odor.

(continued)

261

TABLE 10.1 (continued)

BMP	Pollutants Treated	Type	NRCS Code(s)[5]	Major Concerns
Integrated pest management	Sediment, sediment-bound and soluble pollutants	Source reduction; managerial	None	Increased level of training necessary. Access to specialists. Perception of economic losses by farmers.
Precision farming	Sediment, sediment-bound and soluble pollutants	Source reduction; managerial	None	Costs associated with equipment, increased labor, and information management.
Terraces	Sediment, sediment-bound pollutants	Source reduction; structural	600	Costs associated with construction and maintenance. Cropland taken out of production.
Grass-waterways	Sediment, sediment-bound pollutants	Source reduction; structural	412	Cropland taken out of production.
Diversions	Sediment, sediment-bound and soluble pollutants	Source reduction; structural	362	Construction costs.
Sediment detention basin	Sediment, sediment-bound pollutants	Source reduction; structural	350	Construction and maintenance costs. May not trap fine sediment.
Constructed wetland	Sediment, sediment-bound, biological and soluble pollutants	Transport interruption; structural	657	Land area needed may be large
Fencing and use exclusion	Sediment, sediment-bound, biological and soluble pollutants	Source reduction; structural	528 and 472	Costs associated with construction and maintenance of fence
Off-Stream water sources	Sediment, sediment-bound, biological and soluble pollutants	Source reduction; structural	None	Does not completely exclude livestock from streams
Rotational grazing	Sediment, sediment-bound, biological and soluble pollutants	Source reduction; structural and managerial	528A	Livestock need to be excluded from streams

pathogens, and pesticides. Sediment also acts as a transport vector for pollutants that are attached to soil particles. An example of this problem was presented by Meals[7] who, when addressing the NPS pollution problems in St. Alban's Bay, stated that, even with great reductions in point and nonpoint inputs of phosphorus to the Bay, reductions in phosphorus levels in the Bay were not observed. Meals[7] attributed this lack of improvement to the release of phosphorus from lake sediments. This example demonstrates that the accumulation of pollutants in the environment can contribute to pollution problems for a long time.

Surface runoff is responsible for transport of both sediment-bound and dissolved pollutants. Therefore, BMPs that reduce surface runoff or the availability of pollutants for transport by surface runoff will also reduce the potential for pollution of downstream water bodies. Some BMPs may only reduce surface runoff by increasing infiltration or increasing retention and detention of water on the soil surface. However, BMPs also need to focus on reducing the generation of surface runoff, sediment, and the availability of nutrients and pesticides. When selecting BMPs, it is important to consider the whole system.

The reason for protecting groundwater from pollution is twofold. First, groundwater serves as a drinking water resource for approximately 50% of the U.S. population. Thus, pesticide and nitrate pollution of groundwater is of potential concern in many areas of the U.S. The second reason is that groundwater can pollute surface water resources. Groundwater with high concentrations of dissolved pollutants may discharge to rivers, lakes, and larger water bodies. Effective BMPs for protecting groundwater reduce the potential for the transport of soluble pollutants from the upper soil horizons to groundwater. Therefore, it is imperative to reduce the amount of excess nutrients, manure, or pesticides on fields or pastures. With these issues in mind, some BMPs commonly used for improving water quality are discussed in the following sections.

10.2.2   CONSERVATION TILLAGE

Farmers in the United States started using conservation tillage in the 1930s. Adoption levels of the practice remained low until the widespread availability of herbicides for weed control in the 1970s. There have been steady gains in the adoption of conservation tillage by farmers. In 1983, 23% of all the cropland acres in the United States was under some form of conservation tillage and in 1993 the percentage increased to 37%.[8] Currently, there is a variety of equipment and chemicals available to farmers using conservation tillage practices. Blevins and Frye[9] offer a comprehensive review of the history and methods of conservation tillage.

There are many different forms of conservation tillage. Examples include no-tillage, mulch tillage, and other tillage operations that leave crop residue on the soil surface. Conservation tillage is defined as any production system that leaves at least 30% of the soil surface covered with crop residue after planting to reduce soil erosion by water.[9] Conservation tillage is also defined as any tillage and planting system that maintains at least 1,000 pounds per acre of flat, small-grain residue equivalent on

FIGURE 10.1   Field under conservation tillage (Source NRCS, 1998).

the surface during critical wind erosion periods.[8] An example of a field under conservation tillage is shown in Figure 10.1. The crop residue left on the soil surface protects the soil from rainfall and wind. Other examples of conservation tillage include strip tillage, ridge tillage, slit tillage, and seasonal residue management. Strip, ridge, and slit tillage refer to various methods used to till the field along the rows while minimizing the disturbance of crop residue between the rows. Examples of strip tillage and ridge tillage are shown in Figure 10.2 and Figure 10.3, respectively. For seasonal residue management, the residue is left on the field during the period between harvest and planting. Immediately before planting, most of the residue is tilled over.

The main benefit of conservation tillage is the protection provided to the soil by the crop residue. The crop residue reduces the detachment of soil particles by rainfall impact. Conservation tillage is classified as a source reduction and managerial practice that reduces sheet and rill erosion.[10–15] Researchers have reported reductions of up to 50% with every 9 to 16% increase in crop residue coverage.[16,17] This means that up to a 90% reduction in erosion rates is possible for the minimum amount of residue coverage (30%). Other benefits of conservation tillage include: (1) increased infiltration,[18–21] (2) protection from wind erosion,[9] (3) reduction in evaporation,[5]

FIGURE 10.2   Strip tillage (Source NRCS, 1998).

FIGURE 10.3   Ridge tillage (Source NRCS, 1998).

(4) increased soil organic matter and improved tilth,[22,23] and (5) increased food and habitat for wildlife (Code 329A to 329C and Code 344).[5] There are several economic benefits associated with conservation tillage compared with conventional tillage. These benefits include reduced fuel and labor costs resulting from fewer trips over the field along with a decline in machinery costs because of a smaller machinery complement.[8] One negative aspect of conservation tillage is that new or retrofitted machinery may be needed by the farmer making the transition from conventional tillage.[8]

The main management concern with conservation tillage is to leave sufficient crop residue on the field to protect the soil from erosive forces of rainfall and runoff. In Figure 10.4, residue is left on soil surface after soil has been chisel-plowed. The residue needs to be on the field during the critical periods of the year when the erosion hazard is high (i.e., immediately after harvest when no cover crop exists and the period between primary tillage and crop emergence). If residue is to be harvested via bailing or grazing, care should be taken to ensure sufficient residue remains to provide the desired amount of erosion protection. Finally, the orientation and total amount of crop residue will vary depending on the specific tillage methods used.

FIGURE 10.4   Chisel plowing in residue (Source NRCS, 1998).

The primary effect of conservation tillage on water quality is a reduction of sediment available for transport. Conservation tillage is used to mitigate erosion problems, which in turn contribute to the degradation of water quality.[24] Conservation tillage decreases the erosion potential on cropland and reduces the potential for degradation of receiving waters by sediment-attached pollutants.[11–13,25,26] By keeping the soil in place, soil resources are preserved.

Although conservation tillage is very effective in reducing erosion, there are some concerns that it may increase potential pollution by other transport processes. Conservation tillage increases infiltration and the potential for leaching of dissolved chemicals.[27] Under conventional tillage, fertilizer or manure is incorporated into the soil by direct injection or by tillage operations. Both of these operations incorporate the crop residue. Under conservation tillage, however, the manure or fertilizer is usually applied to the soil surface and not incorporated to minimize residue disruption. Thus, the nutrients tend to accumulate near the soil surface.[28] The increased nutrient level at the soil surface leads to increased nutrient concentrations in surface runoff.[11,12,16,18] Kenimer et al.[10] reported increased pesticide concentrations of sediment-bound atrazine and 2,4-D in runoff from no-till compared with concentrations in runoff from conventionally tilled plots, and concentrations of dissolved atrazine and 2,4-D in runoff increased as residue levels increased. The negative impacts could be addressed through the combination of conservation tillage with other BMPs. Conservation tillage combined with nutrient management would reduce the amount of nutrients in the field, thus reducing the potential for pollution by either surfaceor subsurface routes. The same is true for the combination of integrated pest management (IPM) practices with conservation tillage, which would reduce the amount of pesticides applied to the field, thus reducing the potential for water quality impairment.

Other methods for mitigating the negative impacts of conservation tillage on water resources include the use of innovative chemical application methods that incorporate chemicals without excessive disturbance of the crop residue. Examples of these methods are band-incorporation of fertilizers,[29] spoke-wheel injectors,[30] and other similar approaches.[12,31,32] These methods generally place the fertilizer below the soil surface while minimizing the disturbance of the crop residue. Mostaghimi et al.[12] reported a 33% reduction in total sediment-bound nitrogen (TN_{sed}) losses from no-tillage plots when subsurface application of fertilizer was used instead of surface application. Furthermore, TN_{sed} levels for no-tillage/subsurface application plots were 97% less than the TN_{sed} levels for conventionally tilled/surface application plots and 89% less than the TN_{sed} levels for the conventionally tilled/subsurface application plots.[12]

10.2.3 Contour Farming

Contour farming is an effective erosion control practice on low to moderate sloping land. Contour farming is defined (NRCS Code 330) as farming sloping land in such a way that land preparation, planting, and cultivating are done on the contours.[5] An example of a field under contour farming is shown in Figure 10.5. Contour farming pro-

FIGURE 10.5   Contour farming (Source NRCS, 1998).

vides protection against sheet and rill erosion. The greatest protection is provided against storms of moderate to low intensity on fields with mild slopes. Contour farming is a managerial practice and is an effective source reduction BMP. It is appropriate for situations where sediment is the main pollutant or vector by which other pollutants are transported. Contouring also increases infiltration and reduces surface runoff. Another benefit of contour farming is that soil and associated resources are kept on the field. Thus, contour farming protects receiving waters by conserving the soil resource, which is also critical to crop production.

A shortcoming of contour farming is that it provides minimum protection against high intensity storms on steep slopes. When storm intensity greatly exceeds the infiltration rate, the accumulation of water behind furrows may lead to "overtopping".[33] Overtopping occurs when ponded water overtops the furrow and from one furrow to the next creating a cascade of failures. This failure may result in severe local erosion in the form of gullies. Overtopping can also occur for storms of moderate intensity if contour farming is used on steep fields.[34]

There are also management concerns associated with the implementation of contour farming. Implementation of contour farming requires the development of detailed topographic maps for the fields. An alternative to the development of topographic maps is to directly identify the contour lines on the field. In either case, the farmer uses this information to locate crop rows on the field. The location of crop rows depends on the size of the field and the equipment width. A major concern of the farmer is to minimize the occurrence of point rows. Point rows are areas within the field where the row width is smaller than the equipment width. Point row areas make the navigation through the field laborious and could encourage the farmer to discontinue the practice.

Contour farming is generally used as a component of other practices, such as strip cropping and terraces. Strip cropping on the contour allows for the application of contour farming on steeper slopes. The closely spaced crops used in strip cropping reduce the potential for overtopping. On steeper slopes, terraces may also be used. Contour farming is not effective in situations where soluble pollutants are the main concern. In cases where both soluble and sediment-bound pollutants are of concern, contour farming could be used in combination with nutrient management or IPM.

10.2.4   STRIP CROPPING

Strip cropping is an effective protection against erosion and sediment-bound pollutants. There are two methods for implementing strip cropping. Strip cropping (NRCS Code 585) on the contour is the practice of growing crops in strips along the contours of the field[5] (See Figure 10.6). This type of strip cropping is commonly referred to as contour strip cropping. The strips alternate between close-grown crops, such as small-grain and row crops. The second method is referred to as field strip cropping. Field strip cropping (NRCS Code 586) is defined as growing of crops in strips that are oriented perpendicular to the "general slope" of the field[5] (See Figure 10.7). Both of the strip cropping methods offer protection against soil erosion, although contour strip cropping may offer more protection than field strip cropping. The potential for overtopping is reduced for contour strip cropping compared with contour farming alone. This reduction is related to lower runoff volumes and surface flow velocities asso-ciated with the close grown crops used in strip cropping. Both contour and field strip cropping are classified as managerial and source reduction practices, although both approaches also interrupt the transport of sediment within the field. As with contour farming, point rows are also a concern with contour strip cropping. The problem of point rows could be alleviated by using field strip cropping. The choice between field strip cropping or contour strip cropping heavily depends on site-specific characteristics of the field. When making this choice, one must balance the importance of the erosion protection against the management concerns of the farmer.

Contour and field strip cropping are most effective in situations where sediment is the main pollutant or vector by which other pollutants are transported. Strip cropping farming is commonly used in locations where field slopes are too steep to use contour farming. Strip cropping has the additional benefit of filtering surface runoff from the clean-tilled strips while moving through the close-grown crop strips. Additional sediment may be removed and trapped in the close-grown crop strips. The

FIGURE 10.6   Strip cropping on the contour (Source NRCS, 1998).

FIGURE 10.7   Field strip cropping (Source NRCS, 1998).

most prominent effect of strip cropping is reduced soil erosion. Strip cropping could be used in combination with nutrient management or IPM for cases where losses of both soluble and sediment-bound pollutants are of concern.

10.2.5   BUFFER ZONES

Buffer zones or filter strips are BMPs that reduce the transport of pollutants and are considered structural practices. They are defined as planted or indigenous bands of vegetation that are situated between pollutant source areas and receiving waters to remove pollutants from surface and subsurface runoff. A grass buffer at the edge of a field is shown in Figure 10.8. To varying degrees, filtration, infiltration, absorption, adsorption, uptake, volatilization, and deposition are pollutant removal processes operating in the buffers or filter strips.[5] The most prominent pollutant removal

FIGURE 10.8   Grass buffer at the edge of a field (Source NRCS, 1998).

processes in filter strips tend to be infiltration of dissolved pollutants and deposition of sediment-bound pollutants.[33] The effectiveness of pollutant removal processes is directly related to the changes in surface flow hydraulics that occur in the buffers.[34] Buffers are most effective when shallow overland flow, commonly referred to as sheet flow, passes through the strip. The surface flow passing through the buffer should not be fast moving, concentrated, or channel flow. If concentrated flow occurs, the buffer will be short-circuited and rendered ineffective.[35] Design guidelines (NRCS Code 393A) are available for locating the buffer on the landscape.[5]

Buffers are used for the treatment of surface runoff from cropland or confined animal facilities. Robinson et al.[36] observed that a 3.0-m wide buffer effectively removed up to 70% of the sediment load from cropland runoff. Edwards et al.[37] reported that buffers were effective for removing metals found in runoff from fields treated with poultry litter. Barone et al.[38] reported that buffers were effective for removing nutrients, bacteria, and pesticides from surface runoff. Reductions in *E. coli* (91%), total coliform (86%), and fecal streptococci (94%) were observed for an 8.5-m grass buffer.[38] Other researchers have investigated the effectiveness of buffers for controlling nutrients from surface-applied swine manure[39] and for trapping microbial pollutants.[40] However, these were all short-term studies and did not address the long-term effectiveness of buffers. Dillaha et al.[35] observed that the effectiveness of buffers tended to decrease with time. As stated earlier, it is imperative that flow velocities entering and flowing within the strip remain low and not concentrated for buffers to be effective. Low flow velocities ensure that the travel time through the buffer is long enough for deposition and other pollutant removal processes to take effect. Moreover, the low flow velocities ensure that soil erosion or resuspension of earlier deposits does not occur within the buffer.[5,34]

A modified form of filter strip is used to treat surface runoff or wastewater from animal facilities. This form of filter strip is designed to convey concentrated flow. The wastewater to be treated is routed through a vegetation-lined waterway.[5] This filter-waterway is not a grassed waterway (which is designed to convey water quickly), rather the filter-waterway is designed for slow movement of water to allow for infiltration, deposition, and other pollutant removal processes to take effect. This waterway could be thought of as a very long filter strip (longer than 100 feet) and are generally narrow. The waterways are used to treat wastewater from milk parlors, milking centers, food processing plants, and manure storage structures.[5] Discharge of wastewater into these filter-waterways should be controllable, and storage of wastewater should be included in the design of the treatment system to allow for a recovery time for the filter-waterway.[5]

The direct environmental impacts of buffers are similar to other BMPs that address erosion and sediment problems. Tim and Jolly[41] conducted a modeling study for a watershed in Iowa to evaluate buffers for treating sediment loads. They observed that buffers alone could result in a 41% reduction in sediment loads reaching the outlet of the watershed. These findings and others have made buffers or filter strips a very popular BMP, and many institutional approaches have been used to increase adoption of buffers by landowners.[42] However, filter strips interrupt the transport of pollutants rather than keep these pollutants or resources in place. To the farmer, this trapped sediment is a lost resource. The same is true for nutrients that accumulate in the filter strips.

There are some concerns about the long-term effectiveness of buffers. With proper maintenance, buffers are expected to function for up to 10 years.[43] However, the buffer may become a pollution source without proper maintenance. As sediment accumulates in the buffer over time, large flows from extreme precipitation events may flush (or clean) the buffer of its sediment load. Without "harvesting" of the biomass grown in the buffer, the trapped nutrients will accumulate, thus increasing the risk of groundwater pollution or increasing the nutrient concentrations of waters leaving the buffer. Models have been developed for the design of buffers.[44–46] However, most models do not consider the long-term effects of nutrient accumulation on the effectiveness of the buffers. Médez-Delgado[33] developed a computer simulation model, the Grass Filter Strip Model (GFSM), to investigate the long-term (10 years) effectiveness of buffers. The GFSM simulates the nutrient dynamics, as well as hydraulics and sediment transport, within a buffer.[33] The long-term performance of buffers could be evaluated using a computer model, such as GFSM, to minimize any potential negative environmental impacts.

As with previously mentioned BMPs, buffers may be used in combination with nutrient or pesticide management practices to address both sediment-bound and dissolved pollutants. For example, buffers can be located down-slope of fields under conservation tillage or other soil conservation practices. The addition of buffers at the edge of fields can reduce the transport of fine materials and dissolved pollutants, which are transport processes not addressed by conservation tillage. As for the case of treating wastewater from animal facilities, buffers could be used in combination with sediment basins and constructed wetlands as a complete treatment system. The main function of buffers in this system would be to remove particles too small to be removed by the sediment basin.

Riparian buffers are similar in design and intent to filter strips. A riparian buffer (NRCS Code 391A) is defined as an area consisting of trees and shrubs that are located directly adjacent to permanent or intermittent water bodies.[5] An example of a riparian buffer is shown in Figure 10.9. As with filter strips, riparian buffers are structural practices that interrupt the transport of pollutants to downstream water bodies. Riparian buffers remove sediment and excess nutrients from water flowing across the land surface.[47] Riparian buffers offer environmental benefits in addition to water quality improvements. They also provide esthetic and ecological enhancements, such as increased areas for wildlife habitat.[48] An ideal riparian buffer consists of three zones.[5] Zone 1 starts at the water line and extends a minimum of 4.6 m (15 feet) away from the water line. The vegetation in this zone is primarily trees and shrubs. Zone 1 should remain relatively undisturbed and livestock should be excluded from this zone. Zone 2 is similar to Zone 1, except that selective harvesting of timber or biomass is recommended to remove nutrients collected by the buffer. Zone 3 is a grass filter strip that is intended to disperse the incoming flow and promote more uniform flow through Zones 1 and 2. Zone 3 also traps sediment in an area without trees so the sediment could be more easily collected and moved back to the fields. As with filter strips, the design and location of riparian buffers can dramatically impact their effectiveness.

FIGURE 10.9   Riparian buffer (Source NRCS, 1998).

Riparian buffers have many of the strengths and weaknesses of filter strips. Riparian buffers are useful for interrupting the transport of pollutants (sediment and nutrients) from agricultural lands. Haycock and Pinay[49] observed that the biomass of the riparian buffer enhanced nitrate retention during the winter months. Carbon from the biomass of the riparian buffer allowed soil bacteria to engage in nitrate reduction during winter, when the plants were inactive. The bacterial reduction enhanced the overall nitrate retention efficiency of the buffer.[49] Snyder et al.[50] also observed reductions in nitrate concentrations in groundwater originating from upland agricultural areas. These reductions ranged from 16 to 70%. Snyder et al.[50] reported that the riparian buffers had no effect on orthophosphorus or ammonium concentrations. In fact, increases in orthophosphorus or ammonium concentrations were observed in water passing through the buffer during summer months.[50] Both the water quality and ecological benefits of riparian buffers have led many environmental agencies to advocate their use and provide alternative policy approaches[51] to help increase their adoption as a BMP.

10.2.6 COVER CROPS AND CONSERVATION CROP ROTATIONS

Cover crops are a source reduction managerial practice. They are (Code 340) defined as crops grown during the time period between the harvest and planting of the primary crop.[5] The main purpose of cover crops is to provide soil cover and protection against soil erosion. Cover crops also sequester nutrients over the winter, prevent their loss, and provide a "green" manure source in the spring[52,53] if the cover crop is left in the field or plowed under before planting of the primary crop. Another benefit of cover crops is soil moisture management by reducing soil evaporation when plants are dormant.[54] Cover crops can also provide additional revenue for the farmer. A prime example is winter wheat. Winter wheat is usually planted a few weeks before corn is harvested to ensure that sufficient wheat plant will emerge to protect the soil after the corn is harvested.

Crop rotations that involve cover crops can be used to enhance the economics of the farm and protect the environment. One possible negative impact of the use of cover crops is increased use of herbicides. If the cover crop is not harvested, it needs to be killed before planting of the primary crop. Additional herbicides are needed if cover crops are being used in a conservation tillage system. Furthermore, cover crops may contribute to the loss of some pollutants.[55] Mostaghimi et al.[11] reported that phosphorus losses from experimental plots were greatest for a residue level of 1500 kg/ha versus 750 kg/ha. The elevated phosphorus levels for the 1500 kg/ha residue level were attributed to a lack of sufficient suspended sediment available to bound with the excess phosphorus.[11] Similar findings were reported for nitrogen. Mostaghimi et al.[13] observed that nitrogen yields increased for residue levels greater than 1500 kg/ha. Cover crops can be used in combination with any other BMPs. When using cover crops with nutrient management, the nutrient source or reduction attributed to the cover crop should be accounted for to provide the primary crop with the needed nutrients.

Conservation crop rotations are a source reduction managerial practice. They are (Code 328) defined as the growing of different crops in a specific sequence on the same field.[5] There are several purposes for using conservation crop rotations. Crop rotations are often planned for the reduction of soil erosion, chiefly sheet erosion. Examples of soil conserving crop rotations may include row crops, such as corn, followed by hay. The plants chosen for the rotation need to produce enough above- and below-ground biomass to control soil erosion.[5] Conservation crop rotations can also be used to maintain soil organic matter. As with selecting plants for soil erosion control, plants are selected based on the amount of biomass provided. Another purpose for using conservation crop rotation is to manage excess and deficient plant nutrients. When addressing excess nutrients, the idea is similar to using cover crops. In fact, cover crops may be a part of the conservation rotation. Plants that have the necessary rooting depth and nutrient needs should be selected when addressing nutrient excesses. For the nutrient-deficient case, a plant may provide nutrients for another plant in the rotation. This is commonly used in the case where a plant with high nitrogen demands, such as corn, is put in a rotation with a legume, such as soybeans.

Conservation crop rotations often form the basis of other conservation practices. For instance, plants that produce large amounts of residue may be selected for a crop rotation on a field where conservation tillage is to be implemented. Furthermore, the crop sequence of strip cropping should be consistent with the conservation crop rotation. Finally, the nutrient deficits and excesses produced during a crop rotation are one of the major constraints when developing a nutrient management plan. Although crop rotations are commonly thought of as site conditions, like soil type or topography, alteration of the crop rotation to address these nutrient deficits and excesses could enhance the effectiveness of other BMPs and should be considered.

The environmental and economic impacts of crop rotation are heavily dependent of the types of crops selected. In general, conservation crop rotations reduce runoff and sheet erosion, increase soil organic matter, and reduce pests compared with continuous cultivation of one crop on a field. For a corn-soybean rotation, leaching of pesticides[56] as well as nutrients were reduced compared to continuous corn.[56,57] Crop rotations often reduce economic risk through diversification of farm operations. A drawback of this practice is that the timing of commodity prices and of crops in the rotation may be unfavorable for the farmer. For instance, the price for soybeans may be low during the year that soybeans are being grown. This economic risk could be reduced if the crop rotation is kept out of sequence on different fields within a farm. Conservation crop rotation is a low-cost practice that provides both economic and environmental benefits.

10.2.7 NUTRIENT MANAGEMENT

Nutrient management is one of the most prevalent BMPs used to address NPS pollution from agricultural lands. Many state and local agencies have developed pamphlets, handbooks, and worksheets to assist in the development of nutrient management plans. In addition, some local and state agencies employ nutrient management specialists who develop plans for farmers. Nutrient management is a source reduction managerial practice and is defined (NRCS Code 590) as the optimization of the plant nutrient applications.[5] The objective of this optimization is to enhance forage and crop yields while minimizing the loss of nutrients to surface and groundwater resources. The objective is accomplished by managing the amount, form, placement, and timing of plant nutrient applications. The procedure used to gather information for nutrient management plans depends on the agricultural system where the practice is applied. Beegle and Lanyon[58] defined these systems as crop farms, crop/livestock farms, and intensive livestock farms. Each of these farming systems can be characterized by their respective nutrient status. The nutrient status of a farm could be classified into three categories. A farm can have a nutrient deficit where nutrients inputs needed on the farm exceed on-farm nutrient resources. This nutrient status requires off-farm nutrient inputs to continue production. The farm could be in balance, where nutrient needs on the farm and outputs are equal to on-farm resources and little or no off-farm nutrient inputs are necessary. Finally, a farm could have excess nutrients, where on-farm nutrient resources greatly exceed the on-farm nutrient needs. In practice, the boundaries among these categories may be difficult to define, but these boundaries are useful for the purpose of discussion. Information about the nutrient status of a farm is critical when developing a nutrient management plan. The first important element of any nutrient management plan is to gather information about the nutrient status of the farm.

For any nutrient management plan, the main purpose of the information-gathering process is to determine the amount of nutrients available and needed on the farm. The needs are generally related to type of crops grown on the farm. The crop needs are related to the soil fertility and production goal. Therefore, the first step in the information-gathering process is soil testing. Soil tests are needed to determine residual levels of available nutrients. If possible, crop tissue samples could be collected and analyzed to determine crop nitrogen needs during the growing season. Laboratory analysis may also be needed to determine the nutrient content of plant residues—whether they are left on the fields or harvested. Another component of the information-gathering process is laboratory analysis of manure samples. Manure tests are performed to determine the nutrient content of the manure. Manure tests are especially important when developing nutrient management plans for crop/livestock and intensive livestock farms. The methods used to handle and store the manure influence the natural processes that affect the nutrient content of the manure. This is especially true for nitrogen. Because manure samples are usually collected from storage facilities, handling and storage methods need to be considered when using manure test results in a nutrient management plan. If possible, manure testing should occur immediately before land application to account for the losses. If manure testing is not available, many state and local agencies provide standard nutrient levels for livestock. However, these standard levels are average values observed for a region and vary from farm to farm. When developing a nutrient management plan, the specific procedures used to collect information depend on the characteristics of the farm system.

Crop system farms generally require nutrient inputs from external sources. Judicious use of commercial fertilizers is an essential part of nutrient management for crop system farms.[58] Soil tests every 2 to 3 years and crop tissue samples at critical periods during the growing season should be used to determine how much fertilizer the crops need. Livestock may be present on the farm, but the nutrients provided by the livestock are considered negligible with respect to the nutrient needs of the crops. A nutrient management plan for this type of farm would focus on determining the needs of the crops for specific yield goals. These attainable yields would be based on historical yield levels for the field or farm. In the absence of historical information, yield goals could be based on realistic soil and crop management production levels.

Once the yield goals are determined, the timing of the nutrient applications should be addressed. The ultimate objective of the plan is to ensure that sufficient nutrients are available to satisfy the crop uptake while minimizing the potential loss of nutrients to the environment. There are different ways to approach this objective. One popular method is the use of split application of nitrogen, in which part of the total amount of nutrients needed by the crop is applied before or during planting. The remaining nutrients are applied later in the growing season when they are needed and only at the rates needed for the expected crop yield.

Commercial fertilizers are sometimes modified to reduce pollution potential. One modification is the use of commercial fertilizer formulations that include nitrification inhibitors.[59] These inhibitors slow the bacterial conversion of ammonium to nitrate, which reduces nitrogen leaching. However, the potential for pollution from sediment-bound pollutants could be magnified and because ammonium can be volatilized as ammonia, volatilization losses may increase unless the fertilizer is incorporated. Another modification is to coat solid forms of commercial fertilizers with slowly degradable materials that gradually release nutrients into the soil environment.

When using green manure such as legumes as a nitrogen source for crops, the availability of nitrogen must be determined. Various factors control the nitrogen cycle in the soil, which in turn influences the amount of mineral nitrogen available to crops.[60] These factors include soil pH, soil temperature, and carbon to nitrogen ratio, among others. The availability and reliability of nitrogen from organic sources, such as green manure, manure, or municipal sludge, is also a concern of farmers.

Manure from other farms may be used as a nutrient source for crop farms. The major difficulty in using manure from other farms is that transportation costs are high in comparison with commercial fertilizers. A general concern with using manure as a fertilizer is the consistency of nutrient levels. The nutrients levels, especially nitrogen, depend heavily on the source (animal), along with handling, storage methods, and feed. There is a need for additional testing of the manure to determine the nutrient levels before its application. This additional step adds to the cost, thus reducing the likelihood that manure, instead of cheaper commercial fertilizers, would be used.

Municipal sludge is another organic fertilizer used on crop system farms. Nutrient contents of municipal sludge, commonly referred to as biosolids, are usually determined for the farmer by the biosolids supplier. In addition, use of biosolids as a nutrient source may have some economic advantages over commercial fertilizers. In many regions, farmers are paid for the application of biosolids on their fields. The major concern with biosolids is heavy metals and other industrial pollutants that may be present in the biosolids. However, both federal and state regulations concerning the use of biosolids as a soil amendment address the pollutant-carrying capacity of soils when determining permissible application rates and frequency. The final approach to nutrient management of crop farms addresses the spatial variability of both soil fertility and crop yields within fields. This approach is commonly referred to as precision farming and is discussed later as a separate BMP.

A *crop/livestock farm* may provide enough nutrients supplied by livestock to meet the nutrient needs of the crops.[58] The crops are also used for feed. This is an idealized system and may not be practical on all farms. However, the crop/livestock system does serve as a good discussion model. This type of farm can be considered a closed system with the only nutrient outputs being livestock and some crops. The most important task of any nutrient management plan for a crop/livestock system is the determination of whether there is enough cropland to fully utilize the nutrients from the manure. Soil, manure, and crop tissue tests are all necessary in the development of the nutrient management plan. Soil fertility would need to be assessed and the nutrient content of the manure should be determined. A nutrient management plan may include use of alternative crops that would help utilize excess nutrients. A common problem found in crop/livestock systems is lack of manure storage facilities, which results in daily spreading of the manure. In this case, the construction of a manure storage structure would be critical for the development of successful nutrient management plan. Manure storage structures are discussed as a separate BMP in a subsequent section. The last ingredient of a successful nutrient management plan for a crop/livestock system is proper calibration of manure spreaders. Without precise knowledge of the amount of nutrients being applied, the usefulness of information provided by soil and manure tests is diminished and the possibility of over- and under-application increases.

An *intensive livestock system* is characterized as having excess nutrients generated on the farm.[58] The basic problem is that there is not enough cropland on the farm to utilize the amount of nutrients generated by livestock production. The major focus of a nutrient management plan for this type of system would be to find additional manure utilization options, such as use on other farms (i.e., crop system farms), use as a feed supplement, composting, and resale, among others. The major obstacle to utilization of the manure as a nutrient supplement on other farms is the cost. Some high-nutrient manure, such as poultry litter, can be economically transported up to 75 miles,[58] whereas lower nutrient content (higher moisture content) manure can be economically transported only shorter distances. Processes that would increase the nutrient value of the manure while lowering transportation costs would greatly increase the economic viability of this approach.

A word of caution should be raised when considering how to implement nutrient management plans. In most cases, manure application rates for nutrient management plans are based on the nitrogen needs of the crops.[61] When the amount of manure applied to cropland is based on crop nitrogen needs, over-application of phosphorus may occur because the N content of manure are generally less than the P needed by crops.[61] In the past, it was assumed that excess phosphorus would be held by soil minerals and not be available for transport.[61,62] However, over-application in some regions has resulted in the phosphorus saturation of agricultural soils. Therefore, any phosphorus applied to these soils would increase the potential for degradation of the aquatic habitat in the receiving waters. This is especially true for orthophosphorus P, which is highly mobile by surface runoff and is an essential nutrient in eutrophication process. In areas were excess soil phosphorus levels may be of concern, soil phosphorus tests should be used in the development of nutrient management plans and application rates of manure should be based on the phosphorus needs of the crops.

10.2.8 MANURE STORAGE FACILITIES

Manure storage facilities are an essential part of most nutrient management plans. These facilities are source reduction structural practices. Manure storage facilities are defined (NRCS Code 313) as any impoundment made by constructing an embankment, excavating a pit or dugout, or by fabricating a structure that allows for the storage of manure in an environmentally benign manner.[5] Most facilities typically provide 3 to 6 months of storage. Some examples of manure storage facilities include lagoons, dry-handling structures, and slurry storage tanks.[63] An example of a dry-handling structure is shown in Figure 10.10 and a lagoon facility is show in Figure 10.11. The type of livestock, site characteristics, economics, and requirements of the nutrient management plan determine the type of manure storage facility to be used.[64] For instance, lagoons (NRCS Code 359) provide storage and biological treatment of manure to reduce pollution and protect the environment.[5] The biological treatment reduces the nutrient content of the manure. Thus, if nitrogen is the nutrient limiting land application, less land will be required for application of manure from a lagoon as opposed to other types of storage structures. Manure storage facilities need to be periodically emptied. Ideally, structures are emptied at times when plants can utilize most of the nutrients in the manure. However, the long periods between emptying times require large amounts of storage. As the storage increases, the cost of the facility increases rapidly. Generally, the cost of a manure storage facility is the most serious obstacle in the adoption of animal waste management plan. To encourage the adoption of nutrient management, cost-share funds and tax credits are supplied

FIGURE 10.10   Dry manure handling storage structure (Source NRCS, 1998).

by state and federal agencies to offset the construction costs of manure storage structures.[65]

Manure storage facilities should be designed and constructed by a professional engineer. Failure of these structures could result in severe environmental damage. Some designs of storage facilities are environmentally preferable over others. For instance, the potential for groundwater pollution associated with lagoons is relatively high compared with other manure storage facilities.[66] The lagoons are often lined with an impermeable material, such as a geotextile material or clay, to reduce the potential of groundwater pollution. It has been reported that some types of manure

FIGURE 10.11   Lagoon storage structure (Source NRCS, 1998).

"seal" themselves over time.[66] Lagoons constructed in sandy soils that did not use impermeable linings have been identified as potential sources of groundwater pollution.[66] Dry handling and slurry storage structures greatly reduce this risk, but are not economically feasible for large livestock operations. Facilities also fail when containment walls of the structure rupture. When this happens, liquid manure may contaminate surface and ground waters. There are also odor concerns associated with some types of storage. In areas where farms are close to residential areas, odor can be a major problem. Most odor problems occur when lagoons are stirred or when the manure is applied. Great care should be taken when locating manure storage structures on the landscape to reduce aesthetic degradation as well as environmental hazards.

When manure is stored, organic forms of nitrogen (N) and phosphorus (P) are converted from organic to inorganic forms by bacteria and other microbes. The two main components of N found in manure are organic N and ammonia N.[67] The inorganic portion of N in fresh manure is commonly in the form of ammonia N. Storage of manure, especially in slurry form, generally results in the loss of organic N through ammonification and then volitilization of the ammonia N. Organic N is converted to ammonium N, which then volatilizes as ammonia N. Also, storage of manure at high moisture contents may result in the loss of nitrate N by denitrification.[68] However, the level of nitrate N in manure depends on the presence of nitrifiers, which are microbes commonly found in the soil. There are both benefits and drawbacks to the transformation of N from organic to inorganic forms. The main benefit is that the inorganic forms of N are available to plants, thus nutrient value of the manure may increase. The drawback is that these same inorganic forms of N also promote the growth of aquatic plants and algae, thus increases in the proportions of inorganic N may increase the potential for degradation of the aquatic habitat in the receiving waters. Therefore, great care needs to be taken when applying the manure from the storage structure, and application levels should be based on crop needs to reduce the potential of polluting surface and ground waters. Unlike N, there has not been much research conducted on P transformations in manure storage facilities, but as with N, organic forms of P are converted to inorganic forms by microbial actions during storage. Furthermore, inorganic P is not lost to the atmosphere, but remains in the stored manure until its application. As with N, there are both benefits and drawbacks to the increases in the soluble forms of P. The main benefit is that the soluble forms of P are available to plants, thus nutrient value of the manure may increase. The main drawback is that these same soluble forms of P also promote the growth of aquatic plants and algae, which may increase the potential for degradation of the aquatic habitat in the receiving waters. This is especially true for orthophosphorus P, which is highly mobile by surface runoff and is an essential nutrient for eutrophication.

10.2.9 INTEGRATED PEST MANAGEMENT

Integrated pest management (IPM) is an effective source reduction treatment for water quality impairments by pesticides. It is a managerial practice and is defined as the use of management practices for pest control that result in efficient production of food and fiber using the minimum amount of synthetic pesticides.[34] A basic premise of IPM is that pesticides should be applied only when the costs associated with pest damage exceed the cost of applying the pesticides. This is a radical departure from past pesticide application practices where pesticides are applied as a routine production or prophylactic practice. Important components of IPM are: maximum use of biological and cultural controls, regulatory procedures (certification of applicators), strict adherence to pesticide labels, crop rotation, pest-resistant and pest-tolerant crops and livestock, scouting by IPM specialists and skilled farmers.[34] Significant reductions in pesticide use have been achieved in most IPM programs while agricultural profitability has increased.[34] The most significant factor hindering adoption of IPM is lack of sufficient knowledge on the part of potential users. An excellent overview of IPM principles and practices is given by the Council of Agricultural Science and Technology.[69]

The use of IPM has increased rapidly during the past 2 decades. One study found that more than 80% of New York apple producers use some IPM practices.[70] Producers who use comprehensive IPM practices used 30, 47, and 10% less insecticides, miticides, and fungicides, respectively, with a resulting savings of an annual average of $98.50 /ha over an 11-year period, without significantly affecting fruit quality. Other studies have found that IPM users tend to be younger, better educated, and have less farming experience than nonusers. Significant savings were also reported for celery using IPM in California.[71] Another study found that increased use of IPM with onions led to a 32% reduction in pesticide use between 1980 and 1988.[72] In Indonesia, IPM techniques reduced pesticide use by 60% and increased rice yields by 25%.[73] Apparently, the amount of pesticides required to control the pesticide-resistant organisms was so high that the pesticides had a toxic effect on the rice crop itself.[73]

10.2.10   PRECISION FARMING

Precision farming is an emerging technology with potential environmental and economic benefits. Precision farming can be defined as the site-specific application of variable rates (rather than uniform rates) of farm inputs across agricultural lands.[74] Precision farming is a source reduction managerial practice. This technique considers the spatial-variability of soil and crop over a specific field, and attempts to avoid over- or under-application of farm inputs within the field.[74] The dynamic nature of interactions among soil, crop, management, and environmental factors cause substantial amounts of spatial variability in the physical characteristics of soils. Spatial variability ultimately causes uneven patterns in soil fertility and crop growth, thus reduces the efficiency of fertilizers applied uniformly over an entire field. Research results indicate that the spatially variable characteristics of soil have major effects on the transport of nutrients by surface runoff and leachate through the soil profile.[75,76] In addition, several studies have reported savings in production costs by applying variable rates of fertilizers, compared with costs associated with application of uniform rates over the entire field.[77,78] The principal savings are from reduced fertilizer use, which offsets the additional costs associated with the soil sampling, variable yield monitoring, and variable rate fertilizer application required by precision farming. In the past, it was not possible to apply variable fertilizer rates because of the inaccuracy of soil maps and lack of appropriate technology for applying variable rates of fertilizers to the field. Advancements in geographic information systems (GIS), global positioning systems (GPS), and new farming technologies have now made it possible to develop accurate soil fertility maps and to apply variable rates of fertilizers to agricultural lands.[77,79,80]

It is now widely recognized that application of fertilizer at variable rates has environmental benefits while maintaining or improving crop production. A study conducted in Colorado showed that nitrate–nitrogen leaching from corn grown on coarse-textured soil could be reduced by 53% using precision farming techniques.[81] Eagel and Gaultney[82] reported that a spatially-based decision support system could reduce the agrochemical needs of a 12-acre farm. Mostaghimi et al.[76] used a NPS model to show that 15 to 25% reductions in stream concentrations of dissolved nitrogen could be expected from implementation of precision farming, as opposed to conventional farming practices. In the same study, Mostaghimi et al.[76] used soil sampling on regular grids to investigate the spatial variability of nutrient levels for a 40-acre farm located in the Coastal Plains of Virginia. They observed that P fertilizer requirements varied from 0 to 100 lb/acre compared with 40 lb/acre under conventional systems. Furthermore, K fertilizer inputs varied from to 80 lb/acre for precision farming, compared with 60 lb/acre under conventional farming systems.[76] Studies conducted in Missouri have also shown that the application of variable rates of P fertilizer produced greater returns for corn crops compared with uniform rate application.[83]

10.2.11 TERRACES, VEGETATED WATERWAYS, AND DIVERSIONS

Management practices that address the conveyance of concentrated-surface runoff can be effective in controlling NPS pollution. This is especially true for NPS pollutants associated with sediment. The most common conveyance BMPs are terraces, vegetated waterways, and diversions. All three of these BMPs are considered structural practices. Terraces interrupt the transport of pollutants, whereas grasswaterways and diversion are source reduction practices and, to a lesser extent, affect the transport of pollutants.

Terraces are very effective in reducing NPS pollution in surface runoff.[18] Terraces (NRCS Code 600) are defined as any combination of ridges and channels constructed across the slope.[5] An example of grass-sided terraces is shown in Figure 10.12. Level terraces were reported to reduce soil loss by 94 to 95%, nutrient losses by 56 to 92%, and runoff volume by 73 to 88%.[18] Terraces achieve these reductions by storing water and allowing for sediment deposition and water infiltration. Consequently, terraces would be expected to increase the potential for the movement of soluble pollutants to the groundwater.

There are several drawbacks associated with terraces. Terraces are expensive to install and maintain, and they remove some land from production. The Rock Creek RCWP reported[84] that structural practices reduced sediment loads by 55%. However, their initial capital costs were high, and the annual maintenance costs for sediment

FIGURE 10.12   Grass-sided terraces (Source NRCS, 1998).

retention facilities, which included different types of terraces, were estimated to range from \$22 to \$37/ha.[84] Because of these high costs and the fear that the practice would not be maintained, the project encouraged conservation tillage as an alternative BMP.[84]

Vegetated waterways provide effective control against gully erosion. Vegetated waterways (NRCS Code 412) are defined as channels with established vegetation designed for the stable conveyance of runoff.[5] A vegetated waterway located between two adjacent fields is shown in Figure 10.13. The stable conveyance of runoff reduces gully erosion. The vegetative lining of the waterway helps to control the velocity of the flowing water and reduce channel erosion. The vegetative cover of the waterway is permanent. The vegetation in the waterway should be kept at the height recom-

FIGURE 10.13   Vegetated waterway (Source NRCS, 1998).

mended by the design to ensure optimal protection. The land area used for the waterway is taken out of crop production and may add to the occurrence of point rows. Vegetated waterways may alter farm traffic patterns, but these alterations will be less compared to traffic alterations due to gullies. For areas with slopes that are too steep for vegetated waterways, stone or concrete hydraulic structures may be required.

Diversions redirect surface water away from potential pollutant sources. Some examples of situations where diversions might be used are to divert concentrated flow around cropland to prevent gully erosion or to divert water away from feedlots or agrochemical mixing areas. Diversions (NRCS Code 362) are generally constructed across the slope and have a supporting ridge on the down-slope side.[5] Generally, diversion are vegetated waterways, but other material, such as gravel or concrete, may be used to protect the channel lining if high flow velocities are expected.

10.2.12 SEDIMENT DETENTION STRUCTURES

Sediment detention structures are used to trap and collect sediment.[85,86] Sediment detention structures are transport interruption BMPs. In these structures, deposition by gravity is the principal pollutant removal process. Sediment detention basins are generally large and may collect water from a moderately large area. Smaller structures such as check dams use the same mechanism to trap coarse sediment, but on a much smaller scale. An example of a small check dam with an underground drain is shown in Figure 10.14. The basic design of a sediment detention basin (NRCS Code 350) includes an impoundment (or small dam) placed across a drainage-way perpendicular to the flow.[5]

As the name indicates, problems associated with excess sediment are commonly treated using these basins. Generally, sediment basins are not used as an agricultural

FIGURE 10.14 Small sediment detention structure in a field (Source NRCS, 1998).

BMP. They are mainly used for treatment of disturbed land associated with construction or mining. Care should be taken in the design of any structure that will impound water. Jarret[85] listed three guidelines that should be considered when designing a sediment detention basin. These guidelines are:

1. Failure of the structure should not result in the loss of human life or in large economic losses downstream.
2. The height of the dam should not exceed 35 feet (11 meters).
3. The product of the storage volume (in acre-feet) and the dam height (in feet) should not exceed 3000 acre-ft^2 (113 ha-m^2).

Jarret[85] states that "all structures exceeding these limits should be referred to a consulting engineer." Jarret[85] and Haan et al.[86] discuss the location and design of sediment detention basins. Smaller impoundments, often called sediment traps, are used as a temporary treatment for a sediment problem. Sediment traps are placed in small drainage ways, and multiple traps may be used downstream from the disturbed area. As with sediment detention basins, sediment traps are not commonly used in agricultural areas. They are mainly used on disturbed sites that include construction sites, mining areas, and forest harvesting areas. The most commonly used sediment trap is known as a check dam. The check dam could be composed of many different materials. Some examples of materials used include wood, gravel, gabions, geotextile fabric (silt or filter fence), and straw bales. The check dams are located in areas where concentrated flow occurs. The drainage area of each check dam should not exceed 5 acres.[85] Traps are easier to install than sediment basins but are only temporary structures. The movement of sediment into streams or other water bodies can be minimized using either sediment basins or traps. Sediment basins should be used for larger drainage areas (greater than 5 acres) where the sediment source will be in existence for a relatively long period (greater than 3 months). Some state and local regulations require sediment basins for construction sites and other nonagricultural areas where land will be disturbed. For agricultural areas, sediment basins may be used downstream from animal feedlots or cropland. Check dams and impoundment terraces are the most common forms of sediment traps used on cropland. Impoundments terraces are used not only to address sediment problems but also as a water management BMP.

There are several drawbacks to sediment detention structures. The first drawback is that they address sediment only after it has eroded from the field. As stated earlier, the soil is the most important resource in agriculture and the first objective of any resource manager should be to keep it on the field. Another drawback is the amount of maintenance these structures require. Both basins and traps need to have a certain amount of storage for sediment. A farmer would be required to remove excess sediment or vegetation that may grow in the sediment storage area of the structure. Furthermore, fine particles are rarely removed by sediment detention structures. Sediment-bound nutrient and pesticides attach to these fine particles. For this reason, sediment detention structures offer little protection against the movement of soluble or dissolved nutrients and pesticides to downstream water bodies. In addition, the infiltration of the ponded water in the basin may pose a risk to groundwater. Finally, sediment basins require land to be taken out of production and the cost of construction may be prohibitive.

Sediment basins and traps may be used in combination with other BMPs to enhance pollution control. Source reduction BMPs such as nutrient management and IPM would reduce the amount of excess nutrients and pesticides that are available for transport. Problems of excess sediment should be addressed on the field (the source) with BMPs such as conservation tillage. A potential application of sediment detention structures in agricultural settings is their use as pretreatment structures for water entering a constructed wetland.

10.2.13   CONSTRUCTED WETLAND

Wetlands are a valuable natural resource. Constructed wetlands have also been used to treat municipal, industrial, and more recently, agricultural waste. Because natural wetlands do not always occur in close proximity to where treatment is needed, constructed wetlands have emerged as a treatment alternative. Reed[87] documented the existence of 154 constructed wetlands for treating municipal and industrial effluent in the United States. Thirty six percent were free water surface (FWS) wetlands and are considered to operate as biological reactors, similar to the trickling filters used in conventional wastewater treatment systems.[88,89] However, wetlands offer some advantages over conventional treatment systems. In addition to microbial activity, other contaminant removal mechanisms have been observed in wetlands. Watson[90] listed several mechanisms including sedimentation, filtration, chemical precipitation, adsorption, and uptake by vegetation.

Wetlands are cost-effective, efficient, and suitable for treating a wide range of pollutants. Magmedov and Yakovleva[91] found that sulfates, ammonium, and nitrate-nitrogen concentrations of up to 100 mg/L, chlorides up to 1500 mg/L, suspended solids up to 300 mg/L, fluctuations in pH (of 3 to 5 units), and the presence of heavy metal ions did not suppress wetland biocenosis. The complexity of the processes operating in wetlands makes it difficult to provide general design criteria. Reed and Brown[88] found that nitrogen uptake by vegetation is not a significant removal mechanism in wetlands. Other studies indicate that plant uptake is the main mechanism for nitrogen removal, accounting for up to 90% of the nitrogen removed.[92–94] It appears, therefore, that the complexity of wetland systems may cause wetland behavior to be site- or region-specific.

Hammer and Bastian[95] provide a summary of the advantages and disadvantages associated with wetlands used for effluent treatment. The advantages are: relatively inexpensive to construct, easy to maintain and operate, relatively tolerant of fluctuating hydrologic and contaminant loading rates, and effective and reliable wastewater treatment. Some advantages of wetlands not related to water quality include providing additional green space, wildlife habitats, and recreational/educational areas. There are some disadvantages associated with wetlands. Wetlands require a relatively large land area for advanced treatment. Another disadvantage is that the design and operating criteria are currently imprecise. Wetlands may also serve as a breeding ground for pests. However, pest control strategies have been developed for wetland treatment systems.[96-98] These systems have been successfully used for treating municipal and industrial effluent, and some preliminary performance characteristics have been established. BOD_5 removal rates ranging from 50 to 90%[99], and COD, TKN, and TP removal rates up to 86, 95, and 99%, respectively, have been reported.[92]

Wetlands have been successfully used to treat liquid manure. When used for this purpose, wetlands can be considered a source reduction BMP. The need for treatment arises when there is not enough land available for application of the manure. The nutrient reductions observed in wetlands have been well documented.[100-102] The wetland effluent is applied to the land for final treatment. Other advantages of manure treatment using constructed wetlands are reduced odor problems, increased habitat, and the ability to use conventional irrigation equipment for land application of the effluent. There is guidance available for the design of constructed wetlands for treatment of animal waste.[103-105] Other uses of constructed wetlands include the treatment of agricultural drainage. Allen[106] reported that a constructed wetland effectively removed selenium from agricultural drainage water. This result indicates that wetlands might be suitable for removal of other heavy metals. Indeed, there are several cases of wetlands being used for treatment of effluents from mining operations.[107,108]

10.2.14 STREAM FENCING AND OFF-STREAM WATER SUPPLIES

Stream fencing is defined as the construction of a barrier, usually a wire or electrified fence, along stream corridors that exclude livestock from direct access to streams (NRCS Code 582 or 472).[5] Fencing or use exclusion BMPs are structural practices and focus on source reduction as shown is Figure 10.15. When livestock are allowed access to streams, they may defecate directly into the stream or stir up sediment from the stream bottom because of animal traffic. The stream banks could also be destabilized by animal traffic and grazing. The main drawback of stream fencing are the

FIGURE 10.15 Use exclusion around stream using fencing (Source NRCS, 1998).

installation and maintenance costs. Estimates for implementing fencing along 71,000 miles fishable streams managed by the Bureau of Land Management (BLM) was over $500 million for construction and an additional $100 million for maintenance.[109] Another drawback of stream fencing is the loss of grazing land. This may not be a problem in flat areas, but in mountainous or steeply sloping areas with meandering streams, this loss of grazing land can be significant. In any case, alternative water supplies must be provided when livestock is excluded from streams. Simple troughs are the most common method used to supply water to livestock away from streams.

Off-stream water sources without fencing has been suggested as an alternative to stream fencing. As with stream fencing, off-stream water sources is a structural BMP that focuses on reducing the source of pollutants by reducing the amount of time animals spend in streams. Several researchers have observed that the uses of off-stream water resources reduced the time livestock spent streams from 81 to 90%.[110,111] The costs of off-stream water supplies are less than fencing, but livestock may still spend time in streams for reasons other than drinking, such as wading to escape hot temperatures or to alleviate irritations from parasites or from vegetation.

Some research has been conducted to present alternative methods to stream bank fencing to protect riparian/stream resources. Previous research in Virginia showed that when cattle were given the choice, they drank from the water trough 92% of the time.[112] Coinciding with the cattle's preference for the troughs was reductions in stream bank erosion and in water quality parameters. Sheffield et al.[112] observed a 77% reduction in stream bank erosion and a large reduction in total nitrogen (54%) and phosphorus (81%).[112] Similar reductions were observed in fecal coliform and streptococcus levels in stream sections where water troughs were available.[112] In another study conducted in Oregon,[110] the availability of water troughs was shown to decrease the time cattle spent in the stream by 90%.

10.2.15   ROTATIONAL GRAZING

Rotational or prescribed grazing is defined as the controlled harvest of vegetation with grazing or browsing animals (NRCS Code 528A).[5] Rotational grazing is a source reduction practice that has both managerial and structural components. This practice addresses sediment, nutrient, and biological sources of pollution that may result from high animal densities on the land. There are several variations of the basic practice of rotational grazing that focus on specific livestock systems. Few studies conclusively relate livestock management directly to stream water quality.[113] Most studies relate "land-based" observations of nutrients to stream water quality. Studies that investigate grazing management usually focus on the economics of the practice and possible health improvements of the livestock.[114] A common characteristic of all grazing management approaches is to provide recovery times for the land by moving livestock to other areas. A specific variation of rotational grazing is discussed in the following paragraphs to highlight the methods needed to provide this recovery time for the land.

An example of rotational grazing is referred to as intensive rotational stocking.[114] The practice focuses on dairy production systems and is defined as the

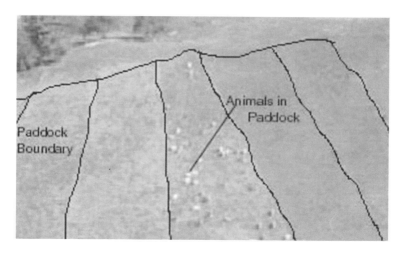

FIGURE 10.16   Cows in paddocks.[114]

rotational grazing of cows among several small pastures or paddocks as opposed to continuous grazing of animals on one large pasture (See Figure 10.16). In Figure 10.16, the borders of the paddocks can be seen as the dark lines running from the top to the bottom of the picture. The cows are rotated among paddocks daily or after each milking. The pasture is separated into the paddocks using inexpensive fencing, such as mobile wire or electrified fencing. Water source must be provided for each paddock. Researchers have found that the intensive rotational stocking practice can increase profits and therefore is the main motivation for adoption of this management system by farmers. Profits increased 72% per acre for intensive rotational stocking compared with profits associated with continuous grazing practices. The profits were derived mainly from lower operational costs. By continuously moving livestock to fresh grazing lands, costs associated with harvesting of feed and management of manure were greatly reduced.[114] The main environmental concern of this practice is that paddocks should not be located near or in watercourses.

10.3   BMP IMPACT ASSESSMENT

Watershed scale assessment of the effects of BMPs on water quality is not an easy task. The complexity of the problem stems from the inherent variability of the natural system and the incremental process of land treatment. Documenting changes in water quality resulting from changes in management practices requires long-term records, rendering the design of an efficient, cost-effective monitoring system especially important. Furthermore, because it is neither economically feasible nor realistically achievable to monitor the watershed for all potential combinations of soils, crops, management practices, and meteorological conditions, mathematical models can be used to aid in the evaluation of BMPs. In the following sections, a framework for the design of monitoring systems for BMP-impact assessment is presented. The essential role of statistics, both for the design of the system and the analysis of the collected data, is emphasized and statistical methods are discussed. The actual techniques and procedures for sampling and laboratory analysis are not provided, but references to the standard manuals are given. The EPA has recently compiled a comprehensive reference on monitoring procedures and data analysis,[115] and this document is recommended to readers who require a more detailed discussion of procedures for assessment of BMP effectiveness than provided in this chapter.

10.3.1 FRAMEWORK FOR THE DESIGN OF A MONITORING SYSTEM FOR BMP IMPACT ASSESSMENT

The time and difficulties involved in the basic water quality monitoring tasks, i.e., sample collection, laboratory analysis, and data management (*analysis and interpretation*), sometimes overshadow the actual purpose of the monitoring effort. To ensure that the end product meets the information expectations of the monitoring system, the development of a well-documented, statistically-based monitoring design is essential. Methods for the design of water quality monitoring systems have been discussed by a number of authors;[116–119] however, little has been published on the design of monitoring systems that focus on detecting changes in water quality resulting from implementation of BMPs. Presented here is a framework for the design of a BMP-assessment monitoring system, based on the design approach of Ward et al.[118]

Ward et al.[118] identified five steps for the design of a water quality monitoring system for assessing BMP impacts:

1. Define the monitoring objectives.
2. Select statistical design and analysis procedures.
3. Design the monitoring network.
4. Develop operating plans and procedures.
5. Develop reporting procedures.

It should be understood that the monitoring system design process is not static. Experience has indicated that once the data collection and analysis have started, additions and modifications in the systems design are often necessary. Therefore, it is essential to thoroughly document all actions taken. The results of the entire design process should be presented in a written report and reviewed by the sponsoring agency. The five steps of the monitoring design process are discussed in detail in the following sections.

10.3.1.1 Step 1: Define the Monitoring Objectives

Watershed monitoring for BMP effectiveness is generally performed to evaluate the impact of a combination of BMPs in achieving water quality standards, rather than the specific effectiveness of individual BMPs. Watershed monitoring studies may be combined with field studies to evaluate the impact of specific BMPs. Results from these monitoring studies can be used to adjust any combination of management practices, guidelines, and management objectives.

The definition of the objectives and goals of the monitoring system is crucial to the success of the system. The entire set of objectives should be comprehensive and non-overlapping. Objectives should be clearly defined and attainable within a realistic time frame. Each objective should focus on a single issue, such that evaluating progress towards one objective will not be contingent upon progress toward another.[120]

Ward et al.[117] diagnosed the disease that plagues many monitoring systems as a "data-rich, but information-poor" syndrome and emphasized that the design of a monitoring system should be guided by the information expectations. An identification of the different types of information that can be produced by the monitoring system is a useful means for the quantification of the information expectations. Ward et al.[118] distinguished the following information types: (1) narrative information, (2) numerical information (raw data), (3) graphical displays, (4) statistical information, and (5) water quality indices.

10.3.1.2   Step 2: Select Statistical Design and Analysis Procedures

After defining the monitoring objectives, the next step is to identify the target population and select a statistical methodology for the design of the monitoring system and the analysis of the collected data. The choice of statistical methodology depends on a number of factors, such as cost-effectiveness, statistical characteristics of the water quality variables, and various practical considerations.[121] However, as noted previously, the selected statistical methodology should have the ability to fulfill the information expectations, as outlined by the defined goals and objectives. Ward et al.[118] noted that it is important to document how conclusions derived from the data analysis will be related to the monitoring objectives.

The target population of a monitoring system has to be selected to provide the information necessary to accomplish the specified monitoring objectives. Cochran[122] conceptualized the statistical sampling plan in terms of three major elements: the target population, the frame, and the sample. The *target population* is used to denote the entire collection of *elements* or units from which the sample is chosen. This is the population about which the information user wishes to make inferences. The *frame* is, in principle, a list of all the *elements* or units in the population. As noted by Cochran,[122] in practice, the frame seldom includes all members of the target population, and often contains elements that are no longer part of the target population. The *statistical sample* is a subset of the frame and also a subset of the target population. The statistical sample is the set of elements available for analysis. In the statistical literature the statistical sample is referred to simply as "the sample", but to avoid confusion with water samples (the volume of water collected from a well or stream), the term statistical sample is used throughout this chapter. The focus of a watershed-scale BMP assessment might be to characterize the water quality of a stream or an aquifer. Nelson and Ward[123] pointed out that if the objective of the study is to monitor the effect of BMPs on the water quality of the entire aquifer, then the target population is made up of all loci within the aquifer. In this case, the target population is infinite. However, if no monitoring wells are to be drilled, only the subpopulation of existing wells is available for sampling. A groundwater well can be considered as a sampling point in a large body of slow moving water in which the chemical composition is spatially variable. However, the composition of water obtained from a well is likely to be influenced by the movement induced by well construction, well development, and pump operation,[124] Spruill.[125] noted that sampling of water supply wells yielded a biased view of the quality of the groundwater because water wells are used for water supply only where the water is usable. This makes it difficult to consider water samples from the subpopulation of existing wells as being representative of an entire aquifer.

10.3.1.2.1 *Statistical Design for BMP Impact Assessment*

Experience from the U.S. Rural Clean Water Program[120] has indicated that the hydrologic variability and nutrient storage in the watershed may mask the impact of BMPs. Nutrients, especially phosphorus, can be stored in the soils of the watershed. Because of this storage, a long period of time (possibly up to several decades) may be required before the water quality impacts of BMPs are observed. Furthermore, the variability of precipitation and stream flow records often overwhelms improvements in water quality because of implementation of BMPs. It has been suggested that 6 to 10 years of monitoring is required, including at least 2 to 3 years of monitoring prior to BMP-implementation. Three general types of BMP evaluation monitoring designs have been suggested.[126]

Paired watershed requires a minimum of two watersheds (control and treatment) and two periods of study (calibration and treatment). The watersheds, which should have similar physical characteristics and land use, are monitored for a number of years to establish pollutant-runoff response relationships for each watershed.[127] During the treatment period, one watershed is treated with a BMP or combination of BMPs and the other remains under the original management. Such an approach is believed to provide the greatest potential for documenting BMP improvements because of ability to control the variabilities in climatic and hydrogeologic factors. However, the cost of monitoring two watersheds might be prohibitive. Furthermore, it might not always be possible to find two watersheds that are sufficiently similar.

Upstream/downstream design uses two water quality monitoring stations, where one station is placed directly upstream from the BMP implementation area and one station directly downstream from that area.[127] Such design is more appropriate for documenting the severity of a problem than for evaluating BMP effectiveness. A similar approach is recommended for groundwater studies. The up-gradient and down-gradient design can account for seasonality and other factors that impact both of the wells. Ward et al.[117] suggested that a tolerance interval approach be used to detect sudden shifts in water quality, such as leaching from pesticide handling sites. The upstream/downstream design requires that water from the BMP implementation area enter along a reach of the stream or river. This design may be difficult to implement, especially if the BMP implementation area is located in the headwaters of a watershed. Therefore, this approach may not be suitable for small watersheds.

Sequential (before-after) design involves monitoring of the same watershed during both pre-BMP and post-BMP phases at a single station downstream from the area of BMP implementation.[127] This design is similar to a paired watershed design without the control watershed. The costs of the single-station design may be significantly less than the paired approach. However, the inability to control the variabilities in climatic and hydrogeologic factors may make it difficult to detect differences in water quality because of the implementation of BMPs.

10.3.1.2.2 Statistical Analysis of the Data

The main objective of any statistical analysis is to extract relevant information from data. The random nature of the observed data is incorporated in the analysis procedures. The randomness of the data may be a result of the processes that generated the data, measurement error, or both. Statistical analysis of data involves two major activities. The first is the estimation of statistical parameters, which are commonly referred to as descriptive statistics. Examples of descriptive statistics are the mean, median, standard deviation, and so on. Descriptive statistics are useful for summarizing data or for focusing on specific data characteristics. The second major activity is statistical inference. Generally, statistical inference implies the application of hypothesis tests to determine the significance of differences observed among data sets. Some statistical inference methods determine the significance of a statistical parameter estimated from a single data set. In either case, statistical inference procedures are used to determine whether differences observed in statistical estimates are the result of some process other than the random fluctuation inherent to the data.

Estimation of Descriptive Statistics is an essential part of any statistical analysis.[115,121,128] The descriptive statistics allows for (1) an initial overview of the results, (2) a visual interpretation of the data, (3) the selection of categorical variables to be used as explanatory poststratification variables in subsequent analyses of variance, and (4) an additional control for faulty entries. Many software packages are available to chart and tabulate frequencies (counts) and relative frequencies (percentages) of the qualitative categorical variables (e.g., land use, management practice). For the quantitative variables (e.g., sediment, nutrient concentrations, pH), a variety of sample statistics describing location, dispersion, and shape of the empirical distribution can be computed for each stratum. Histograms, stem-and-leaf plots, and boxplots can also be generated using statistical software packages.

Parametric and Nonparametric Tests differ in their assumptions about the distribution of the data being analyzed. The assumption of normality is required for many parametric statistical tests, such as the chi-square test, *t*-test, regression, and analysis of variance. Departures from normality (e.g., data that exhibit skewness or lack of symmetry) can invalidate the results of these tests. After an extensive literature review and statistical analysis of existing groundwater data, Montgomery et al.[129] reported that groundwater quality variables were often not normally distributed; exhibited seasonal patterns, especially in shallow or highly permeable aquifers; and exhibited significant serial correlation when data were collected on a quarterly basis. Nonparametric procedures afford significant advantages over their parametric counterparts. Nonparametric procedures generally reproduce the empirical structure of multivariate data sets, yet do not require assumptions about statistical distribution of the data. Nonparametric tests tend to be more robust compared with parametric tests when there are outliers in the data. Also, missing and censored data are easily dealt with when using nonparametric tests. Because of these characteristics, nonparametric statistical procedures have been favored for water quality applications.[130,131]

Statistical Inferences can be made by way of contingency table analysis, estimation of proportions (point estimates and exact confidence limits), multivariate and univariate analysis of variance, and principal component analysis. Documentation of how statistical inferences are to be formulated allows for confirmation that the monitoring system will actually provide data in sufficient quantities for the precision and sensitivity required.

Relationships among selected variables (land use and management practices) and the measured continuous response variables (sediment and chemical concentrations and loadings) can be modeled and tested using a variety of procedures,[115] such as;

- Chi-square test of homogeneity of proportions and independence in contingency tables to test relationships among categorical variables,
- Multivariate analysis of variance to test the effect of explanatory categorical variables on the measured continuous response variables,
- Univariate analysis of variance to test the effects of explanatory categorical variables on continuous response variables in the light of the multivariate results.
- Principle component analysis to test the relationship among the continuous variables.

Monotonic Trends tests can be used to detect slow changes in water quality over time. Monotonic trend techniques would be appropriate for watersheds where the implementation of BMPs occurs gradually and data are collected continuously during implementation.[132] An excellent review of statistical methods and computational procedures for trend detection and estimation was presented by Gilbert.[121] Trend analysis preferably starts with graphical display of the data (i.e., by plotting the measured water quality parameter values over time). If the plots suggest a linear relation and if the data were normally distributed and neither serially correlated nor affected by seasonality, a simple linear regression analysis can be conducted. To analyze if the slope of the linear regression line is statistically significant, a *t*-test is applied.[121] Alternatively, the nonparametric Mann-Kendall test for trend or Sen's nonparametric estimator of slope can be used. Both tests require equal time intervals for the data but accommodate missing data, ties, or data below the detection limit. To test for the homogeneity of trend direction at different sampling stations, a chi-square test of the Mann-Kendall statistics can be conducted. If the chi-square value is not significant, trends should be tested for each station separately.

Water quality data often exhibit seasonality, which obscures the detection of long-term trends. The seasonal Kendall test and Sen's nonparametric trend test are unaffected by seasonal cycles. The seasonal Kendall test was proposed by Hirsch et al.[130] for analysis of seasonality using monthly data (i.e., 12 seasons). Because the normal approximation is used to test the null hypothesis of no-trend, the test requires a minimum of 3 years of data for each of the 12 seasons (a total of 36 observations). A technique for computing the exact distribution of the test statistic for different numbers of seasons and years can be found in Hirsch et al.[130] These authors also defined an unbiased estimator for the magnitude of the trend (i.e., the seasonal Kendall slope estimator). If there are no missing data, the use of Sen's nonparametric trend test, which is more likely to detect monotonic trends, is recommended.[133] Homogeneity of trend directions in different seasons and different stations can again be tested with the chi-square statistic.[133] If the direction of the trends varies among the seasons, the seasonal statistics are not meaningful, and individual Mann-Kendall and Sen's slope estimators should be computed for each season.

Step Trend tests should be used when BMP implementation in the watershed was immediate or when there is a gap in the data record between the pre- and post-BMP phase.[133] If the data are normally distributed, a t-test can be used to analyze if the water quality data from the pre- and post-BMP period are significantly different from each other. The nonparametric tests for this analysis is the Mann-Whitney test or the almost identical Hodges-Lehman test.[133] The Mann-Whitney test was modified for seasonal correlation by Lettenmaier.[134] A seasonal Hodges-Lehman estimator for determining the magnitude of the step trend was introduced by Hirsch et al.[130]

10.3.1.3   Step 3: Design of the Monitoring Network

With the monitoring objectives and hypothesis defined (step 1), and the appropriate statistical methods identified (step 2), the third step takes the design process to the watershed. This step answers questions about where, what, and when to sample. These questions are outlined in the following three tasks: (1) identification of the sampling locations, (2) selection of the variables to measure, and (3) scheduling of the water sampling.

10.3.1.3.1   *Identification of the Sampling Locations*

This task uses the selected statistical design to identify sampling locations and, if relevant, the determination of the size of the statistical sample. The size of the sample is dependent upon many factors such as the size of the target population, the degree of precision desired for the study, the variance of the data, and the cost of obtaining a sample. A Geographic Information System (GIS) can be used to effectively identify the location of the sampling sites. The use of Global Positioning Systems (GPS) would greatly facilitate implementation of this task.

10.3.1.3.2   *Selection of Water Quality Variables*

The water quality variables to be monitored should be the same as the variables targeted by the BMPs and should reflect the water quality problem. These variables may include various forms of sediment, nutrients, pesticides, bacteria, and so on. For example, because it is impractical to test all possible pesticides, a careful selection needs to be made. The selection depends on the use, toxicity, and physical and chemical properties of the chemicals; the sampling area; and the economics and availability of the chemical analysis. A procedure for screening pesticides for their inclusion in a monitoring program was presented by Shukla et al.[135]

Background variables, such as pH, conductivity, and temperature of the water at the time of sampling, are generally included in water quality studies to provide a data-quality check.[125,136] Tests for these variables are typically easy to conduct and require minimum expense, equipment, and training. Finally, information on the land use and management practices, which is essential for establishing relationships between water quality parameters and land use, should be collected.

10.3.1.3.3   Scheduling of Sampling

After the identification of the monitoring network, the next step is the determination of sampling frequency. When sampling is too frequent, serial correlation causes the information to be redundant and wasteful. On the other hand, infrequent sampling may miss critical information, thus rendering the results of the BMP impact assessment inconclusive. An important consideration when selecting a sampling scheme to evaluate water quality impacts of agricultural nonpoint source pollution is the temporal variability of the surface water flow, ground-water recharge, and agricultural practices. Loftis and Ward[137] emphasized that statistical data analysis procedures should match the sample frequencies. They listed three factors that affect the sample statistics; (1) random changes from precipitation events, (2) seasonal changes from climatic variations, and (3) serial correlation of samples that are closely spaced in time. To address seasonal variability, the sample population needs to be monitored at regular intervals during the year. Finally, chemical monitoring requires careful consideration of the maximum holding time for the samples, laboratory capacity and storage space, the duration of analytical procedures, and the availability of staff and resources.

10.3.1.4   Step 4: Develop Operating Plans and Procedures

To ensure that the data obtained are valid and comparable, all samples should be collected and analyzed according to documented standardized methods.[138,139] The development of sound *Quality Assurance/Quality Control (QA/QC)* procedures can help ensure control and documentation of data quality.[140] *Quality Assurance (QA)* refers to the overall management activities conducted to ensure that a project meets the agreed-upon quality standards and to ensure compliance with standard operating procedures. *Quality Control (QC)* refers to the operation-level management activities conducted to ensure that these standards are met.[141] To facilitate the incorporation of QA/QC practices, the U.S. EPA[142] presented guidelines for the development of a *Quality Assurance Project Plan (QAPjP)*. QAPjPs are required by many federal, state, and private organizations, although the exact requirements of the different agencies and organizations may vary. All the sample collection and analytical procedures that are routine in nature should be described by *Standard Operating Procedures (SOPs)*. The U.S. EPA defined an SOP as a written document which details an operation, analysis, or action whose mechanisms are thoroughly prescribed and which is commonly accepted as the method for performing routine or repetitive tasks.[143] A review of the requirements of an EPA QAPjP was presented by Brossman et al.[140] and Mostaghimi et al.[144,145]

10.3.1.5   Step 5: Develop Reporting and Information Utilization Procedures

The last step in the monitoring system design process is to specify the frequency, type, and format for reporting. Assessment studies generally prepare seasonal or annual data summaries. All reports need to state progress toward achieving the stated programs objectives and goals. Interaction between the designers and information users or sponsoring agency will be needed to achieve consensus about the type and format of the information to be reported.

Ward et al.[118] stressed the importance of identifying the receiving party and what they will do with the information. The language (e.g., technical or layman) and information of each report need to be tuned to the specific audience. Thus, it might be necessary to prepare documents with highly varying layout and contents. Reports are essential for providing feedback to funding agencies, project staff, and the general public and may play an important role in motivating their long-term involvement in the monitoring program.

REFERENCES

1. Magleby, R., et al., Soil erosion and conservation in the United States: an overview, vol. Agriculture Information Bulletin 718, Washington D.C., USDA, Economic Research Service, 1996.
2. Wolkomir, R., Bringing ancient ways to our farmers' fields, *Smithsonian,* 26(8), 99, 1995.
3. SWCS, Resource conservation glossary, 3 ed, Ankeny, IA, Soil Conservation Society of America, 1982.
4. Novotny, V. and O. H. Olem, *Water Quality: Prevention, Identification, and Management of Diffuse Pollution,* New York, Van Nostrand Reinhold, 1994.
5. NRCS, National handbook of conservation practices, NRCS-USDA: Washington D.C., 1998.
6. Wayland, R., What progress in improving water quality?, *Journal of Soil and Water Conservation,* 48(4), 261, 1993.
7. Meals, D. W. Water quality trends in the St. Albans Bay watershed, Vermont, following RCWP land treatment. in *The national rural clean water program: 10 years of controlling agricultural nonpoint source pollution,* Washington D.C., USEPA, 1992.
8. Bull, L. and C. Sandretto, Crop residue management and tillage system trends, Vol. Statistical Bulletin 930, Washington D.C., USDA, Economic Research Service, 1996.
9. Blevins, R. L. and W. W. Frye, Conservation tillage: an ecological approach to soil management, *Advances in Agronomy,* 51, 33, 1993.
10. Kenimer, A. L., et al., Effects of residue cover on pesticide losses from conventional and no-tillage systems, *Transactions of American Society of Agricultural Engineers,* 30(4), 953, 1987.
11. Mostaghimi, S., T. A. Dillaha, and V. O. Shanholtz, Influence of tillage systems and residue levels on runoff, sediment and phosphorus losses, *Transactions of American Society of Agricultural Engineers*, 31(1), 128, 1988.

12. Mostaghimi, S., T. M. Younos, and T. M. Tim, The impact of fertilizer application techniques on nitrogen yield from two tillage systems, *Agricultural Ecosystems and the Environment,* 36, 12, 1991.

13. Mostaghimi, S., T. M. Younos, and U. S. Tim, Crop residue effects on nitrogen yields in water and sediment runoff from two tillage systems, *Agricultural Ecosystems and the Environment,* 39, 187, 1992.

14. King, L. D., Reduced chemical input cropping systems in the southeastern United States. II. effects of moderate rates of N fertilizer and herbicides, tillage, and delayed cover crop plow-down on crop yields, *American Journal of Alternative Agriculture,* 9(4), 162, 1994.

15. Morrison J. E., Jr, et al., Rill erosion of a Vertisol with extended time since tillage, *Transactions of American Society of Agricultural Engineers,* 37(4), 1187, 1994.

16. Baker, J. L. and J. M. Laflen, Effects of corn residue and fertilizer management on soluble nutrient losses, *Transactions of American Society of Agricultural Engineers,* 25(2), 344, 1982.

17. Dillaha, T. A., et al., Rainfall simulation: A tool for best management practice education, *Journal of Soil and Water Conservation,* 43(4), 288, 1988.

18. Baker, J. L. and H. P. Johnson. Evaluating the effectiveness of BMPs from field studies. in *Agricultural Management and Water Quality,* Ann Arbor, MI, Ann Arbor Scientific Publications, Inc., 1983.

19. Baumhardt, R. L. and R. J. Lascano, Rain infiltration as affected by wheat residue amount and distribution in ridged tillage, *Soil Science Society of America Journal,* 60(6), 1908, 1996.

20. Dao, T. H., Tillage and winter wheat residue management effects on water infiltration and storage, *Soil Science Society of America Journal,* 57(6), 1586, 1993.

21. Pikul J. L. Jr. and J. K. Aase, Infiltration and soil properties as affected by annual cropping in the Northern Great Plains, *Agronomy Journal,* 87(4), 656, 1996.

22. Hubbard, V. C. and D. Jordan, Nitrogen recovery by corn from nitrogen-15 labeled wheat residues and intact roots, *Soil Science Society of America Journal,* 60(5), 1405, 1996.

23. Singh, H. and K. P. Singh, Nitrogen and phosphorus availability and mineralization in dryland reduced tillage cultivation: effects of residue placement and chemical fertilizer, *Soil Biology & Biochemistry,* 26(6), 695, 1994.

24. Edwards, W. M., et al., Tillage studies with a corn-soybean rotation: hydrology and sediment loss., *Soil Science Society of America Journal,* 57(4), 1051, 1993.

25. Sharpley, A. N., et al., The transport of bioavailable phosphorus in agricultural runoff, *Journal of Environmental Quality,* 21(1), 30, 1992.

26. Gaynor, J. D., D. C. MacTavish, and W. I. Findlay, Atrazine and Metolachlor loss in surface and subsurface runoff from three tillage treatments in corn, *Journal of Environmental Quality,* 24(2), 246, 1995.

27. Drury, C. F., et al., Influence of tillage on nitrate loss in surface runoff and tile drainage, *Soil Science Society of America Journal,* 57(3), 797, 1993.

28. Erbach, D. C., Tillage for continuous corn and corn-soybean rotations, *Transactions of American Society of Agricultural Engineers,* 25(4), 906, 1982.

29. Mueller, D. H., et al., The effect of conservation tillage on the quality of runoff water, American Society Agricultural Engineers, St. Joseph, MO, ASAE Paper No. 82-2022,), 1982.

30. Blaylock, A. D. and R. M. Cruse, Ridge-tillage corn response to point-injected nitrogen fertilizer, *Soil Science Society of America Journal,* 56(2), 591, 1992.

31. Chichester, F. W. and J. E. Morrison, Jr., Agronomic evaluation of fertilizer placement methods for no-tillage sorghum in vertisol clays, *Journal of Production Agriculture,* 5(3), 378, 1992.

32. Reeves, D. W., C. W. Wood, and J. T. Touchton, Timing nitrogen applications for corn in a winter legume conservation-tillage system, *Agronomy Journal,* 85(1), 98, 1993.

33. Médez-Delgado, A., Nitrogen transport and dynamics in grass filter strips in *Biological Systems Engineering Department.* Virginia Polytechnic Institute and State University: Blacksburg, VA, 1996.

34. Heatwole, C., T. A. Dillaha, and S. Mostaghimi, Agricultural BMPs applicable to Virginia, Bulletin 169, Blacksburg, VA, Water Resources Research Center, Virginia Polytechnic Institute and State University, 1991.

35. Dillaha, T. A., J. H. Sherrard, and D. Lee, Long-term effectiveness and maintenance of vegetative filter strips, Bulletin 153, Blacksburg, VA, Water Resources Research Center, Virginia Polytechnic Institute and State University, 1986.

36. Robinson, C. A., M. Ghaffarzadeh, and R. M. Cruse, Vegetative filter strip effects on sediment concentration in cropland runoff, *Journal of Soil and Water Conservation,* 51(3), 227, 1996.

37. Edwards, D. R., et al., Vegetative filter strip removal of metals in runoff from poultry litter-amended fescue grass plots, *Transactions of American Society of Agricultural Engineers,* 40(1), 121, 1997.

38. Barone, V. A., et al., Effectiveness of vegetative filter strips in reducing NPS pollutant losses from agricultural lands: sediment, nutrients, bacteria, and pesticides, American Society Agricultural Engineers, St. Joseph, MO, ASAE Paper No. 982037/982038,), 1998.

39. Chaubey, I., et al., Effectiveness of Vegetative Filter Strips in Retaining Surface-Applied Swine Manure Constituents, *Transactions of American Society of Agricultural Engineers,* 37(3), 845, 1994.

40. Coyne, M. S., et al., Soil and fecal coliform trapping by grass filter strips during simulated rain, *Journal of Soil and Water Conservation,* 50(4), 405, 1995.

41. Tim, U. S. and R. Jolly, Evaluating agricultural nonpoint-source pollution using integrated geographic information systems and hydrologic/water quality model, *Journal Environmental Quality,* 23), 1, 1994.

42. Lant, C. L., S. E. Kraft, and K. R. Gillman, Enrollment of filter strips and recharge areas in the CRP and USDA easement programs, *Journal of Soil and Water Conservation,* 50(2), 193, 1995.

43. Dillaha, T. A. and J. C. Hayes, A procedure for the design of vegetative filter strips, Final Report, Prepared for the USDA-SCS, 1991.

44. Hayes, J. C., J. Barfield, and R. I. Barnhisel, Filtration of sediment by simulated vegetation II. unsteady flow with non-homogeneous sediment, *Transactions of American Society of Agricultural Engineers,* 22(5), 1063, 1979.

45. Lee, D., Simulation of phosphorus transport in vegetative filter strips, in Biological Systems Engineering Department. Virginia Polytechnic Institute and State University: Blacksburg, VA, 1987.

46. Muñoz-Carpena, R. J., Modeling hydrology and sediment transport in VFS, Biological and Agricultural Engineering Department. North Carolina State University: Raleigh, NC, 1993.

47. Gilliam, J. W., Riparian wetlands and water quality, *Journal of Environmental Quality,* 23(5), 896, 1994.

48. Triquet, A. M., G. A. McPeek, and W. C. McComb, Songbird diversity in clear cuts with and without a riparian buffer strip, *Journal Soil and Water Conservation*, 45(4), 500, 1990.

49. Haycock, N. E. and G. Pinay, Groundwater nitrate dynamics in grass and poplar vegetated riparian buffer strips during the winter, *Journal Environmental Quality*, 22(2), 273, 1993.

50. Snyder, N. J., et al., Impact of riparian forest buffers on agricultural nonpoint source pollution, *Journal of the American Water Resources Association*, 34(2), 385, 1998.

51. Dellapenna, J. W., The regulated riparian version of the ASCE model water code: the third way to allocate water, *Water Resources Bulletin*, 30(2), 197, 1994.

52. Brandi-Dohrn, F. M., et al., Nitrate leaching under a cereal rye cover crop, *Journal Environmental Quality*, 26(1), 181, 1997.

53. Wyland, L. J., et al., Winter cover crops in a vegetable cropping system: impacts on nitrate leaching, soil water, crop yield, pests and management costs, *Agricultural Ecosystems and the Environment*, 59, 1, 1996.

54. Ewing, R. P., M. G. Wagger, and H. P. Denton, Tillage and cover crop management effects on soil water and corn yield, *Soil Science Society American Journal*, 55(4), 1081, 1991.

55. Miller, M. H., E. G. Beauchamp, and L. J. D., Leaching of nitrogen and phosphorus from the biomass of three cover crop species, *Journal of Environmental Quality*, 23(2), 267, 1994.

56. Kanwar, R. S., T. S. Colvin, and D. L. Karlen, Ridge, moldboard, chisel, and no-till effects on tile water quality beneath two cropping systems, *Journal of Production Agriculture*, 10(2), 227, 1997.

57. Owens, L. B., W. M. Edwards, and M. J. Shipitalo, Nitrate leaching through lysimeters in a corn-soybean rotation, *Soil Science Society of America Journal*, 59(3), 902, 1995.

58. Beegle, D. B. and L. E. Lanyon, Understanding the nutrient management process, *Journal of Soil and Water Conservation*, 42(9), 23, 1994.

59. Darst, B. C. and L. S. Murphy, Keeping agriculture viable: industry's viewpoint, *Journal of Soil and Water Conservation*, 42(9), 8, 1994.

60. Sander, D. H., D. T. Walters, and K. Frank, Nitrogen testing for Optimum Management, *Journal of Soil and Water Conservation*, 42(9), 46, 1994.

61. Sharpley, A. N., et al., Managing agricultural phosphorus for protection of surface waters: Issues and options, *Journal of Environmental Quality*, 23(3), 437, 1994.

62. Sims, J. T., Characteristics of animal wastes and waste-amended soils: an overview of the agricultural and environmental issues, in *Animal Waste and the Land-Water Interface*, K. Steele, Editor, CRC/Lewis, New York, p. 589, 1995.

63. NRCS, National Engineering Handbook: Agricultural Wastewater Management Field Handbook, Washington D.C, USDA, 1992.

64. Safley, J., L. M., et al., Lagoon management, Extension Technical Bulletin, Fayetteville, AK, University of Arkansas Cooperative Extension Service, 1994.

65. Virginia DCR, Virginia Agricultural BMP Manual, Richmond, VA, Department of Conservation and Recreation, Division of Soil and Water Conservation, 1997.

66. Westerman, P. W., R. L. Huffman, and J. S. Feng, Swine-lagoon seepage in sandy soil, *Transactions of American Society of Agricultural Engineers*, 90(1), 139, 1995.

67. Collins, E. R., J. D. Jordan, and T. A. Dillaha, Nutrient values of dairy manure and poultry litter as affected by storage and handling, in *Animal Waste and the Land-Water Interface*, K. Steele, Editor, CRC/Lewis, New York, p. 589, 1995.

68. Cabrera, M. L. and R. M. Gordillo, Nitrogen release from land-applied animal manures, in *Animal Waste and the Land-Water Interface,* K. Steele, Editor, CRC/Lewis, New York, p. 589, 1995.

69. Council of Agricultural Science and Technology, Integrated pest management, Report No. 93, Ames, IA, Council of Agricultural Science and Technology, 1982.

70. Kovach, J. and J. P. Tette, A survey of the use of IPM by New York apple producers, *Agricultural Ecosystems and the Environment,* 29, 101, 1988.

71. Trumble, J. T., W. G. Carson, and G. S. Kund, Economics and environmental impact of a sustainable integrated pest management program in celery, *Journal of Economic Entomology,* 90(1), 139, 1997.

72. Environmental Science and Technology, ES&T currents: technology, *Environmental Science Technology,* 23(7), 758, 1989.

73. Bouwer, H., Agriculture and groundwater quality, *Civil Engineering,* 59(7), 60, 1989.

74. Carr, P. M., et al., Farming soils, not fields: a strategy for increasing fertilizer profitability, *Journal of Production Agriculture,* 4(1), 57, 1991.

75. Wallach, R., W. A. Jury, and W. F. Spencer, The concept of convective mass transfer for prediction of surface-runoff pollution by soil surface applied chemicals, *Transactions of American Society of Agricultural Engineers,* 32(3), 906, 1989.

76. Mostaghimi, S., et al., Suitability of precision farming technology in the Virginia's coastal resources management area, Grant No.NA570Z0561-01, Blacksburg, VA, Department of Biological Systems Engineering, Virginia Polytechnic Institute and State University, 1997.

77. Hammond, D. Cost analysis of variable fertility management of phosphorus and potassium for potato production in central Washington, in *Soil Specific Crop Management, First Workshop on Research and Development Issues,* American Society of Agronomy, Crop Science Society of America, and Soil Science Society of America, 1993.

78. Mulla, D. J. and M. Hammond, Mapping soil test results from large irrigation circles, in *Far west regional fertilizer conference,* Bozeman, MT, Far West Fertilizer AgChem Association, Pasco, WA, 1988.

79. Mulla, D. J. Mapping and managing spatial patterns in soil fertility and crop yield, in *Soil-specific crop management, first workshop on research and development issues,* American Society of Agronomy, Crop Science Society of America, and Soil Science Society of America, 1993.

80. Murphy, D. P. Yield mapping—a guide to improved techniques and strategies, in *Site-specific management of agricultural systems, Second Conference,* 677 S. Segoe Rd., Madison, WI, 1995.

81. Shaffer, M. J., B. K. Wylie, and M. D. Hall, Identification and mitigation of nitrate leaching hot spots using NLEAP-GIS technology, *Journal of Contaminant Hydrology,* 20(3–4), 253, 1995.

82. Eagel, B. A. and L. D. Gaultney, Environmentally sound agricultural production systems through site-specific farming, Society Agricultural Engineers, St. Joseph, MO, ASAE Paper No. 902566), 1990.

83. Wollenhaupt, N. C. and D. D. Buchholz, Profitability of farming by soils, in Proc. Soil-Specific Crop Management, in *First Workshop on Research and Development Issues,* American Society of Agronomy, Crop Science Society of America, and Soil Science Society of America, 1993.

84. Little, C. E., Rural clean water: the economy of rain and the Tillamook imperative, *Journal Soil and Water Conservation,* 44(3), 199, 1989.

85. Jarret, A. R., *Water management,* Dubuque, IA, Kendall/Hunt Publishing Co., 1995.

86. Haan, C. T., B. J. Barkfield, and J. C. Hayes, *Design Hydrology and Sedimentology for Small Catchments,* New York, Academic Press, 1994.

87. Reed, S. C., Constructed wetlands for wastewater treatment, *Biocycle,* 23(1), 44, 1991.

88. Reed, S. C. and D. S. Brown, Constructed wetland design—the first generation, *Water Environmental Research,* 4(6), 6, 1992.

89. Benham, B. L. and C. R. Mote, Treating dairy waste utilizing laboratory-scale constructed wetlands, American Society Agricultural Engineers, St. Joseph, MO, ASAE Paper No. 932576,), 1993.

90. Watson, J. T., et al. Performance expectations and loading rates for constructed wetlands, in *Constructed Wetlands for Wastewater Treatment: Municipal, Industrial, and Agricultural,* Lewis Publishing, Chelsea, MI, 1989.

91. Magmedov, V. G. and L. I. Yakovleva, Ecological features of an artificial wetlands area, in *International Symposium on the Hydrology of Wetlands in Temperate and Cold Regions,* Joensuu, Finland, The Academy of Finland, 1988.

92. Breen, P. F., A mass balance method for assessing the potential of artificial wetlands for wastewater treatment, *Water Resources Research,* 24(6), 689, 1990.

93. McIntyre, B. D. and S. J. Riha, Hydraulic conductivity and nitrogen removal in an artificial wetland system, *Journal Environmental Quality,* 20(1), 259, 1992.

94. Rogers, K. H., P. F. Breen, and A. J. Chick, Nitrogen removal in experimental wetland treatment systems: evidence for the role of aquatic plants, *Research Journal of the Water Pollution Control Federation,* 63(7), 934, 1991.

95. Hammer, D. A. and R. K. Bastian. Wetlands ecosystems: natural water purifiers, in *Constructed Wetlands for Wastewater Treatment: Municipal, Industrial, and Agricultural,* Lewis, Chelsea, MI, 1989.

96. USEPA, Design manual constructed wetlands and aquatic plant systems for municipal wastewater treatment, EPA/625/1-88/022, Cincinnati, OH, Center for Environmental Research Information, 1988.

97. Martin, C. V. and B. F. Eldridge. California's experience with mosquitoes in aquatic wastewater treatment systems, in *Constructed Wetlands for Wastewater Treatment: Municipal, Industrial, and Agricultural,* Lewis Publishing, Chelsea, MI, 1989.

98. Snoddy, E. L. and J. C. Cooney. Insecticides for pest control in constructed wetlands for wastewater treatment: a dilemma, in *Pesticides in terrestrial and aquatic environments: proceedings of a national research conference,* Blacksburg, VA, Virginia Water Resources Research Center, Virginia Polytechnic Institute and State University, 1989.

99. Sereico, P. and C. Larneo. Use of wetlands for wastewater treatment, in *Civil engineering practice,* Water Resources/Environmental Technomic Publication Co., Inc., Lancaster, PA, 1988.

100. Costello, C. J. Wetlands treatment of dairy animal wastes in Irish drumlin landscape, in *Constructed Wetlands for Wastewater Treatment: Municipal, Industrial, and Agricultural,* Lewis Publishing, Chelsea, MI, 1989.

101. Strong, L., et al. Establishment of a constructed wetland to treat wastewater from a confined animal operation, in *Proceedings, twenty-first Mississippi water resources conference,* Soil Conservation Service, Jackson, MS, 1991.

102. Hunt, P. G., et al., Constructed wetland treatment of swine wastewater, American Society Agricultural Engineers, St. Joseph, MO, ASAE Paper No. 932601/933510), 1993.

103. Payne, W. E. and R. L. Knight, New design procedures for animal waste constructed wet-lands, American Society Agricultural Engineers, St. Joseph, MO, ASAE Paper No. 982089), 1998.

104. CH2M and Payne Engineering, Constructed wetlands for animal water treatment: a manual on performance, design, and operation with case histories, Montgomery, AL., Prepared for the Gulf of Mexico Program under contract with ASWCC and NCASI, 1997.

105. NRCS, Constructed wetlands for agricultural waste treatment: technical requirements, Washington, D.C., USDA-NRCS, 1991.

106. Allen, K. N., Seasonal variation of selenium in outdoor experimental stream-wetland systems, *Journal of Environmental Quality,* 20(4), 856, 1991.

107. Brodie, G. A., D. A. Hammer, and D. A. Tomljanovich. Constructed wetlands for treatment of ash pond seepage, in *Constructed Wetlands for Wastewater Treatment: Municipal, Industrial, and Agricultural,* Lewis, Chelsea, MI, 1989.

108. Wildeman, T.R. and L.S. Laudon. Use of Wetlands for Treatment of Environmental Problems in Mining: non-coal-mining applications, in *Constructed Wetlands for Wastewater Treatment: Municipal, Industrial, and Agricultural,* Lewis Publishing, Chelsea, MI, 1989.

109. Platts, W. S. and F. J. Wagstaff, Fencing to control livestock on riparian habitats along streams: is there a viable alternative?, *North American Journal of American Fisheries,* 4(2), 266, 1984.

110. Miner, J. R., J. C. Buckhouse, and J. A. Moore, Will a water trough reduce the amount of time hay-fed livestock spend in the stream (and therefore improve water quality)?, *14,* 1(35), 1992.

111. Clawson, J. E., The use of off-stream water developments and various water gap config-urations to modify the watering behavior of grazing cattle, Oregon State University: Corvallis, OR, 1993.

112. Sheffield, R. E., et al, Off-stream water sources for grazing cattle as a stream bank stabilization and water quality BMP., *Transactions of the ASAE,* 40(3), 595–604., 1997.

113. Correll, D. L., T. E. Jordan, and D. E. Weller, Livestock and pasture land effects on the water quality of Chesapeake Bay watershed streams, in *Animal Waste and the Land-Water Interface,* CRC/Lewis, New York, p. 107, 1995.

114. NRCS, Dairy farmer profitability using intensive rotational stocking, Fort Worth, TX, NRCS, Grazing Lands Technology Institute, 1996.

115. USEPA, Monitoring guidance for determining the effectiveness of nonpoint source controls, EPA 841-B-96-004, Washington D.C., U.S. EPA, Office of Water, 1997.

116. Sanders, T. G., et al., Design of networks for monitoring water quality, Littleton, CO, Water Resources Publications, 1983.

117. Ward, R. C., J. C. Loftis, and G. B. McBride, The data-rich but information-poor syndrome, *Water Quality Monitoring Environmental Management,* 10(3), 291, 1986.

118. Ward, R. C., J. C. Loftis, and G. B. McBride, *Design of Water Quality Monitoring Systems,* New York, Van Nostrand Reinhold Co., 1990.

119. Ward, R. C. and J. C. Loftis, Monitoring systems for water quality, *Critical Reviews in Environmental control,* 19(2), 101, 1989.

120. USEPA. Evaluation of the experimental rural clean water program, in *National Water Quality Evaluation Project,* NCSU Water Quality Group, EPA 841-R-93-005, U.S. Environmental Protection Agency, Office of Water, Washington, D.C., 1993.

121. Gilbert, R. O., *Statistical Methods for Environmental Pollution Monitoring,* New York, Van Nostrand Reinhold, 1987.

122. Cochran, W. G., *Sampling Techniques 3rd ed,* New York, John Wiley & Sons, Inc., 1977.

123. Nelson, J. D. and R. C. Ward, Statistical considerations and sampling techniques for ground-water quality monitoring, *Groundwater,* 19(6), 617, 1981.

124. Brown, E., M. W. Skougstad, and M. J. Fishman, Methods for collection and analysis of water samples for dissolved minerals and gases, in *Techniques of water-resources investigations of the U.S. Geological Survey,* U.S. Government Printing Office, Washington D. C, 1970.

125. Spruill, T. B., Monitoring regional ground-water quality, statistical considerations and description of a monitoring network in Kansas, Water Resources Investigation Report 90-4159, Lawrence, KA, U.S. Geological Survey, 1990.

126. USEPA, Quality assurance project plan for the national pesticide survey of drinking water wells, EPA 810-B-92-001, Cincinnati, OH, U.S. EPA, Technical Support Division, Office of Drinking Water, 1992.

127. Spooner, J., et al. Appropriate designs for documenting water quality improvements from agricultural NPS control programs, in *Perspectives on nonpoint source pollution,* Washington D.C., U.S. EPA, 1985.

128. Ott, L., *An Introduction to Statistical Methods and Data Analysis,* Boston, PWS-Kent Publishing Co. 835, 1984.

129. Montgomery, R. H., J. C. Loftis, and J. Harris, Statistical characteristics of groundwater quality variables, *Groundwater,* 25(2), 176, 1987.

130. Hirsch, R. M., A. G. Scott, and T. Wyant, Investigation of trends in flooding in the Tug Fork Basin of Kentucky, Virginia, and West Virginia, Water-supply Paper 2203, Washington D.C., United States. Geological Survey, 1982.

131. Cooke, R. A., S. Mostaghimi, and P. W. McClellan, Application of robust regression to the analysis of BMP effects in paired watersheds, *Transactions of American Society Agricultural Engineers,* 38(1), 93, 1995.

132. Walker, J. F., Statistical techniques for assessing water-quality effects of BMPs, *Journal Irrigation and Drainage Engineering,* 120(2), 334, 1994.

133. Weaver, A. B. and D. A. Hughes. Continuous monitoring of rainfall, streamflow and suspended sediment concentration in semiarid environments, in *Challenges in African hydrology and water resources, proceedings, Harare Symposium,* Wallingford, Oxfordshire, International Association of Hydrological Sciences, 1984.

134. Lettenmaier, D. P., Detection of trends in water quality data from records with dependent observations, *Water Resources Research,* 12(5), 1037, 1976.

135. Shukla, S., S. Mostaghimi, and A. C. Bruggeman, A risk-based approach for selecting priority pesticides for groundwater monitoring programs, *Transactions of American Society Agricultural Engineers,* 39(4), 1379, 1997.

136. Vitale, R. J., O. Braids, and R. Schuller, *Groundwater Sample Analysis, Practical Handbook of Ground-Water Monitoring,* Ed. D.M. Nielsen, Chelsea, MI, Lewis, Inc., 1991.

137. Loftis, J. C. and R. C. Ward, Water quality monitoring—some practical sampling frequency considerations, *Environmental Management,* 4(6), 521, 1980.

138. Brakensiek, D. L., H. B. Osborn, and W. J. Rawls, Field manual for research in agricultural hydrology, Chapter 10, Agriculture handbook; no. 224., ed. USDA, Washington, D.C., Science and Education Administration, 1979.

139. USEPA, Handbook of sampling and sample preservation of water and wastewater, EPA 600/4-82/-029, Cincinnati, OH, U.S. EPA, Environmental Monitoring and Support Laboratory, Office of Research and Development, 1982.

140. Brossman, M. W., T. J. Hoogheem, and R. C. Splinter, Quality assurance project plans— a key to effective cooperative monitoring programs, in Quality Assurance for Environmental Measurements, Philadelphia, PA, American Society for Testing and Materials, 1985.

141. USEPA, National survey of pesticides in drinking water wells, Phase I report, EPA 570/9-90-015, Washington, D.C., U.S. EPA, 1990.

142. USEPA, Interim guidelines and specifications for preparing water quality assurance project plans, QAMS-005/80, Washington, D.C., U.S. EPA, 1980.

143. Stanley, T. W. and S. S. Verner, The U.S. Environmental Protection Agency's quality assurance program, in Quality Assurance for Environmental Measurements, Philadelphia, PA, American Society for Testing and Materials, 1985.

144. Mostaghimi, S., Watershed/water quality monitoring for evaluating BMP effectiveness— Nominal Creek Watershed. Quality Assurance/Quality Control Project Plan. Report No. N-QA3-8906, Blacksburg, VA, Virginia Polytechnic Institute and State University, 1989.

145. Mostaghimi, S., Watershed/water quality monitoring for evaluating BMP effectiveness—Owl Run Watershed. Quality Assurance/Quality Control Project Plan. Report No. O-QA3-8906, Blacksburg, VA, Virginia Polytechnic Institute and State University, 1989.

11 Monitoring

William L. Magette

CONTENTS

11.1 INTRODUCTION

Generically, "monitoring" can be described as the process of making observations for purposes of control or decision making. Although nonspecific, this is a useful

1-56670-222-4/01/$0.00+$.50
© 2001 by CRC Press LLC

conceptual definition of monitoring within the context of agricultural nonpoint source (i.e., diffuse) pollution identification and assessment. Indeed, "control" or "decision making" is the purpose of virtually every (nonresearch) diffuse pollution monitoring program.

Monitoring of agricultural nonpoint source pollution (NPSP) is conducted for several reasons (e.g., regulation, policy development, resource assessment, evaluation of managerial practices, research, and other purposes). Regardless of purpose, all monitoring programs involve making observations (i.e., specific measurements) somewhere in a watershed (catchment) and evaluating the meaning of such observations. The specific purpose of a monitoring program modifies the detail in which monitoring is conducted and the types of measurements that are made.

Diffuse pollution results from the interaction between uncontrollable (and largely unpredictable) weather events and the landscape. The landscape is itself a patchwork of areas that differ in topography, geology, vegetative cover, soils, management, and other factors, all of which influence hydrologic and pollutant response. Given so many variables, monitoring agricultural nonpoint source pollution is anything but straightforward. Compared with monitoring point sources of pollution, the challenges of monitoring diffuse pollution are exceedingly more difficult and typically result in greater costs. This is because of the inherent causes of nonpoint source pollution, but also because the "system" (i.e., a watershed or portion thereof) being monitored is large in area and spatially and temporally variable.

Another difficulty is finding ideal (representative, readily accessible, and reliable) monitoring sites at which to make necessary measurements. Except at the outlet of a watershed, there is generally no singular point in a catchment that is comparable with the final discharge point for effluent from a point source of pollution. Yet, at best, measurements made at the outlet of a catchment reflect the cumulative effect of all weather-landscape-human activity interactions in the catchment, as modified by various transport and attenuation processes. Such integrated measurements make it difficult to discern the impact of a specific situation within the watershed. Depending on the monitoring program objectives, monitoring might be necessary at other scales, requiring measurement points instead of (or in addition to) the catchment outlet. These can include the edges of fields, the bottom of a root zone, the water table or points below, a drain outlet, a spring, a stream, or anywhere water moves either continuously or intermittently.

As suggested by the previous statement, this chapter concentrates on monitoring water-borne diffuse pollutants from agricultural land. Although some attention also is afforded to soil monitoring and collection of agricultural management data, air quality monitoring is not addressed. The intent of this chapter is to give the reader an overview of diffuse pollution monitoring from the perspective of a practitioner. Excellent texts devoted solely to this subject are available (e.g., Dressing,[1] Bartram and Ballance,[2] Kunkle et al.,[3] Gibbons,[4] Ward et al.,[5] Chapman[6]). Readers are encouraged to consult such references for more detail than is possible or appropriate to give here.

11.2   DESIGNING AN EFFECTIVE MONITORING PROGRAM

An effective NPSP monitoring program is one that produces desired information at an acceptable level of effort and cost. Such a program results from good planning, careful execution, and continuous review and evaluation. These three elements are embodied in the monitoring system design, which evolves as the culmination of numerous discussions between various groups of professionals.

Essential groups involved in designing a NPSP monitoring program are those ultimately responsible for implementing the program and those who will use the resulting information. However, the latter group may be represented via a project brief, such as a solicitation for services or a regulatory stipulation. Regardless of whether end users of program results physically participate in the discussions, developing the monitoring design is very much an iterative process. Once a design is agreed, however, it represents a blueprint for the monitoring program that reconciles users' needs for information against the technical, financial, and temporal considerations that invariably constrain a program. Among many other things, the design describes:

- how, when and where samples will be collected
- how the samples will be analyzed
- how the resulting data will be stored, retrieved, analyzed, and interpreted
- how the program results will be reported

Typically, financial resources for a NPSP monitoring program will be defined by a solicitation for services (as in the case of a fixed-term investigation) or by a government-based budgetary process (as in the case of long-term, ambient monitoring efforts). The challenge is to develop a monitoring program that will deliver credible and useful information within these financial constraints.

11.2.1   PLANNING

Poor planning is more frequently the cause for failures of nonpoint source pollution monitoring programs than are deficiencies in implementation. Contrary to intuition, the success of an NPSP monitoring program is not necessarily proportional to the size of the monitoring budget. Rather, success is a function of the amount of effort devoted to planning the endeavor. In the absence of proper planning, large expenditures will not produce acceptable monitoring results. By contrast, well-planned programs with only modest budgets are capable of producing data that can be interpreted to yield useful information. (Yet, despite the power of planning, even outstanding planning cannot overcome the constraints caused by hopelessly inadequate budgets.)

In short, good planning is essential to a successful monitoring project. The key to good planning is having clear goals and identifying precise objectives to achieve those goals. If the human, financial, and time resources available for the monitoring program are inadequate to achieve the objectives, either the objectives or the resources must be modified. Otherwise, the monitoring program will not be a success. Finding the balance between objectives and resources is partly what makes the design of monitoring programs an iterative process.

11.2.2 GOALS AND OBJECTIVES

Planning a NPSP monitoring program is understandably difficult because, like any planning process, it involves making projections into the future. The uncertainty of anticipating what might happen over the lifetime of a monitoring program can be overcome somewhat through experience. However, even for experienced personnel, nothing can improve planning so much as having clear monitoring goals and objectives.

In broad terms, goals identify the reasons for conducting a monitoring program. The generation of useful information should be an overriding goal of every monitoring program. However, specificity must be added to this goal by appending the reason(s) for which the information is needed (e.g., to provide a "snapshot" of regional water quality, or to assess the suitability of a water resource for human consumption). A group other than those responsible for implementing the NPSP monitoring program often sets the goal or goals of a monitoring program.

In contrast, those responsible for conducting the monitoring program typically define the monitoring objectives. Objectives are the precise pathways by which the program goal or goals are satisfied. Objectives must be articulated in very specific language and committed to writing, as these are the guiding forces for all other aspects of the monitoring program.

In setting objectives, it is insufficient only to answer the question "what is to be accomplished by monitoring?" Instead, objectives must be defined in measurable terms. For example, an objective of a field-scale monitoring project may be "to demonstrate the environmental benefits of nutrient management planning." Although descriptive, this objective lacks specificity in terms of measurability and would be better considered as an overarching monitoring goal. Focus and measurability could be given by making a simple change in the original statement (e.g., "to determine annual field-scale losses of nitrogen and phosphorus as a result of implementing nutrient management planning"). An equally acceptable objective could be "to quantify changes in in-stream concentrations of phosphorus as a result of implementing nutrient management planning, compared with those concentrations resulting from traditional management of nutrients."

Of course, the monitoring approach would be entirely different to achieve each of the two restated objectives given above. That is precisely why goals and objectives must be specific and why objectives must be measurable. There is little hope of devising a monitoring scheme capable of satisfying an intent that is not clearly articulated. Likewise, there is no way of knowing if the purpose of a monitoring scheme has been achieved unless there is a measurable standard (i.e., objectives) against which results can be evaluated. Thus, a lack of clearly stated goals and measurable objectives undermines the NPSP monitoring program from the outset, as well as threatens the credibility of those implementing the program when the achievements of the effort are reviewed.

11.2.3 DATA NEEDS AND DATA COLLECTION

Once monitoring objectives are agreed, it is possible to plan how to achieve them. This aspect of planning should begin by assessing what data and information are needed to achieve the objectives, and what already exist. A reconnaissance of all possible data sources should be among the first tasks undertaken. Are there ambient water quality monitoring schemes in place; if so, are the results applicable and available? Are there regulations dictating that certain discharges of pollutants be monitored; if so, are the data part of the public record? Are sales data available that could help define the quantity of potential pollutants (e.g., pesticides or fertilizer nutrients) in a given geographic area? What research is available from universities, other research organizations, or regulatory agencies? Are there trade associations, such as farmers organizations, that have useful information about the implementation of specific agricultural management practices? Most important, is there an adequate definition of the hydrologic behavior (i.e., specific pollutant transport pathways) of the study area?

It is easy to underestimate the effort (cost and time) required to gather, collate, and interpret existing information. Even when seemingly useful data already exist and are available, much time might be needed to assess the true relevance of the data to the monitoring program. For example, ambient water quality data are readily available in developed countries. However, in the context of diffuse pollution assessments, a serious deficiency with these data sets is that synchronous measurements of terrestrial data (such as land management practices) rarely exist. Thus, much effort might be needed just to ascertain if the available data could be used for, say, identifying baseline cause-and-effect conditions. Typically, it is relatively easy to characterize ambient water quality in a catchment, but far more difficult to ascribe reasons for the ambient conditions. Other obstacles that complicate the use of existing data are questions about data quality, and the effects that changing measurement methods might have had on data comparability over time.

Once all available data have been assembled, assessed for usability, and interpreted, it is possible to identify data "gaps." Then, using the monitoring program objectives as determinants, program directors can draft a list of data needs. With such a list, the directors can debate and agree on the most effective strategies to satisfy the data needs, and therefore, the monitoring program objectives.

An "existing conditions" report is an effective way to summarize what is known (and what is not known) about water quality and the terrestrially based impacts influencing it in the area to be monitored. A report of this type, or extracts from it, also can be a useful way to communicate information to those who will ultimately use the results of the monitoring program, and to other important stakeholders (e.g., the general public, farmers, local government representatives). As well, the very process of assembling an existing conditions report helps those implementing the monitoring

program identify genuine data needs and rank those needs according to their importance in satisfying program objectives. This, in turn, helps rationalize choices
between competing ways to expend limited monitoring resources.

11.2.4 IMPLEMENTATION STRATEGIES

11.2.4.1 Data Interpretation

Goals and objectives define the targets for a monitoring program; the implementation
strategy defines how the program will be conducted. As such, it is a comprehensive
description of all facets of data collection, data analysis and interpretation, and data
reporting, including the important elements of quality assurance and quality control
associated with each of these aspects.

An effective way to begin developing a monitoring implementation strategy is to
first specify how data will be interpreted. This may seem surprising given that data
analysis and interpretation are tasks usually considered only after data have been collected. In reality, data collection should be guided by the data interpretation techniques to be used and, of course, by the monitoring objectives. By first deciding how
data will be handled (e.g., the specific statistical tests), those responsible for data collection can avoid or minimize two common and costly monitoring mistakes: (1) collecting data that are not needed, and (2) failing to collect data that are needed.

Waiting to select data interpretation tools and procedures until after sampling
stations have been located, sampling frequencies adopted, and data collection begun,
invites disaster. Environmental data are prone to being correlated (especially when
taken over a short period of time), flow dependent (for water-based data), and subject
to seasonal variability. "Missing data," either resulting from laboratory error or
mechanical failure of sampling equipment, and "censored data" resulting from a
finite detection limit of analytical instruments add to the problems in statistically
evaluating environmental measures. The net effect of these complications is that environmental data typically violate most, if not all, assumptions required for analysis by
classical (i.e., parametric) statistics.

In devising an implementation strategy for monitoring NPSP, it is imperative to
anticipate these complications before data are collected, not after. Guidance on
selecting data analysis and interpretation procedures is readily available (Gibbons,[4]
Gilbert,[7] Schweitzer and Santolucito,[8] Hipel[9]). Given the critical role of data interpretation in the success of a monitoring program, the importance of having data interpretation specialists as members of the monitoring program development team
should be obvious. Further, these persons should be involved at the early stages of
planning and throughout the development of the implementation strategy.

11.2.4.2 Sampling Sites

Both the monitoring objectives and the data interpretation techniques will guide the
frequency of sample collection. Likewise, the monitoring objectives will dictate the
kinds of sampling sites that will be required. As mentioned previously, NPSP monitoring can take place at a variety of scales from field to watershed.

The first priority of any monitoring scheme should be to produce valid data from which rational decisions can be made. In this context, monitoring sites must yield representative samples. However, from a logistical standpoint, sites also must be readily accessible and reliable. Of these three criteria, the degree to which a site is representative is the most difficult to assess. Clear monitoring objectives facilitate this assessment. Under no circumstances should a site be selected for its accessibility in preference to its ability to yield representative samples.

When monitoring diffuse agricultural pollution, the question often arises as to which environmental medium (soil, water, or air) should be monitored. The farm-yards and farm fields in which agricultural production (and therefore, potentially pol-luting activity) takes place can be relatively far removed from surface water resources. Likewise, groundwater below pollutant sources may occur at great depth or be relatively insulated from terrestrial activities by impermeable layers of geologic material. One could argue, then, that soil is the most proximate environmental medium to agricultural sources of pollutants. However, soil is tremendously variable in virtually all of its characteristics. And, monitoring the soil by itself does not pro-duce information about the water quality impacts (if any) of diffuse agricultural pollution.

This dilemma reinforces the importance of precise monitoring objectives in determining sampling sites. If the objective is, for example, to quantify pollutant losses from a specific agricultural practice or from a particular combination of site characteristics and management practices, sampling should take place as close to the potential origin of pollutants as possible. Thus, edges of fields, the root zone, or the soil itself are appropriate monitoring sites for such an objective. This is because the ability to relate pollutant losses to a specific (geographic) source in a watershed is inversely proportional to the distance from the source at which monitoring takes place. In addition, tremendous attenuation of diffuse pollutants can take place as a function of transport distance and intervening conditions between the pollutant source and monitoring point.

In contrast, if the monitoring objectives include determining water quality impacts of diffuse agricultural pollution, there is no alternative to sampling surface water, ground water, or spring discharges. As mentioned previously, the challenge is to select sites that are representative and accessible. Given that pollutant attenuation is a function of transport distance, it is essential that monitoring sites in an aqueous medium fit into a statistical design capable of detecting changes in water quality and also relating those changes to activities on the landscape.

11.2.4.3 Quality Assurance/Quality Control

Just as the strength of a chain is determined by its weakest link, the effectiveness of a diffuse pollution monitoring program is limited by its least robust component. By its very nature, diffuse pollution monitoring is prone to errors. Sample collection typ-ically occurs during or just after inclement weather that creates adverse working con-ditions. Large numbers of samples create logistical challenges in terms of logging, transport, analysis, and data reporting. Automated equipment can fail, resulting in the loss of critical samples. The list of difficulties and error sources is lengthy indeed.

A quality assurance/quality control (QA/QC) plan can help minimize these difficulties by controlling sources of error and assuring confidence in the results. It is, therefore, a critically important component of a diffuse pollution monitoring program as well as an influence over all other components. In short, a QA/QC plan describes the set of practices that will be followed to ensure that the output from every component of the monitoring program will be credible.

An acceptable QA/QC plan addresses both field and laboratory activities. Elements of the plan focus on all aspects of

- management (e.g., staffing and management hierarchy)
- training of field and laboratory staff
- standard operating procedures (field and laboratory)
- facilities (e.g., mobile and fixed laboratories)
- equipment maintenance and calibration
- sample collection
- sample handling (including logging, preservation, transport, chain of custody, and storage)
- reporting of laboratory results (including data checking)
- analysis and interpretation of data

"Large" monitoring programs benefit from having a designated quality assurance officer to oversee implementation of the QA/QC plan and monitor compliance. Some sources of funding may require accreditation of laboratories before providing money for monitoring programs. Another approach to implementing QA/QC is to have the monitoring program conform to quality standards such as ISO 9000.

Numerous sources of information are available for laboratory quality assurance schemes (e.g., Bartram and Balance,[2] Keith[10]). The principles embodied in such guidance can be adapted for application to field activities.

11.2.4   RECONCILING RHETORIC AND REALITY

The "rhetoric" of planning a NPSP monitoring program is the initial identification of objectives, sampling sites, and operational procedures. These must fit within the "reality" of available budgetary and other resources. Reconciling rhetoric and reality is the process of finalizing the monitoring program so that objectives can be realized within available time and financial constraints. This reconciliation process may dictate that objectives or other initial decisions change, but under no circumstances should QA/QC procedures be compromised. In diffuse pollution monitoring, it is always preferable to do a limited number of things well than to do many things poorly.

It is useful, if not essential, that a trial of the monitoring program be conducted before finalizing the plan. Ideally, this trial would include going through the entire program step-by-step in as realistic a "simulation" as possible (except for installing costly monitoring facilities). Going through the process of collecting and transporting samples is especially useful in highlighting potential logistical problems and can serve as a valuable training exercise. "Desktop" simulations of the monitoring program are useful as well. It may be possible to use existing data sets with the proposed data analysis procedures. It is certainly feasible to forecast staffing availability allowing for both anticipated changes (vacations) and unanticipated changes (e.g., illness). It is also relatively easy to visualize the potential effect (and appropriate responses) of malfunctions in key equipment and other critical aspects in the monitoring program.

11.3 MONITORING TECHNIQUES

As described previously, a variety of sampling scales can be used in an NPSP monitoring scheme, depending on the monitoring objectives. Each of these has particular demands (and constraints) in terms of monitoring techniques. The process of collecting samples is quite simple compared with deciding monitoring objectives and a logistical plan that assures the objectives will be satisfied within budget limits.

Nevertheless, detecting changes in the quality of an uncontrolled environment is fraught with difficulty not only because of the unpredictable nature of weather events (which are the driving forces that transport pollutants to receiving waters), but also the natural variability of the system (topography, stream density, vegetative cover, soil characteristics, and other factors). Unlike man-made systems, such as sewerage works or industrial wastewater treatment systems, the hydraulic linkage between points in a natural system is not obvious. As regards free-flowing streams, it may be a relatively simple matter to separate stream flow into its component parts of stormwater runoff (i.e., overland flow and shallow subsurface interflow) and base flow (i.e., groundwater discharge), but it is not at all simple to identify where specifically within a catchment a particular contribution of water to the stream originated. The same obstacle exists concerning ground water and other types of surface water (i.e., lakes and estuaries). The problem of flow path identification is all but insurmountable in areas having complex hydrogeography, such as regions dominated by karstified limestone.

In contrast, installing the necessary equipment to allow representative samples to be collected is relatively straightforward. There are relatively few places at which and ways that samples can be retrieved from an aqueous medium, as listed below.

Sampling point	Sampling technique
Edges of fields	Flumes or other constructed device (e.g., flow splitter, Coshocton wheels)
Bottom of root zone	Suction cups, plates or candles, gravity, lysimeters
Groundwater	Wells (boreholes)
Drainage pipes and springs	Flumes or other constructed device
Surface water	Weirs, flumes, or other stable cross section

11.3.1 EDGE-OF-FIELD OVERLAND FLOW

Surface runoff, or overland flow, results from two processes. Hortonian overland flow occurs after precipitation has filled all surface depressions on the soil surface and either continues to fall at a rate faster than it can be absorbed into the soil, or continues to fall in an amount that exceeds the storage capacity in the soil profile. Saturation (or apparent) overland flow can occur at the bottom of some hill slopes, where topography changes from convex to concave or where an impeding subsurface layer intersects the soil surface. Saturation overland flow also can occur where a rising water table reduces the water storage capacity of the soil profile to such an extent that even low intensity rainfall cannot infiltrate.

Regardless of the causative mechanism, overland flow can transport pollutants in particulate form (e.g., soil particles, organic material) and in dissolved form (e.g., soluble nutrients). Overland flow is most conveniently measured and sampled at places in the landscape where it naturally becomes concentrated (such as in drainage ways) or where the natural topography can be modified to force the overland flow to concentrate.

11.3.1.1 Flow Measurement

Whereas simple grab samples of overland flow can be collected and analyzed to yield the concentrations (mg L^{-1}) of pollutants contained therein, only by simultaneously measuring flow rate (and therefore volume) can these concentrations be translated into mass losses (e.g., kg or kg ha^{-1}). In general, both mass and concentration data are needed in a diffuse pollution monitoring program. Concentration data are useful in evaluating habitat impacts because these tend to be specified in terms of concentrations; mass data are useful in evaluating the efficiencies of management practices to control pollutant losses. Unfortunately, as mentioned previously, the relevance of edge-of-field data in assessing water resource impacts decreases as the distance between the source area and receiving water increases.

Until it concentrates because of natural topographic features or artificial means, overland flow occurs at a relatively shallow depth spread over a broad area. Sample collection is therefore dependent on forcing the overland flow through a constricted flow path that causes flow depth to increase. If the constriction is chosen carefully so that a unique relationship between flow volume and flow depth can be determined, then the constriction serves a dual role of facilitating both flow measurement and sample collection. Flumes are particularly useful for this purpose.

Flumes used for edge-of-field monitoring tend to be either of the H design (including HS and HL) or Parshall design. H flumes are particularly useful where floating material is likely to be transported in the runoff as these flumes have a self-cleaning critical section that generally prevents clogging by debris. Standard designs can be modified where sedimentation is anticipated to be a problem. Flumes are typically constructed of stainless steel or fiberglass, depending on the pollutants expected to be encountered. Assuming flumes are carefully constructed and put in place, flow measurement is exceptionally accurate. Their theoretical calibration should, nevertheless, be checked following installation. Numerous sources give design and construction details for flumes (Brakensiek et al.,[11] Bos et al.,[12] Leupold & Stevens,[13] Grant[14]).

Flow measurement in flumes is accomplished by measuring the depth of flow in the control section. This can be accomplished by traditional float-and-pulley systems connected to a recording device (paper chart, punched paper tape, or data logger). Alternatively, bubbler systems, pressure transducers, and ultrasonic sensors, each connected to a data logger, may be used. Each of these techniques has particular advantages and disadvantages. Whereas the use of electronic measurement techniques is very much the norm because of the obvious benefits these offer in terms of data handling and remote sensing, care needs to be exercised in selecting the particular sensor. Manufacturers provide guidance on equipment selection. Obviously, a source of power (batteries, solar cells, or line electricity) is required for electronic devices.

11.3.1.2 Sample Collection

Integral to the process of flow measurement is the collection of samples for subsequent analysis. Automated (discrete and composite) samplers are very much the norm for this application, particularly where electronic flow measurement is used, as the samplers integrate with the flow recorders. Nevertheless, float-and-pulley systems for flow measurement can be modified to operate automated samplers. Both types of samplers can be set to collect samples on a timed basis or on a flow basis. Flow proportional sampling is usually preferred because pollutant transport is typically a function of runoff rate.

Likewise, discrete automatic samplers are preferred to composite samplers when it is important to know when, during a runoff event, pollutants are transported. This information is particularly useful when devising pollutant control strategies and when gathering data for ultimate use in mathematical models. If it is important only to know mass losses of pollutants for an entire event, then composite samplers are satisfactory.

Like electronic flow recorders, automatic samplers require a source of power. Refrigerated samplers are available for use at monitoring sites where it is not feasible to retrieve samples immediately after they have been collected. Nevertheless, sample holding times (including the time needed to transport samples to the laboratory) must not exceed the recommended maximums for the analytical tests to be used.

An alternative to automated sampling is hand sampling, but this is almost always impractical because of the unpredictable nature of runoff events and the high labor requirements. Nevertheless, it is a reliable, and often preferred, sample collection technique during plot-scale intensive studies, as with rainfall simulation. Other alternatives include flow-splitting devices such as multislot divisors and Coschocton wheels (Brakensiek et al.[11]). These instruments operate by diverting some fraction of the total flow into a collection vessel. Thus, they provide flow-proportional composite samples and a crude estimate of total flow volume. If composite samples are acceptable, flow splitting devices offer some advantage because of their low cost (compared to automated samplers) and freedom from power requirements.

11.3.2 BOTTOM OF ROOT ZONE

Water enters the soil profile by infiltration. If at any time the quantity of water in the profile exceeds the demands exerted by plants (the amount lost by evaporation and the amount that the soil can retain naturally), the water will move downward through the profile in response to gravity. Traditionally, the bottom of the root zone (i.e., the deepest extent of most roots for a given type of plant) has been used as a convenient hypothetical boundary for measuring vertical losses of pollutants from agricultural fields. The rationale for this selection has been that once pollutants, which include valuable plant nutrients, exit the root zone, there is little (especially plant uptake) to impede their delivery to groundwater. Although this rationale is not strictly true, the extent to which physical, chemical, or biological processes below the root zone can attenuate pollutants is significantly lower than in the root zone itself.

Regardless of whether sampling of water leaching through the soil profile is attempted at the bottom of the root zone or deeper in the profile, the collection techniques are basically the same. Ceramic (or fritted glass or Teflon®*) samplers (also called suction or tension lysimeters) can be inserted into the soil profile and fitted with a vacuum to extract soil water from the soil (Morrison,[15] Wilson[16]). Alternatively, so-called "zero tension" samplers can be used to collect soil water when the profile at the point of measurement is saturated. This liquid can then be analyzed for pollutant concentrations.

However, because it is not possible to determine the specific origin of the extracted water, it is usually impossible to translate concentration data from either suction or zero-tension lysimeters into mass data. The ceramic samplers also pose many practical problems: intimate contact between the sampler and the bulk soil is essential, yet difficult to achieve; the samplers are difficult to install in stony or gravely soils; and some pollutants adhere to the ceramic used to manufacture the samplers. Fritted glass and Teflon® can be substituted for ceramic to overcome the latter problem.

Naturally draining lysimeters offer some improvement over ceramic samplers, but they tend to be even more difficult or expensive to install. Lysimeters are typically of three types: column lysimeters, monolithic lysimeters, and pipe lysimeters. In general, pipe lysimeters perform exactly like subsurface drains (discussed below). Column lysimeters usually are constructed of PVC or concrete pipes ranging from 15–60 or 90 cm in diameter that encase either disturbed or undisturbed soil profiles. Monolithic lysimeters are much larger structures capable of supporting full-sized agricultural machinery that encase undisturbed soil profiles. Undisturbed profiles are regarded as being superior to disturbed ones in mimicking natural conditions.

Both column and monolithic lysimeters are placed into the bulk soil or into a purpose-built excavated site. These lysimeters permit all drainage to be collected and therefore facilitate both concentration and mass data to be accumulated. However because their bases are no longer a part of the soil mass, these devices tend to create artificial water tables within the lysimeter that may not exist in a natural setting. Lysimeters are typically best suited for research applications.

---

*Registered Trademark of E.I. du Pont de Nemours and Company, Inc., Wilmington, Delaware.

11.3.3 GROUNDWATER

Groundwater is that resource existing at variable depths below the soil surface in zones called aquifers. Groundwater is often a source of drinking supplies for rural inhabitants; it also usually makes its way toward and eventually becomes surface water. It is replenished naturally by precipitation that percolates downward through the soil profile. In so doing, this percolating water can also transport unwanted pollutants, such as nitrate nitrogen. The area of land surface that precipitation enters as it makes its way to replenish groundwater is called a recharge zone.

Groundwater is divided into two categories, confined and unconfined, depending on whether the aquifer in which it is contained is confined by restricting layers of geologic material or not. Because such restricting layers consist of highly impermeable material, such as clay, that do not transmit water or pollutants readily, they tend to insulate confined aquifers from the downward movement of water through the soil profile directly above the aquifer. Thus, the replenishment of confined ground water typically occurs from recharge areas that may be tens to hundreds of kilometers away from where the aquifer is monitored. In contrast, unconfined aquifers lack impermeable layers above them. These aquifers are thus most susceptible to contamination by pollutants originating from human activity directly above them.

Regardless of whether an aquifer is confined or unconfined, movement of groundwater within aquifers has both a horizontal and vertical component. The rate of movement is extremely slow compared with surface water, except perhaps in karstified limestone aquifers. Also compared with surface water, which is generally well mixed by turbulent flow, groundwater moves slowly along flow lines. Under laminar flow conditions, a theoretical droplet of water moving along one flow line mixes relatively little with neighboring droplets. Generally, the deeper flow lines in an aquifer transmit the oldest water, which has traveled the farthest distance.

For all these reasons, monitoring groundwater to detect the influence of human activity above it is not straightforward. In general, except for unconfined aquifers, a thorough hydrogeologic investigation must be completed before the locations of bore holes for monitoring can be determined. Yet, considering that the recharge area for a confined aquifer can be quite distant from the area to be monitored, the sampling of a confined aquifer can be quite irrelevant in many cases. Even for unconfined aquifers, bore holes must be carefully constructed to make certain only the top or uppermost region of the aquifer is sampled (to detect the influence of land use directly above). The farther into the depth of an unconfined aquifer a monitoring point is inserted, the farther from that point will be the recharge area from which the water at that point originated.

Thus, before land management activities can be accurately monitored by examining groundwater, a thorough geohydrologic investigation should be performed by appropriately trained professionals to identify groundwater flow paths. This assessment should tell where to establish bore holes. However, it likely will not tell how many to establish. In practice, statistical rigor (Gibbons[4]) is difficult to achieve because of the costs of constructing monitoring bore holes. Nevertheless, every attempt to achieve a statistically sound distribution of monitoring sites should be made. The use of some arbitrary rule of thumb, such as one bore hole "up-gradient" and two holes "down-gradient" of the site of interest, yields only minimal useful information. Nevertheless, even when using such a simple monitoring design, it is imperative that the groundwater being monitored at the down-gradient sites is water that has actually been (or likely to have been) impacted by the area of interest.

Improperly constructed monitoring wells can themselves be sources of groundwater contamination. Thus, it is imperative that wells be installed by trained professionals. The choice of drilling technique depends more on the geologic conditions than on the ultimate use of the well as a monitoring device. Regardless of the drilling procedure, the resulting annulus around the well casing must be carefully sealed with a grouting material to prevent surface water from traveling down the casing and into the water table. Likewise, if a monitoring well penetrates one or more confined aquifers, care must be taken to assure that the casing is firmly set in the confining layer to prevent a hydraulic cross-connection between aquifers.

In general, it is more useful to collect "depth-discrete" groundwater samples, than depth integrated. Depth-discrete samples are obtained using short (0.6 m) screens placed at strategic depths within an aquifer, usually in a collection called a "nest," and provide insight into both the vertical and horizontal movement of pollutants. In principle, a depth-discrete sample can also be retrieved using multiple wells set at different depths in a single bore hole; however, some literature suggests that the hydraulic seals separating the well screens are not always effective. If the objective is to monitor the vertical contribution of pollutants from land use directly above a monitoring well, a single screen long enough to span the anticipated variation in water table level (and providing a depth integrated sample) may be acceptable. Regardless of the type of screen used, it is critical to accurately locate its elevation and that of the water table and soil surface. Also, the well must be properly developed to remove drilling debris and fine sediments from around the screen so that representative samples of native groundwater can be retrieved.

Sample retrieval can be accomplished by a variety of means, ranging from a simple, hand-operated bailer to a mechanically powered pump. Care must be taken to remove stagnant groundwater from the casing before collecting a sample for analysis. The sample retrieval process must not introduce contamination into the well, nor must it alter the intrinsic composition of the native groundwater. The latter can be of particular concern if volatile compounds are the pollutants of interest.

Because of the concern about groundwater contamination over recent years (at least in the U.S.), there is ample guidance available regarding all aspects of groundwater monitoring (Barcelona et al.,[17] Scalf et al.,[18] USEPA,[19] Gibbons,[4] Nelson and Dowdy[20]).

11.3.4 DRAINAGE PIPES AND SPRINGS

In some respects, subsurface drainage pipes offer the best opportunity to monitor the vertical losses of pollutants from agricultural fields. This assumes that the drainage tiles were designed correctly, are working properly (i.e., are not blocked), and that their discharge points are easily accessible. If these conditions are met, it is possible to measure flows and collect samples for analysis. Thus, both pollutant concentrations and mass losses can be determined. In addition, because drainage theory is well advanced, it is possible to calculate fairly accurately what area of a field is contributing flow to an individual drain. This calculation permits mass losses to be expressed on an areal basis (e.g., kg ha^{-1}). Further, because drainage pipes typically discharge to flowing water, pollutant losses measured at the discharges of these devices are equivalent to those delivered to a surface water resource.

Likewise, springs and seeps can provide a location for collection of water samples and sometimes for flow measurement. As a minimum, data about the concentrations of pollutants in this flow can be obtained; in some cases data about masses of pollutants lost in these flows can be developed also. However, it is usually not possible to express mass losses on an areal basis because the drainage area contributing flow to the spring or seep is difficult to define.

Except when flowing full, drainage lines are basically open channels. Thus, open channel flow measuring techniques (e.g., flumes and weirs) can be applied to drainage pipes if the discharge can be appropriately directed through the measuring device. The same is true for spring and seepage discharges. In addition, depending on the pipe diameter, it is possible to measure flow using Doppler technology (flow area/velocity) and to insert weirs or flumes into the drainage pipe itself. Flow recording and sample retrieval are accomplished using the same techniques described previously in the "Edge-of-Field Overland Flow" section.

11.3.5 SURFACE WATER

When attempting to measure the impacts of a particular land management practice (or land use) on diffuse pollutant losses, surface water as a possible sampling point is most relevant when a stream or other open conveyance borders the agricultural field under evaluation. In any event, when the monitoring objective is to evaluate the water quality impacts of diffuse pollutants, surface water sampling is unavoidable (unless, of course, the focus of the monitoring program is solely on groundwater).

Despite its appeal as an accessible environmental medium, surface water presents many monitoring challenges. For example, the diversity of surface water is large, ranging from ditches, drains, and minor channels that flow intermittently, to large rivers, lakes, estuaries, and oceans. Another challenge is the fact that, in general, free-flowing streams and rivers contain a mixture of groundwater and direct surface runoff. Only by judiciously choosing the time(s) when sampling occurs is it possible to determine the relative contributions of pollutants from the two separate pathways. However, flows (both surface runoff and groundwater discharge) enter a stream/river coming from both sides of the channel. If the land areas bordering each side of the stream/river are not more or less identical and subjected to equivalent managerial practices, attributing water quality impacts to land management on either side will be difficult at best.

Surface water monitoring is best suited to catchment-scale evaluations of land use impacts on water resources. Catchments can range from large to small, being defined simultaneously by topography (for surface runoff) and hydrogeology (for groundwater contributions). In unit-source catchments (those in which land use and land management is the same throughout), surface water monitoring offers a reasonable means of assessing the cumulative impact of management over the entire catchment. Otherwise, surface water monitoring is a generally less straightforward means of evaluating land management impacts at a particular point in the catchment than is edge-of-field monitoring.

In contrast to other forms of open channel flow (e.g., ditches, springs, and overland flow), streams and rivers do not lend themselves well to flow measurement by flumes. However, many streams are amenable to flow measurement using weirs. Weirs are low-profile obstructions of specific cross sections built across open channels. A unique head discharge relationship allows flow volume to be measured by monitoring the depth of flow over the crest of the weir. Flow depth (and sample collection) can be measured by any of the techniques described previously in section 11.3.1. Brakensiek et al.[11] include helpful guidance in selecting an appropriate weir design based on a variety of site specific considerations.

Weirs are not appropriate for flow measurement on large rivers and streams. Instead, a stable cross section must be found at which the relationship between depths of flow and cross sectional areas of flow can be determined. In addition, the average velocity of flow at each depth of flow must be determined, from which a rating curve (flow depth versus flow volume) can be developed. This is a time-consuming process, but the technique is well established (e.g., Brakensiek et al.[11]).

Once a rating curve has been established for a stream or river, samples can be collected automatically by equipment described previously in section 11.3.1, or by hand. Regardless of the retrieval methods, particular attention must be given to making sure that representative samples are collected. As flow volume increases, the proportion of total flow represented by a single discrete sample decreases. Pollutant concentrations, particularly of suspended sediment, are known to vary considerably as a function of depth below the surface of a river and distance from each shore. These variations must be determined by repeated point sampling prior to the start of the monitoring program.

11.3.6 SOIL

For adhering to the principle of monitoring as close as possible to the source of agricultural nonpoint source pollutants, bulk soil is itself a relevant sampling point. In land-based agricultural production systems, it is soil that is the recipient of inputs (nutrients, lime, organic, and other amendments) that can become pollutants.

Bulk soil is, in fact, the medium that is sampled and analyzed to determine soil fertility status so that crop nutrition recommendations can be formulated. Soil testing laboratories typically have a standardized protocol for the collection and analysis of soil samples from which these recommendations are derived. These soil sampling procedures and the associated nutrient application recommendations have been experimentally tested and validated to take into account the tremendous variability inherent in the soil medium.

In tandem, these techniques produce scientifically valid results for fulfilling plant nutrition needs. Used separately, however, neither procedure is likely to produce equally good results. In general, the precision of statements that can be made about soil properties at a given point depends largely on how variable the area being sampled is; for a fixed number of samples, as heterogeneity increases, precision decreases.

As emphasized previously, when monitoring the environmental impact of agricultural best management practices, care must be taken to assure that sampling is reflective of these impacts. Although soil offers convenient and relatively inexpensive sampling opportunities, the sampling strategy must recognize and accommodate the spatially variable nature of soil properties. The sampling strategies that are sufficient for collecting soil samples from which to make agronomic recommendations may not be sufficient for documenting pollutant movement. In general, the intensity of sampling depends on the desired accuracy of the result and on the variability of soil population. Peterson and Calvin[21] provide a discussion of soil sampling strategies specifically for the soil medium. Gilbert[7] and Keith[10] provide more generic discussions of environmental sampling.

11.4 DETERMINING CHANGES IN ENVIRONMENTAL MEASURES

An essential part of every scientist's job is to determine changes resulting from an imposed experimental treatment. Scientists must continually ask themselves if one observation they make is actually different from another. Until they are sure they can make realistic measurements and determine true differences between measurements, they are helpless in assessing the results of the perturbations they deliberately cause through their experimental treatments. This assessment is accomplished using appropriate measuring techniques combined with proper statistical control. In the context of environmental management, one must be just as rigorous in asking: 1) if we can make representative measures, and 2) if two or more measures are actually different.

11.4.1 STATISTICAL CONTROL

It is impossible to disregard statistical control when discussing the monitoring of agricultural nonpoint source pollution or the evaluation of agricultural best management practices. Adhering to accepted monitoring protocol is but half of the requirement for credible monitoring and evaluation. Only when good statistical control accompanies an appropriate sampling and analysis protocol can differences between measurements be detected with confidence.

The variability in the natural environment is large, as noted previously. Because of this variability, it is not uncommon to find that measures of environmental quality (such as water samples or soil samples) differ quite dramatically from place to place, as well as from time to time at the same place. Regarding agricultural nonpoint source pollution control, the challenge for environmental managers and scientists is to determine if these differences are caused by natural variability (random effects) or by changes in agricultural management practice or land use (treatment effects). A specific example would be collecting soil samples from a given field on two separate occasions to determine if a farmer had followed a nutrient management plan. If the samples were different, one would have to ask if the differences occurred because the soil is naturally variable or because the person followed (or failed to follow) nutrient application guidelines.

Only statistical analysis can determine if the differences in separate environmental measures are caused by treatment effects. In a given monitored system, there will be some minimum detectable change (MDC) in a given measure below which it is impossible to determine if a change (or difference) in the measure is statistically significant (i.e., due to more than natural variability). For purposes of nonpoint source pollution monitoring, a system is a combination of size of the area being examined, monitoring program design, duration of monitoring program, the media being monitored, weather, and other factors (Spooner et al.[22]). Because many of the factors in a system are very variable, measures of the system performance will also be very variable, meaning that any differences in measures will have to be very large to have statistical significance.

Large MDCs make it difficult to determine treatment effects. To detect treatment effects on environmental measures, all sources of uncontrolled variability should be minimized as a way to reduce MDC. Although it is usually difficult to control natural variability, this can be accomplished to some extent by restricting the size of the system being examined. For example, in the previous example of soil sampling, one could confine the system being monitored to a particular part of a farm field, or by segregating sampling according to soil type or some other feature (stratified sampling). Alternatively, MDC can be reduced by collecting more samples, increasing the period of monitoring, and by using more sophisticated (and restrictive) statistical techniques (Spooner et al.[22]).

11.4.2 SURFACE WATER

Spooner et al.[23,24] have described several statistical designs for improving the ability to detect changes in surface water quality. These include (1) before and after testing (time trends or time series analyses), (2) above and below testing (upstream and downstream analyses), (3) paired catchments testing (treated–untreated catchment analyses).

Each of the designs has particular strengths, weaknesses, and economic costs, but all improve the ability to detect true changes in surface water quality beyond simple collection and analysis of grab samples by helping to reduce MDC. Depending on the parameter in question and the number of samples collected per year, changes in magnitude on the order of 30–60% can be required (Spooner et al.[22]) for differences to be statistically significant (due to treatment effects). The above designs can help improve the sensitivity of monitoring so that smaller impacts can be detected.

If surface water is monitored, it is imperative that the monitoring scheme be designed to measure both base flow and storm runoff events to adequately determine both pollutant concentrations and mass losses (Blevin et al.[25]). Water quality parameters of the type that would be of interest in nonpoint source studies are distinctively non-normal and positively skewed (Hirsch and Slack[26]). In particular, the magnitudes of these parameters are very much streamflow dependent. Thus, the collection of grab samples at occasional times during the year can result in overestimating impacts, underestimating them, or failing to detect any change.

Biological monitoring is becoming increasingly popular as a complement to traditional chemical analyses of surface water (e.g., to determine nutrient content, dissolved oxygen, etc.). This type of monitoring is based on the observation that the numbers and types of aquatic organisms (especially benthos) at any point in a given body of surface water are reflective of the quality of water at that point. Studies around the world (e.g., Cairns and Dickson[27]) have documented that certain species typically tolerate only good water quality, whereas other species characterize polluted water. The results of these studies have been collated into guidelines (Terrell and Perfetti[28]) for making water quality assessments without need for physical or chemical measurements. Because biological monitoring tends to detect cumulative impacts on water quality, sampling times are not as critical as for sampling the water column. On the other hand, results from biological monitoring are qualitative instead of quantitative for water quality and should therefore be used with, rather than exclusive of, chemical and physical measurements (Chapman et al.[29]). The problem remains to relate the results of biological monitoring to agricultural practices conducted at a discrete location within a catchment.

11.4.3 GROUNDWATER

Monitoring of groundwater is subject to the same constraints (in terms of obtaining statistically valid data) as is monitoring of surface water. In contrast to surface water, however, groundwater quality tends to change more slowly. Thus, monthly sampling is commonly used as a sample frequency. However, this is a general rule of thumb that may require modification under specific geohydrologic conditions (such as depth to water table, overlying material, and aquifer characteristics), which can speed the delivery of dissolved pollutants to the water table (Smith et al.[30]). Collecting samples on a strict time schedule, such as monthly, can fail to detect groundwater impacts that occur at a frequency different than that of sampling. This problem would be expected where preferential (or macropore) flow through the soil profile is prevalent (such as in karstified limestone areas with shallow top soils).

Like surface water quality data, groundwater data can be non-normally distributed and exhibit seasonality, autocorrelation, and flow dependence (Gibbons[4]). Consequently, non-parametric statistical analyses encompassing trend detection are typically required to properly analyze ground water data. Of the available nonparametric tests, a variation (Gilbert[7]) of the Mann-Kendall test is particularly well suited to groundwater data analysis because it requires less than 40 measures, has no distributional assumptions, can accommodate missing data (nondetects), and does not require that measurements be equally spaced in time (Gilbert[7]). As with surface water monitoring and data analysis, it is theoretically possible to detect groundwater impacts using appropriate monitoring designs and statistical analyses.

11.4.4 SOIL

As is the case for surface and groundwater, classical statistics fail to describe the variability in soil quality characteristics (Trangmar et al.[31]), because the random component of soil variability often is spatially dependent. Soil properties are continuous variables whose values at any location vary according to direction and distance of separation from neighboring samples (Burgess and Webster[32]). The smaller the distance between samples, the smaller will be the difference in the values of soil parameters at the two points (Trangmar et al.[31]).

Gilbert[7] described a variety of sampling approaches suitable for application to soils as well as other media. These range from "haphazard" sampling (guessing where samples should be collected) to rigorous probability-based sampling capable of detecting statistically significant changes in soil parameters. The probabilistic sampling designs include:

- simple random sampling—not as rigorous as other statistical designs but easy to apply
- stratified random sampling—useful when homogeneous regions can be created from heterogeneous population
- systematic sampling—for use to estimate spatial trends or patterns
- double sampling—useful when a strong linear relationship exists between a parameter of interest and one that is easier/cheaper to collect/analyze

As with other environmental measures, soil monitoring is subject to problems related to pollution studies in general: seasonality effects on data, correlated data, changes in protocol during the period of monitoring, and other confounding effects. Thus, having the numbers of samples on which to make valid statistical inferences about changes (or lack thereof) in pollutant levels is as critical for soil as for other media.

11.5 SUMMARY

Water quality impairments are caused both by point and nonpoint sources of pollution. Point sources include easily defined sites from which pollutants are discharged; in contrast, nonpoint sources are of a diffuse nature and difficult to pinpoint. The predominant nonpoint pollution source is land-based agricultural activity, although road construction, forestry, and other land-based enterprises also contribute pollutants.

To control the losses of agricultural pollutants, farmers might improve physical facilities around farmyards, such as providing increased manure storage capability. As well, they could implement better managerial practices, such as nutrient management planning, for tasks occurring on the landscape. Collectively, these improvements are called best management practices (BMPs) and are site-specific measures believed to be the most cost-effective and practical techniques by which farmers can control nonpoint source pollution from agriculture.

As with other pollution control strategies, it is often desirable to define precisely how well BMPs do protect or improve water or soil quality in a specific situation. This knowledge could be useful for purposes of managing environmental quality on a catchment basis, for optimally managing farm resources, and for documenting compliance with environmental mandates. Likewise, assessing the relative contribution of point and nonpoint pollutant sources to water quality is an essential step to managing water on a catchment basis.

Monitoring of water quality can, in short, be conducted for a variety of reasons associated with diffuse agricultural pollution. The process of monitoring diffuse pollution is difficult, time consuming, and expensive. A monitoring program that produces useful information requires good planning with measurable objectives, as well as a major commitment of resources (both time and money).

Although the precise purposes for conducting a diffuse pollution monitoring program can be varied, these can be broadly classified into either measuring pollutant losses or measuring pollutant impacts. The techniques used for each of these broad objectives are similar, but the locations selected at which to monitor are generally different for the two objectives. Measurement of diffuse pollutant impacts necessitates monitoring ground and surface water; monitoring pollutant losses may involve neither resource. When a particular location of diffuse pollution is of interest, monitoring should be conducted as close to the pollutant source as practicable, consistent with program objectives, and regardless of whether pollutant impacts or losses are being measured.

In designing a diffuse pollution monitoring program, there is no substitute for thorough planning, with particular emphasis on quality control and quality assurance. The planning process is iterative, and should involve a multidisciplinary implementation team.

ACKNOWLEDGMENT

Portions of this chapter were developed from the author's personal lecture notes from the University of Maryland at College Park, and were expanded while he was a Research Officer at Teagasc, Environmental Research Centre, Johnstown Castle, Wexford, Ireland.

REFERENCES

1. Dressing, S. A., Ed., Monitoring Guidance for Determining the Effectiveness of Nonpoint Source Controls (EPA 841-B-96-004), U.S. Environmental Protection Agency, Office of Water, Washington, D.C., 1997.
2. Bartram, J. and Balance, R., *Water Quality Monitoring,* E & FN Spon, London, 1996.
3. Kunkle, S., Johnson, W. S., and Flora, M., Monitoring Stream Water for Land-use Impacts: A Training Manual for Natural Resource Management Specialists. U.S. Department of Agriculture, Forest Service, Washington, D.C., 1987.
4. Gibbons, R. D., *Statistical Methods for Groundwater Monitoring,* John Wiley & Sons, Inc., New York, 1994.
5. Ward, R. C., Loftis, J. C., and McBride, G. B., *Design of Water Quality Monitoring Systems,* Van Nostrand Reinhold, New York, 1990.
6. Chapman, D., *Water Quality Assessments,* E & FN Spon, London, 1997.

7. Gilbert, R. O., *Statistical Methods for Environmental Pollution Monitoring,* Van Nostrand Reinhold, New York, 1987.

8. Schweitzer, G. E. and Santolucito, J. A., Environmental Sampling for Hazardous Wastes, American Chemical Society, Washington, D.C., 1984.

9. Hipel, K. W., Ed., Nonparametric Approaches to Environmental Impact Assessment (AWRA Monograph No. 10), American Water Resources Association, Herndon, Virginia, 1988.

10. Keith, L. H., Ed., Principles of Environmental Sampling, American Chemical Society, Washington, D.C., 1988.

11. Brakensiek, D. L., Osborn, H. B., and Rawls, W. J., Field Manual for Research in Agricultural Hydrology (Agriculture Handbook 224), U. S. Department of Agriculture, Washington, D.C., 1979.

12. Bos, M. G., Replogle, J. A., and Clemmens, A. J., *Flow Measuring Flumes for Open Channel Systems,* John Wiley & Sons, Inc., New York, 1984.

13. Leupold & Stevens, Stevens Water Resources Data Book, 3rd edition, Leupold & Stevens, Inc., Beaverton, Oregon, 1978.

14. Grant, D. M., ISCO Open Channel Flow Measurement Handbook, 2nd edition, ISCO, Inc., Lincoln, Nebraska, 1981.

15. Morrison, R. D., Ground Water Monitoring Technology: Procedures, Equipment and Applications, TIMCO Manufacturing, Inc., Prairie du Sac, Wisconsin, 1983.

16. Wilson, N., *Soil Water and Ground Water Sampling,* CRC Press, Inc., Boca Raton, Florida, 1995.

17. Barcelona, M. J., Gibb, J. P., Helfrich, J. A., and Garske, E. E., Practical Guide for Ground-Water Sampling, Illinois State Water Survey, Champaign, Illinois, 1985.

18. Scalf, M. R., McNabb, J. F., Dunlap, W. J., Cosby, R. L., and Fryberger, J., Manual of Ground-Water Sampling Procedures, National Water Well Association, Worthington, Ohio, 1981.

19. USEPA, Ground Water, Volume II: Methodology (EPA/625/6-90/016b). U.S. Environmental Protection Agency, Office of Research and Development, Washington, D.C., 1991.

20. Nelson, D. W. and Dowdy, R. H., Methods for Ground Water Quality Studies. Agricultural Research Division, University of Nebraska-Lincoln, Lincoln, Nebraska, 1988.

21. Peterson, R. G. and Calvin, L. D., Sampling, in Methods of Soil Analysis, Part 1. Physical and Mineralogical Methods, Agronomy Monograph no 9, 2nd Edition, Klute, A. (Ed.), American Society of Agronomy, Soil Science Society of America, Madison, Wisconsin, 1986, Chapter 2.

22. Spooner, J., Jamieson, C. J., Maas, R. P., and Smolen, M. D., Determining statistically significant changes in water pollutant concentrations, *Journal of Lake and Reservoir Management,* 3, 195, 1987.

23. Spooner, J., Maas, R. P., Dressing, S. A., Smolen, M. D., and Humenik, F. J., Appropriate designs for documenting water quality improvements from agricultural NPS control programs, in *Perspectives on Nonpoint Source Pollution,* EPA/440/5-85-001, U.S. Environmental Protection Agency, Washington, D.C., 1985, 30–34.

24. Spooner, J., Maas, R. P., Smolen, M. D., and Jamieson, C. A., Increasing the sensitivity of nonpoint source control monitoring programs, in *Proceedings, Symposium on Monitoring, Modeling and Mediating Water Quality,* American Water Resources Association, Herndon, Virginia, 1987, 243–257.

25. Blevin, L. F., Humenik, F. J., Koehler, F. A., and Overcash, M. R., Dynamics of rural nonpoint source water quality in a southeastern watershed, *Transactions of the American Society of Agricultural Engineers,* 23, 1450, 1980.

26. Hirsch, R. M. and Slack, J. R., A nonparametric trend test for seasonal data with serial dependence, *Water Resources Research,* 20, 727, 1984.

27. Cairns, J. and Dickson, K. L., A simple method for the biological assessment of the effects of waste discharges on aquatic bottom dwelling organisms, *Journal of the Water Pollution Control Federation,* 43, 755, 1971.

28. Terrell, C. R. and Perfetti, P. B., Water Quality Indicators Guide: Surface Waters, SCS-TP-161, U. S. Department of Agriculture, Soil Conservation Service, Washington, D.C., 1989.

29. Chapman, D., Jackson, J., and Krebs, F., Biological monitoring, in *Water Quality Monitoring,* Bartram, J. and Ballance, R., Eds., E & FN Spon, London, 1996, Chapter 11.

30. Smith, M. C., Thomas, D. L., Bottcher, A. B., and Campbell, K. L., Measurement of pesticide transport to shallow groundwater, *Transactions of the American Society of Agricultural Engineers,* 33, 1573, 1990.

31. Trangmar, B. B., Yost, R. S., and Uehara, G., Application of geostatistics to spatial studies of soil properties, *Advances in Agronomy,* 38, 45, 1985.

32. Burgess, T. M. and Webster, R., Optimal interpolation and isarithmic mapping of soil properties: I. the semi-variogram and punctual kriging, II. block kriging, *Journal of Soil Science,* 31, 315, 1980.

Index

A

Acetanilide herbicides, 128
AF, see Attenuation Factor
AFO, see Animal feeding operation
Agricultural best management practices,
    259
Agricultural drainage and water quality,
    207–231
  history of drainage in United States,
    208–209
  impact of drainage of surface water
    quality, 225
  institutional and social constraints,
    225–226
  materials and methods for subsurface
    drainage, 209–210
  soil and crop management aspects of
    water-table management,
    215–216
  types of drainage systems, 210–211
    conventional subsurface drainage,
     210–211
    surface drainage, 210
    water-table management, 211
  water quality impacts, 216–225
    hydrology, 216
    nutrients, 216–222
    pesticides, 222–225
  water table management design,
    211–215
    design computations, 214
    detailed field investigations, 213
    operations and system management,
     214–215
    preliminary evaluation and feasibility
     of site, 211–213
    system layout and installation, 214
Alachlor, 115, 120
Aldicarb, 115
Alfalfa
  as nutrient scavenging crop, 159
  phosphorus uptake of, 101
  yield, evapotranspiration and, 185
Aluminum sulfate, addition of to litter, 159
AMC, see Antecedent moisture condition
Ametryne, 115
Ammonia volatilization, 65
Animal feeding operation (AFO), 146
Animal manures
  introduction of phosphorus into
    ecosystem by, 93
  salts in, 145
Animal wastes, 72
Antecedent moisture condition (AMC), 7
Antecedent soil moisture (ASM), 7
Aquatic systems, phosphorus loading to,
    101, 105
Aquifer(s), 22
  confined, 21
  exchange of water between streams and,
    75
  Iowa, 67
Aquitards, 21, 22
Arylanilines, 112
ASM, see Antecedent soil moisture
Atrazine, 115, 122, 170
  loss, 125, 126
  detection of in watershed studies, 120
  effects of tillage on, 223
  rainfall simulation on reduction of, 127
Attenuation Factor (AF), 247
Azinphos-methyl, 116
Azoles, 112

B

Bank storage, in streams, 74
Barley, phosphorus uptake of, 101
Benomyl, 116
Benzimidazoles, 112
Benzoic acids, 112